Pure and Applied Mathematics

A Series of Texts and Monographs (continued)

D1189476

Algebraic Numbers

PURE AND APPLIED MATHEMATICS

A Series of Texts and Monographs

Edited by: R. Courant · L. Bers · J. J. Stoker

VOLUME XXVII

ALGEBRAIC NUMBERS

PAULO RIBENBOIM
Queen's University
Kingston, Ontario, Canada

WILEY-INTERSCIENCE
A Division of John Wiley & Sons, Inc.
New York · London · Sydney · Toronto

Library of Congress Catalogue Card Number: 74-37174
ISBN 0-471-71804-1

Printed in the United States of America.

10 9 8 7 6 5 4 3 2 1

*A Jean Dieudonné
maître et ami*

Preface

This book is intended as a text for a first course in the theory of algebraic numbers. Accordingly, we have decided to maintain a very unsophisticated point of view and a rather elementary level throughout. It is our hope that the students will appreciate this beautiful branch of mathematics and then turn to more advanced books on the subject, either in the classical or in the more modern presentations.

The material included in this volume has been recently taught at Queen's University, for senior undergraduate and beginning graduate students. From our experience, this approach proves very successful toward a well-motivated introduction of many concepts in algebra.

As main prerequisites for this course, we assume a general knowledge of basic concepts in algebra, including Galois theory, and the elementary properties of rings, ideals, and fields. For convenience, we have preceded the text by an Introduction, where the main definitions and pertinent facts are included.

For the theory of algebraic numbers, we require results from ring theory and also the structure of finitely generated modules over principal ideal domains. We have chosen to develop these theorems as the necessity arises; this has the advantage of not making the Introduction any longer. Besides, in each case, some applications of the property in question are immediately derived. Even though the development of the theory itself has to suffer interruptions, we think that this approach justifies the consideration of several fundamental topics in Algebra.

The principal subject of the book is dealt with in Part Two. The reader may be familiar with some of the results in Part One, especially if he has had a course in elementary number theory. However, we believe that for much of this material our approach is more algebraic in character than is customary in a traditional course. Also, much of Part One is extended and generalized in Part Two, so a full development seemed desirable.

As the subject unfolds, we consider deeper topics and we devote extensive sections to the theory of ramification and the different, culminating with a proof of Kronecker and Weber's theorem, independently of class field theory.

The last section is devoted to the study of many interesting special cases and numerical examples. It is of a great importance to understand the theory and the difficulties usually met in specific computations. The student should be encouraged to solve the numerous exercises.

In order to confine the book to a reasonable size, we had to exclude essential topics. The analytical methods are absent, thus we have not proved Dirichlet's theorem on primes in arithmetic progressions. Similarly, we did not indicate how to compute the class number of quadratic and cyclotomic fields. The connections between ideal theory and quadratic forms have not been dealt with, nor has the theory of valuations.

In an elementary text such as this one, there is no question of including original results and new techniques. We have limited our task to a compilation of results from the best existing books on the subject; their influences are patent and easily recognizable at different places. We hope however that our presentation, using a more up-to-date language, is fit to the present taste and appropriate for the use of the book as a text. In this respect, it may serve as a stimulating method for learning algebra *in vivo*, as concepts become necessary to develop number theory. The instructor is advised to begin the presentation with Part One and to develop more fully the topics in the Introduction at the moment they are actually required. At this level, it is expected not to go beyond Chapter 8 in a one-semester course.

On the other hand, the more mature students may work through the book without any great hindrance, if they are well acquainted with algebra.

Kingston, Ontario   P. RIBENBOIM

Acknowledgements

G. Edwards and T. Gulliksen have carefully read a preliminary version of the text. My colleagues N. Sankaran and I. Hughes have helped in testing the book with the students. It is a pleasure to express my gratitude for their very valuable suggestions and constant encouragement.

P. R.

Contents

xi

Introduction

CHAPTER 1

Principal Ideal Domains and Unique Factorization Domains

1. Let A be a domain, that is, a commutative ring with unit element (different from 0), having no zero-divisors (except 0). Let K be its field of quotients.

If $a,b \in K$ we say that a *divides* b (with respect to the domain A) when there exists an element $c \in A$ such that $a \cdot c = b$. We write $a \mid b$ to express this fact. Thus, every element of K divides 0 and $1 \mid a$ for every $a \in A$.

We have $a \mid a$, for every $a \in K$ and if $a,b,c \in K$, from $a \mid b$, $b \mid c$ we deduce that $a \mid c$. However, we may well have $a \neq b$, but $a \mid b$ and $b \mid a$. For example, a divides $-a$ and $-a$ divides a.

We say that a,b are *associated* whenever it is true that $a \mid b$ and $b \mid a$; we write $a \sim b$ to express this fact.

The set $U = \{a \in K \mid a \sim 1\}$ is a subgroup of $K^{\cdot}$ (multiplicative group of non-zero elements of K), and $U \subseteq A$. The elements of U are therefore *invertible in A*, and they are called the *units* of the ring A. For example, if $A = \mathbb{Z}$ then the units are precisely 1, -1. If $A = K[X]$ (ring of polynomials with coefficients in a field K) then the units are the non-zero elements of K.

If $a \in K$ then the set $Aa = \{xa \mid x \in A\}$ is called a *principal fractional ideal*; if $a \in A$ we simply say that Aa is a *principal ideal*.

Two elements $a,b \in K$ generate the same principal fractional ideal ($Aa = Ab$) if and only if a,b are associated. Similarly, if $a,b \in K$ then $Aa \subseteq Ab$ if and only if b divides a.

The set of non-zero principal fractional ideals of K (relative to A) forms a multiplicative group, with the operation: $Aa \cdot Ab = Aab$. This group is isomorphic to the quotient group $K^{\cdot}/U$.

2. A domain such that every ideal is principal, is called a *principal ideal domain*. The ring $\mathbb{Z}$ of integers and $K[X]$ of polynomials with coefficients in a field are well-known examples of principal ideal domains. The arithmetic of principal ideal domains is very similar to that of the ring of integers.

An element p of a domain A is said to be a *prime element of A* if p is not a unit of A and it satisfies the following condition: if $a \in A$, $a \mid p$ then either $a \sim 1$ or $a \sim p$.

3

If A is *a* principal ideal domain, then $p \in A$ is a prime element if and only if the principal ideal Ap is a *maximal ideal* of A (among those distinct from the unit ideal of A); it is equivalent to say that the quotient ring A/Ap is a field.

For example, if $A = \mathbb{Z}$, the prime elements are p, $-p$ where p is any prime number; then the ideal $\mathbb{Z}p$ is maximal, and the quotient ring $\mathbb{Z}/\mathbb{Z}p$ is the field $\mathbb{F}_p$ with p elements.

If $A = K[X]$, where K is a field, the prime elements are the irreducible polynomials.

We now state the *fundamental theorem for principal ideal domains*:

Let A be a principal ideal domain, let S be a set of prime elements of A, such that

(a) Every prime element of A is associated with some prime element in S.

(b) No two prime elements in S are associated.

Then, every element $x \in K$, $x \neq 0$, may be written uniquely in the form

$$x = u \prod_{p \in S} p^{n(p)}$$

where u is a unit, $n(p) \in \mathbb{Z}$ and $n(p) \neq 0$ at most for a finite number of prime elements $p \in S$.

Any domain A for which the above theorem is true is called a *unique factorization domain*. Thus, every principal ideal domain is a unique factorization domain. However, for example, $K[X,Y]$ is a unique factorization domain, but not a principal ideal domain.

If A is a unique factorization domain, if a prime element p divides a product ab, with $a,b \in A$, then either $p \mid a$ or $p \mid b$.

If A is a unique factorization domain, any two elements $a,b \in A$ have a *greatest common divisor d* (which is unique up to unit elements); by definition d satisfies the following properties: $d \mid a$, $d \mid b$; if $d' \in A$ and $d' \mid a$, $d' \mid b$ then $d' \mid d$. We write $d = \gcd(a,b)$.

Similarly, $a,b \in A$ have a *least common multiple m* (unique up to a unit element): $a \mid m$, $b \mid m$; if $m' \in A$, $a \mid m'$, $b \mid m'$ then $m \mid m'$. We write $m = \operatorname{lcm}(a,b)$.

The elements $a,b \in A$ are said to be *relatively prime* when 1 is a greatest common divisor of a,b.

If A is a principal ideal domain, $a,b \in A$ and $d = \gcd(a,b)$, then there exist elements $x,y \in A$ such that $d = xa + yb$. In particular, if $a,b \in A$ are relatively prime, we may write $1 = xa + yb$, with $x,y \in A$. If A is a principal ideal domain, if $a,b,c \in A$ and $a \mid bc$ but $\gcd(a,b) = 1$ then necessarily $a \mid c$.

3. In our study we shall encounter more general types of commutative rings than the unique factorization domains.

Let A be a commutative ring with unit element 1. A subset J of A is called an *ideal* of A when it satisfies the following properties:

(a) If $a,b \in J$, then $a + b \in J$.

(b) If $b \in J$, $a \in A$ then $ab \in J$.

In particular, J is also an additive subgroup of A.

Among the ideals of A we have the zero ideal 0 and the unit ideal A. These are among the *principal ideals* Aa (where $a \in A$). In general, the ring A may have ideals which are not principal.

Every ideal J of A gives rise to the ring A/J, whose elements are the cosets $a + J = \{a + b \mid b \in J\}$ for every $a \in A$. We have $a + J = a' + J$ if and only if $a - a' \in J$. The operations of A/J are defined by

$$(a + J) + (a' + J) = (a + a') + J, \ (a + J).(a' + J) = (a . a') + J.$$

The mapping $\varphi: A \to A/J$, $\varphi(a) = a + J$ for every $a \in A$, is the canonical ring homomorphism from A onto A/J.

An ideal P of A is said to be a *prime ideal* when it satisfies the following conditions:

(a) $P \neq A$.

(b) If $a,b \in A$, $a,b \in P$ then either $a \in P$ or $b \in P$.

Thus P is a prime ideal of A if and only if A/P is a domain.

An ideal P of A is said to be a *maximal ideal* when:

(a) $P \neq A$.

(b) There exists no ideal J of A such that $P \subset J \subset A$.

It is easily seen that P is a maximal ideal of A if and only if A/P is a field.

Hence every maximal ideal is a prime ideal.

We define operations of *addition* and *multiplication* between ideals in the following way:

$$J + J' = \{a + a' \mid a \in J, a' \in J'\}$$
$$J . J' = \left\{\sum_{i=1}^{n} a_i a_i' \mid n > 0, a_i \in J, a_i' \in J'\right\}.$$

Then $J + J'$ and $J . J'$ are ideals of A and we have the following properties: $J + J'$ contains J and J' and if any ideal of A contains J and J' then it contains $J + J'$; $J . J' \subseteq J \cap J'$; if J'' is another ideal of A then $(J + J') + J'' = J + (J' + J'')$, and $(J . J') . J'' = J . (J' . J'')$; $A . J = J$ for every ideal J; $Aa . Ab = Aab$ for any elements $a,b \in A$.

We note also that if P is a prime ideal, if J, J' are ideals of A such that $J . J' \subseteq P$, then $J \subseteq P$ or $J' \subseteq P$. In fact, otherwise there would exist elements $a \in J$, $a \notin P$, $a' \in J'$, $a' \notin P$, such that $aa' \in J . J' \subseteq P$, which is impossible since P is a prime ideal.

Any intersection of ideals of the ring A is still an ideal of A. Thus, given any subset S of A the intersection of all ideals of A containing S is the smallest ideal of A containing S; it is called *the ideal generated by* S. If $S = \{a\}$ then the ideal generated by S is the principal ideal Aa, which is also denoted by (a). If $S = \{a_1, \ldots, a_n\}$ then the ideal generated by S is $\Sigma_{i=1}^n Aa_i$, also denoted by $(a_1, \ldots, a_n)$.

EXERCISES

1. Show that every integer $n \geqslant 1$ has only finitely many divisors.

2. Show that there exists infinitely many prime numbers.

3. Prove: If $a,m,n \in \mathbb{Z}$, if a divides m,n and a is relatively prime to m, then a divides n.

4. Let p be a prime number dividing m,n, and not dividing m; show that p divides n.

5. Show that every natural number $m > 0$ may be written in a unique way as a product of powers of prime numbers.

6. If m,n are non-zero integers, show that there exists a greatest common divisor d of m and n. Show that if d,d' are greatest common divisors of m,n then $d = d'$ or $d = -d'$.

7. Determine a computational procedure to find the greatest common divisor d of integers $m,n \geqslant 1$, by means of the Euclidean algorithm.

Hint: Divide m by n (when $m \geqslant n$) and n by the remainder (if not zero) and so on.

8. Show that if m,n are non-zero integers, d a greatest common divisor of m,n, then there exist integers a_0,b_0 such that $d = a_0 m + b_0 n$. Conversely, every number of the form $am + bn$ (with $a,b \in \mathbb{Z}$) is a multiple of d (this is known as the *Bézout property* of the integers).

9. Let d be a greatest common divisor of the integers $m,n \geqslant 1$. Show that m/d, n/d are relatively prime.

10. If m,n are non-zero integers, show that there exists a least common multiple l of m,n; if l' is any other one then $l = l'$ or $l = -l'$.

11. Generalize Exercises 6 to 10 for the case of non-zero integers $m_1, \ldots, m_r$.

12. Let m,n be non-zero integers, let d be a greatest common divisor and l a least common multiple of m, n. Show that $m \cdot n = \pm l \cdot d$.

13. Let d,l be positive integers. Show that there exist integers a,b such that $gcd(a,b) = d$ and $lcm(a,b) = l$ if and only if d divides l. In this case, show that the number of possible pairs of positive integers a,b with above properties is 2^r, where r is the number of distinct prime factors of b/a.

14. Determine the highest power of the prime number p which divides $n!$.

15. Show that the product of n successive integers is a multiple of $n!$.

16. If p is a prime number, prove that the binomial coefficients $\binom{p}{k}$, with $1 \leqslant k < p$, are multiples of p. More generally, if $1 \leqslant k < p^m$, $m \geqslant 1$, if p^r divides k but p^{r+1} does not divide k, then p^{m-r} is the exact power of p dividing $\binom{p^m}{k}$.

17. Let p be a prime number. Prove that every natural number a may be written in unique way in the form

$$a = a_0 + a_1 p + a_2 p^2 + \cdots + a_n p^n$$

where $0 \leqslant a_i \leqslant p - 1$ (for every i). This is called the *p-adic development of a*. Find explicitly the 11-adic development of 32356.

18. Let p be a prime number. Prove that every rational number a/b may be written in a unique way in the form

$$a/b = p^{-m}(c_0 + c_1 p + c_2 p^2 + \cdots + c_n p^n + \cdots)$$

where $m \geqslant 0$, $0 \leqslant c_i \leqslant p - 1$ (for every i) and there exists $n_0 \geqslant 0$ and $r \geqslant 1$ such that $c_{n+r} = c_n$ for $n \geqslant n_0$. This is called the *p-adic development of a/b*. Find the 5-adic development of the following numbers -1, $1/2$, -4, $3/7$.

19. The rational number $x = a/b$ (where a,b are relatively prime integers) is said to be *p-integral* when p does not divide b. Let $\mathbb{Z}_p$ be the set of p-integral rational numbers. Show that $\mathbb{Z}_p$ is a ring containing $\mathbb{Z}$. If p,p' are different primes, then $\mathbb{Z}_p, \mathbb{Z}_{p'}$ are distinct; neither contains the other.

20. Let p be any prime number. For every non-zero integer a let $v_p(a) = n \geqslant 0$ when p^n divides a, but p^{n+1} does not divide a; let also $v_p(0) = \infty$. If $x = a/b$ is any rational number (with $b \neq 0$) let $v_p(x) = v_p(a) - v_p(b)$. Show that v_p is well-defined and that it satisfies the following properties:

$$v_p(x + y) \geqslant \min \{v_p(x), v_p(y)\},$$
$$v_p(xy) = v_p(x) + v_p(y).$$

Moreover, if $v_p(x) \neq v_p(y)$ then $v_p(x + y) = \min\{v_p(x), v_p(y)\}$, and $v_p(x) \geqslant 0$

if and only if x is a p-integral rational number. Show that if $p \neq p'$ then $v_p \neq v_{p'}$.

21. If p is a prime number, define $|0|_p = 0$ and $|x|_p = p^{-v_p(x)}$, when x is a non-zero rational number (where v_p has been defined in the previous exercise). Show that

$$|x + y|_p \leqslant \max\{|x|_p, |y|_p\} \leqslant |x|_p + |y|_p,$$
$$|xy|_p = |x|_p \cdot |y|_p.$$

Also prove that if $|x|_p \neq |y|_p$ then $|x + y|_p = \max\{|x|_p, |y|_p\}$ and $|x|_p \leqslant 1$ if and only if x is a p-integral rational number. If $p \neq p'$ prove that $|\ |_p \neq |\ |_{p'}$.

22. The *Fibonacci numbers* 1, 1, 2, 3, 5, 8, 13, 21, 34, ... are defined inductively by the relation

$$a_n = a_{n-1} + a_{n-2} \text{ for } n \geqslant 2 \text{ (with } a_1 = 1, a_2 = 1).$$

Prove: (a) Two consecutive Fibonacci numbers are relatively prime.
(b) $a_n^2 - a_{n-1}a_{n+1} = (-1)^{n-1}$ for every $n \geqslant 1$.

23. The *Lucas numbers* 1, 3, 4, 7, 11, 18, 29, 47, ... , are defined inductively by the relation

$$b_n = b_{n-1} + b_{n-2} \text{ for } n \geqslant 2 \text{ (with } a_1 = 1, a_2 = 3).$$

Prove: (a) Two consecutive Lucas numbers are relatively prime.
(b) $b_n^2 - b_{n-1}b_{n+1} = (-1)^n \cdot 5$ for every $n \geqslant 1$.

24. Let a_n denote the nth Fibonacci number and b_n the nth Lucas number. Prove:
(a) $2a_{m+n} = a_m b_n + a_n b_m$.
(b) $b_n^2 - 5a_n^2 = 4(-1)^n$ and $b_{n+1}b_n - 5a_{n+1}a_n = 2(-1)^n$.
(c) a_n and b_n have the same parity.
(d) If they are both odd then $gcd(a_n, b_n) = 1$; otherwise, $gcd(a_n, b_n) = 2$.
(e) If m divides n then a_m divides a_n.
(f) If $gcd(m,n) = d$ then $gcd(a_m, a_n) = a_d$.
(g) If $gcd(m,n) = 1$ then $a_m a_n$ divides a_{mn}.

25. Let K be a field. Prove that $K[X]$ is a principal ideal domain.

26. Let R be a domain; we say that R is a *Euclidean ring* when for every element $a \in R$, $a \neq 0$, it is associated a natural number $\delta(a)$ and the following properties are satisfied:
(a) If $a, b \in R$ are non-zero elements then $\delta(ab) \geqslant \delta(a)$.
(b) If $a, b \in R$, $b \neq 0$, there exist elements $q, r \in R$ such that $a = bq + r$ and $r = 0$ or $\delta(r) < \delta(q)$.

Show: (1) $\mathbb{Z}$ is a Euclidean ring.

(2) If K is a field then $K[X]$ is a Euclidean ring.

(3) If R is a Euclidean ring then it is a principal ideal domain.

27. Show that every principal ideal domain is a unique factorization domain.

28. If $f = a_n X^n + a_{n-1} X^{n-1} + \cdots + a_0 \in \mathbb{Z}[X]$, let

$$\gamma(f) = gcd(a_0, a_1, \ldots, a_n)$$

be the *content of f*. We say that f is *primitive* when $\gamma(f) = 1$. Prove *Gauss' lemma*: If $f, g \in \mathbb{Z}[X]$ are primitive polynomials then fg is also primitive.

29. Show that if $f, g \in \mathbb{Z}[X]$ then their contents satisfy $\gamma(fg) = \gamma(f) \cdot \gamma(g)$.

30. Show that if $f \in \mathbb{Z}[X]$ is a primitive polynomial then it is the product of primitive polynomials which are irreducible; this decomposition is unique (up to the order of the factors and the sign).

31. Show that if $f \in \mathbb{Z}[X]$ then it may be written in unique way (up to the order of the factors) in the form:

$$f = \pm p_1 p_2 \ldots p_r h_1 h_2 \ldots h_s$$

where each $p_i \in \mathbb{Z}$ is a prime number and each $h_j \in \mathbb{Z}[X]$ is a primitive irreducible polynomial.

32. Show that if $f \in \mathbb{Z}[X]$ is irreducible in $\mathbb{Z}[X]$ then it is also irreducible in $\mathbb{Q}[X]$.

33. Generalize Exercises 28 through 32, replacing $\mathbb{Z}$ by any unique factorization domain R. Conclude that if R is a unique factorization domain then $R[X]$ is also a unique factorization domain.

34. Prove that if R is a unique factorization domain and if $(X_i)_{i \in I}$ is any family of indeterminates, then $R[X_i]_{i \in I}$ is a unique factorization domain.

CHAPTER 2

Commutative Fields

For the convenience of the reader we recall some definitions and facts about commutative fields.

1. Let L be a field, K a subfield of L. The element $x \in L$ is said to be *algebraic over* K when there exists a non-zero polynomial f, with coefficients in K such that $f(x) = 0$; dividing by the leading coefficient, we may assume that f is monic. In other words, there exist elements $a_1, \ldots, a_n \in K$ (with $n > 0$) such that $x^n + a_1 x^{n-1} + \cdots + a_n = 0$.

If $x \in L$ but x is not algebraic over K, we say that it is *transcendental over* K.

We shall be mainly concerned with algebraic elements.

If $x \in L$, $K[x]$ shall denote the subring of L of all elements of the form $\Sigma_{i=0}^{m} c_i x^i$, where $m \geqslant 0$, $c_i \in K$ (for $i = 0, 1, \ldots, m$).

If $x \in L$ is algebraic over K let $J = \{ f \in K[X] \mid f(x) = 0 \}$. J contains a unique monic polynomial $f_0 = X^n + a_1 X^{n-1} + \cdots + a_n$ of smallest degree; f_0 is irreducible over K and $a_n \neq 0$. A polynomial f belongs to J if and only if f is a multiple of f_0. Thus J is the principal ideal of $K[X]$ generated by f_0 and since f_0 is irreducible, J is a maximal ideal. The polynomial f_0 defined above is called the *minimal polynomial of x over K*. Its degree is called the *degree of x over K*.

The set of elements $\{1, x, \ldots, x^{n-1}\}$ is a basis of the K-vector space $K[x]$. Indeed, x^n is expressible in terms of the lower powers of x because $x^n + a_1 x^{n-1} + \cdots + a_n = 0$. On the other hand, no relation with coefficients in K may exist between $1, x, \ldots, x^{n-1}$, because f_0 is a polynomial of minimal degree in the ideal J.

The mapping $\varphi \colon K[X] \to K[x]$, defined by $\varphi(f) = f(x)$ for every polynomial $f \in K[X]$, is a ring-homomorphism with kernel J, and $K[X]/J \cong K[x]$. Since J is a maximal ideal then $K[x]$ is a field. This may also be seen by noting that x is invertible in $K[x]$, since

$$x \cdot [-a_n^{-1}(x^{n-1} + a_1 x^{n-2} + \cdots + a_{n-1})] = 1.$$

2. Let L be a field, K a subfield of L. If every element of L is algebraic over K, then L is said to be an *algebraic extension* of K. Otherwise, L is called a *transcendental extension* of K.

11

If the K-vector space L is of finite dimension $n = [L:K]$, we say that n is the *degree of L over K*.

If K is a subfield of L, if S is any subset of L, we denote by $K(S)$ the smallest subfield of L which contains K and S. If $S = \{x_1, \ldots, x_n\}$, we write $K(x_1, \ldots, x_n)$. L is said to be *finitely generated over K* if there exist elements $x_1, \ldots, x_n \in L$ such that $L = K(x_1, \ldots, x_n)$.

Every extension of finite degree must be algebraic and finitely generated; the converse is also true.

If K is a subfield of L, and S a set of elements of L which are algebraic over K, then $K(S)$ is an algebraic extension of K. In particular, the set of all elements of L which are algebraic over K is itself a field, called the *algebraic closure of K in L*.

More generally, Steinitz' theorem states: given any field K there exists a field $\bar{K}$ with the following properties:

(a) $\bar{K}$ is an algebraic extension of K.

(b) If L is any algebraic extension of $\bar{K}$ then $L = \bar{K}$.

(c) If $\tilde{K}$ is any field satisfying properties (a), (b) above, then there exists a K-isomorphism between $\tilde{K}, \bar{K}$ (that is an isomorphism leaving invariant all the elements of the subfield K).

$\bar{K}$ is called the *algebraic closure* of K.

Any field satisfying property (b) only of $\bar{K}$ is called an *algebraically closed field*. This implies that if f is a polynomial with coefficients in $K \subseteq \bar{K}$, then all its roots belong to $\bar{K}$; also, if L is any algebraic extension of K, then there exists a K-isomorphism from L onto a subfield of $\bar{K}$. Thus, for the purpose of studying algebraic extensions of K we may restrict our attention to the subfields of an algebraic closure $\bar{K}$. We note also that if L is an algebraically closed field containing K, if $\bar{K}$ is the subfield of all elements of L which are algebraic over K, then $\bar{K}$ is an algebraic closure of K.

3.   A very important theorem (sometimes called the Fundamental Theorem of Algebra) states that the field $\mathbb{C}$ of complex numbers is algebraically closed.

We say that a complex number x is an *algebraic number* if it is algebraic over the field $\mathbb{Q}$ of rational numbers. The set $\mathbb{A}$ of all algebraic numbers is a field which is an algebraic closure of $\mathbb{Q}$.

Let us note that $[\mathbb{A}:\mathbb{Q}] = \infty$, since by Eisenstein's irreducibility criterion there exist irreducible polynomials of arbitrary degree over $\mathbb{Q}$ (for example $X^n - p$, where p is any prime number). Every algebraic extension of finite degree over $\mathbb{Q}$ is called an *algebraic number field*.*

* In a more advanced level one considers also algebraic number fields of infinite degree over $\mathbb{Q}$.

It is an easy matter to show that the field A must be countable. Since $\mathbb{C}$ is not countable, there exist uncountably many transcendental complex numbers (for examples π, e). Hermite showed in 1873 that e is transcendental, while Lindemann proved the transcendency of π in 1882. More recent proofs were given by Hilbert, Gelfond, and Schneider (see Lang [14]).

4. Let K be a field, let 1 denote its unit element; for every integer $n > 0$ let $n \cdot 1 = 1 + \cdots + 1$ (n times). If $n \cdot 1 \neq 0$ for every $n > 0$, we say that K has characteristic 0. If $n > 0$ is the smallest integer such that $n \cdot 1 = 0$, then we say that K has characteristic n; it is easily seen that n must be a prime number $n = p$.

The field $\mathbb{Q}$ of rational numbers has characteristic 0. For every prime p, the field $\mathbb{F}_p$ of residue classes modulo p, namely $\mathbb{F}_p = \{\bar{0}, \bar{1}, \bar{2}, \dots, \overline{p-1}\}$ has characteristic p (see Chapter 3).

If K is any field, the intersection of all its subfields is again a subfield, called the *prime field of K*. It is isomorphic to $\mathbb{Q}$ when K has characteristic 0, or to $\mathbb{F}_p$ when K has characteristic p.

5. Let K be a field and $\bar{K}$ an algebraic closure of K. The elements $x, x' \in \bar{K}$ are *conjugate over K* whenever they have the same minimal polynomial over K. This happens if and only if there exists a K-isomorphism σ from $K(x)$ onto $K(x')$ such that $\sigma(x) = x'$.

Let L, L' be extensions of K contained in $\bar{K}$. We say that L, L' are *conjugate over K* if there exists a K-isomorphism σ from L onto L'.

If $[L:K]$ is finite the number of K-isomorphisms from L into an algebraic closure $\bar{K}$ of K is at most equal to the degree $[L:K]$.

If L is equal to all its conjugates over K, then L is said to be a *normal extension* of K. For example, $\bar{K}$ is a normal extension of K. Every extension of degree 2 is also normal.

A typical normal extension of K is obtained by considering an arbitrary polynomial $f \in K[X]$ and adjoining to K the roots of f. The resulting field L is equal to all its conjugates over K, hence it is a normal extension of K. It is called the *splitting field* of f over K.

Conversely, if $L \mid K$ is a normal extension of finite degree, then $L = K(x_1, \dots, x_n)$, where $x_1, \dots, x_n$ are algebraic elements. If f_i is the minimal polynomial of x_i over K and $f = f_1 f_2, \dots, f_n$ then all the roots of f are still in L, so L is the splitting field of f.

Given any field $L, K \subseteq L \subseteq \bar{K}$, the intersection L' of all normal extensions of K between L and $\bar{K}$ is the smallest normal extension of K containing L and contained in $\bar{K}$. If L has finite degree over K then L' has also finite degree over K.

If L is a normal extension of K, the set of K-automorphisms of L forms a group (under composition), which we denote by $G(L \mid K)$.

6. Let $K \subseteq L \subseteq \bar{K}$ where $[L:K] < \infty$; we say that L is *separable over K* if the number of distinct K-isomorphisms from L into $\bar{K}$ is exactly equal to the degree $[L:K]$. More generally, if L is any field, $K \subseteq L \subseteq \bar{K}$, we say that L is separable over K if every subfield F of L, such that F contains K and $[F:K] < \infty$, is separable over K (in the sense just defined).

In particular, the algebraic extension $L = K(x)$ is separable over K whenever x has $n = [L:K]$ distinct conjugates over K. In this case, we say that x is a *separable element over K*.

It follows that an algebraic extension L of K is separable if and only if every element x of L is separable over K.

If K is a field of characteristic 0, all the roots of any irreducible polynomial f over K are necessarily distinct. Hence every algebraic extension of a field of characteristic 0 is separable.

If L is a separable extension of finite degree of a field K, then there exists an element t in L such that $L = K(t)$; t is called a *primitive element* of L over K. In particular, this theorem holds when $K = \mathbb{Q}$.

If L is an algebraic number field of degree n over $\mathbb{Q}$, we have $L = \mathbb{Q}(t)$, where t has a minimal polynomial f with coefficients in $\mathbb{Q}$ and degree n. Let $t_1 = t, t_2, \ldots, t_n$ be the roots of f, which belong to $\mathbb{C}$. If t_i is not a real number, then its complex conjugate $\bar{t}_i$ is also a root of f; thus the non-real roots of f occur in pairs.

Let r_1 denote the number of real roots of f, let $2r_2$ denote the number of non-real roots of f; then

$$n = r_1 + 2r_2.$$

7. An algebraic extension L over K is said to be a *Galois extension* whenever L is normal and separable over K. If K has characteristic 0, this means that L is a normal extension over K. The group $G(L \mid K)$ of all K-automorphisms of a Galois extension L of K is called the *Galois group of L over K*. If $[L:K] = n$ then $G(L \mid K)$ has precisely n elements. L is called an *Abelian extension* (respectively a *cyclic extension*) of K when its Galois group is abelian (respectively cyclic).

Let L be a Galois extension of finite degree of K, let $G = G(L \mid K)$. If G' is any subgroup of G, let us consider the subfield $fi(G')$ of all elements $x \in L$ which are invariant under G' (that is $\sigma(x) = x$ for every $\sigma \in G'$). Similarly, if K' is any field, $K \subseteq K' \subseteq L$, then L is a Galois extension of K' and we may consider the Galois group $gr(K') = G(L \mid K')$ of all K'-automorphisms of L.

The *fundamental theorem of Galois theory* states:

(a) G' is equal to the Galois group $G(L \mid K')$ where $K' = fi(G')$

(b) K' is the field of invariants of the group $G' = G(L \mid K')$.

Moreover, G' is a normal subgroup of G if and only if the corresponding field K' is a Galois extension of K. Then $G(K' \mid K) \cong G/G'$.

Let $\tilde{K}$ be an extension of K, let L, L' be subfields of $\tilde{K}$ containing K. The *compositum* of L, L' in $\tilde{K}$ is the smallest subfield of $\tilde{K}$ containing L and L'. We denote it by LL' or $L'L$.

If $L \mid K$ is a Galois extension then the compositum LL' is a Galois extension of L' and $G(LL' \mid L') \cong G(L \mid L \cap L')$. Explicitly, every $(L \cap L')$-automorphism of L may be uniquely extended to a L'-automorphism of LL'.

If moreover $L' \mid K$ is also a Galois extension then the compositum LL' is a Galois extension of $L \cap L'$ and

$$G(LL' \mid L \cap L') \cong G(L \mid L \cap L') \times G(L' \mid L \cap L').$$

In particular, if $L \cap L' = K$ we have

$$G(LL' \mid K) \cong G(L \mid K) \times G(L' \mid K).$$

8. Let K be a field and n an integer, $n > 0$. An element $x \in K$ such that $x^n = 1$ is called a nth *root of unity*. The set $W_{n,K}$ of all nth rooths of unity in K is a multiplicative group. If K is a subfield of L then $W_{n,K}$ is a subgroup of $W_{n,L}$. If m divides n then $W_{m,K} \subseteq W_{n,K}$.

We denote by W_K the set of all roots of unity in K; that is,

$$W_K = \bigcup_{n \geqslant 1} W_{n,K}.$$

Let $x \in W_K$ and let n be its order in the multiplicative group W_K, that is, the smallest integer such that $x^n = 1$. Then we say that x is a *primitive nth root of unity*. It is easily seen that if K has characteristic p and $x \in W_K$ then its order n is not a multiple of p.

To deal simultaneously with fields of any characteristic, it is customary to say that K has *characteristic exponent* 1 when its characteristic is 0, and characteristic exponent p, when its characteristic is p.

Let K be an algebraically closed field and let n be a natural number relatively prime to the characteristic exponent of K. Thus $X^n - 1$ has exactly n distinct roots in K, which are the elements of the group $W_{n,K}$. This group is cyclic, each generator being a primitive nth root of unity.

If ζ is such a generator then $W_{n,K} = \{\zeta, \zeta^2, \zeta^3, \ldots, \zeta^{n-1}, 1\}$. ζ^a is a primitive nth root of unity if and only if a, n are relatively prime. Thus the number of primitive nth roots of unity is equal to $\varphi(n)$, where φ indicates the *Euler function* (see Chapter 3, definition 2). It follows that if L is a subfield of K then $W_{n,L}$ is also a cyclic group and its order divides n.

If ζ is a primitive nth root of unity in K, then its conjugates over the prime subfield K_0 of K are also primitive nth roots of unity. The polynomial

$$\Phi_{n,K_0} = \prod_{\substack{gcd(a,n)=1 \\ 1 \leqslant a \leqslant n}} (X - \zeta^a)$$

is called the nth *cyclotomic polynomial* (over K_0). It has degree $\varphi(n)$. The coefficients of Φ_{n,K_0} belong to K_0, since they are invariant by conjugation.

If K has characteristic 0 it is known that $\Phi_{n,\mathbb{Q}} = \Phi_n$ is irreducible over $\mathbb{Q}$; that is, $\mathbb{Q}(\zeta)$ has degree $\varphi(n)$ over $\mathbb{Q}$ (see Chapter 13, (**3B**)). For example,

$$\Phi_1 = X - 1$$
$$\Phi_p = X^{p-1} + X^{p-2} + \cdots + X + 1,$$
$$\Phi_{p^r} = X^{p^{r-1}(p-1)} + X^{p^{r-1}(p-2)} + \cdots + X^{p^{r-1}} + 1.$$

More generally

$$X^n - 1 = \prod_{d \mid n} \Phi_d$$

since both sides are equal to the product of all linear factors $X - \zeta^a$ for $a = 1, 2, \ldots, n$. From this relation it is possible to compute Φ_n by recurrence.

$\mathbb{Q}(\zeta)$ is a Galois extension of $\mathbb{Q}$, having Galois group isomorphic to the multiplicative group $P(n)$ of prime residue classes modulo n (see Chapter 3). $\mathbb{Q}(\zeta)$ is therefore an Abelian extension of $\mathbb{Q}$.

It is worth mentioning that if p does not divide n it may occur that $\Phi_{n,\mathbb{F}_p}$ is reducible over $\mathbb{F}_p$.

9.   We describe now the finite fields. If K is a finite field, it has characteristic p (for some prime p), and the number of its elements is a power p^n, $n > 0$. Precisely, n is the degree of K over the prime field $\mathbb{F}_p$. For every n there exists exactly one field with p^n elements, which we denote by $\mathbb{F}_{p^n}$. Its nonzero elements are the roots of $X^{p^n-1} - 1$ and $\mathbb{F}_{p^n} = \mathbb{F}_p(\zeta)$, where ζ is a primitive $(p^n - 1)$th root of unity.

$\mathbb{F}_{p^n}$ is a Galois extension of $\mathbb{F}_p$ with cyclic Galois group. A canonical generator of $G(\mathbb{F}_{p^n} \mid \mathbb{F}_p)$ is the *Frobenius automorphism* σ, defined by $\sigma(x) = x^p$ for every $x \in \mathbb{F}_{p^n}$.

In particular, if m divides n, then $\mathbb{F}_{p^n}$ is a Galois extension of $\mathbb{F}_{p^m}$ with cyclic Galois group generated by σ^m, where $\sigma^m(x) = x^{p^m}$ for every $x \in \mathbb{F}_{p^n}$.

10.   Let K be a field, and let L be a separable extension of degree n. Thus, there exist n distinct K-isomorphisms $\sigma_1 = \varepsilon$ (identity), $\sigma_2, \ldots, \sigma_n$ of L into an algebraic closure $\bar{K}$ of K (which may be assumed to contain L).

If $x \in L$ the *trace of* x (relative to K) is defined as

$$\mathrm{Tr}_{L|K}(x) = \sum_{i=1}^{n} \sigma_i(x)$$

and the *norm of* x (relative to K) is

$$N_{L|K}(x) = \prod_{i=1}^{n} \sigma_i(x).$$

If $f = X^n + a_1 X^{n-1} + \cdots + a_n$ is the minimal polynomial of x over K, then $\mathrm{Tr}_{L|K}(x) = -a_1$, $N_{L|K}(x) = (-1)^n a_n$.

Among the properties of the trace and norm, we note:

$$\mathrm{Tr}_{L|K}(x + y) = \mathrm{Tr}_{L|K}(x) + \mathrm{Tr}_{L|K}(y);$$
$$N_{L|K}(xy) = N_{L|K}(x) \cdot N_{L|K}(y).$$

If $x \in K$ then $\mathrm{Tr}_{L|K}(x) = nx$, $N_{L|K}(x) = x^n$. If $K \subseteq L \subseteq L'$ are fields and L' is separable over K, then we have the transitivity of the trace and the norm:

$$\mathrm{Tr}_{L|K}(\mathrm{Tr}_{L'|L}(x)) = \mathrm{Tr}_{L'|K}(x)$$

and

$$N_{L|K}(N_{L'|L}(x)) = N_{L'|K}(x).$$

If $L \mid K$ is a separable extension, there exists always an element $x \in L$ such that $\mathrm{Tr}_{L|K}(x) \neq 0$.

Indeed, since $\mathrm{Tr}_{L|K}(1) = [L:K] \cdot 1$, if K has characteristic 0 then $\mathrm{Tr}_{L|K}(1) \neq 0$. Similarly, if K has characteristic p and p does not divide $[L:K]$ then $\mathrm{Tr}_{L|K}(1) \neq 0$. The remaining case, where K has characteristic p and $[L:K]$ is a multiple of p, requires a non-trivial proof based on Dedekind's theorem on independence of the K-isomorphisms of L:

If $\sigma_1, \ldots, \sigma_n$ are the K-isomorphisms from L into an algebraic closure $\bar{K}$ of K, if $x_1, \ldots, x_n \in \bar{K}$ and $\Sigma_{i=1}^{n} x_i \sigma_i = 0$ (null mapping from L to $\bar{K}$) then necessarily $x_1 = \cdots = x_n = 0$.

It is also useful to know that if $L \mid K$ is a separable extension of finite degree n and the characteristic of K does not divide n, then there exists a primitive element t such that $\mathrm{Tr}_{L|K}(t) = 0$.

11. We shall now consider the notion of discriminant.

Let $L \mid K$ be a separable extension of degree n, let $(x_1, \ldots, x_n)$ be a n-tuple of elements in L. We define its *discriminant* (in $L \mid K$) to be

$$\mathrm{discr}_{L|K}(x_1, \ldots, x_n) = \det(\mathrm{Tr}_{L|K}(x_i x_j)),$$

that is, the determinant of the matrix, whose (i,j)-entry is equal to $\mathrm{Tr}_{L|K}(x_i x_j)$ (for $i, j = 1, \ldots, n$). Thus, the discriminant belongs to K.

If $(x_1', \ldots, x_n')$ is another n-tuple of elements in L, and $x_j' = \Sigma_{i=1}^n a_{ij}x$ (for all $j = 1, \ldots, n$), where $a_{ij} \in K$, then

$$\text{discr}_{L|K}(x_1', \ldots, x_n') = [\det(a_{ij})]^2 \cdot \text{discr}_{L|K}(x_1, \ldots, x_n).$$

Another expression of the discriminant may be obtained in terms of the K-isomorphisms $\sigma_1, \ldots, \sigma_n$ of L:

$$\text{discr}_{L|K}(x_1, \ldots, x_n) = [\det(\sigma_i(x_j))]^2.$$

In order that $x_1, \ldots, x_n \in L$ be linearly independent over K it is necessary and sufficient that $\text{discr}_{L|K}(x_1, \ldots, x_n) \neq 0$.

Consider the special basis $\{1, t, \ldots, t^{n-1}\}$ where t is a primitive element, $L = K(t)$, we obtain

$$\text{discr}_{L|K}(1, t, \ldots, t^{n-1}) = \prod_{i<j} (t_i - t_j)^2$$

where $t_1 = t, t_2, \ldots, t_n$ are the conjugates of t.

In this situation, it is customary to call the above expression the *discriminant of t* (over K). Thus, all the conjugates of t have the same discriminant.

If $f \in K[X]$ is a monic polynomial of degree n, with roots $x_1, \ldots, x_n$, we define the *discriminant of f* by:

$$\text{discr}(f) = \prod_{i<j} (x_i - x_j)^2 = (-1)^{n(n-1)/2} \prod_{i \neq j} (x_i - x_j).$$

Thus $\text{discr}(f) \in K$ and $\text{discr}(f) \neq 0$ if and only if f has distinct roots. If t is a separable element over K, if f is its minimal polynomial, then $\text{discr}_{L|K}(t) = \text{discr}(f)$.

It is important to compute the discriminant of an irreducible polynomial $f = X^n + a_1 X^{n-1} + \cdots + a_n$ without knowing a priori its roots. If $p_k = x_1^k + \cdots + x_n^k$ (where $x_1, \ldots, x_n$ are the roots of f) for $k = 0, 1, 2, \ldots$, then $p_0 = n$, $p_1 = -a_1$, and $p_2, p_3, \ldots$ may be computed recursively by the well-known Newton formulas (see Exercise 17). Then

$$\text{discr}(f) = \begin{vmatrix} p_0 & p_1 & \cdots & p_{n-1} \\ p_1 & p_2 & \cdots & p_n \\ \cdot & \cdot & & \cdot \\ \cdot & \cdot & & \cdot \\ \cdot & \cdot & & \cdot \\ p_{n-1} & p_n & \cdots & p_{2n-2} \end{vmatrix}$$

In some cases, the computation of the above determinant is rather awkward. However, it is possible to compute the discriminant of f by the more direct formula:

$$\text{discr}(f) = (-1)^{n(n-1)/2} N_{K(x)|K}(f'(x))$$

where x is any root of f and f' is the derivative of the polynomial f.

EXERCISES

1. Show that $\sqrt{2}$ is not a rational number.

2. Let $m,n \geqslant 1$ be integers. Show that $\sqrt[n]{m}$ is a rational number if and only if m is an nth power of a natural number.

3. Prove *Eisenstein's irreducibility criterion*: The polynomial

$$f = X^n + a_{n-1}X^{n-1} + \cdots + a_1 X + a_0$$

with coefficients in $\mathbb{Z}$ is irreducible over $\mathbb{Q}$ provided there exists a prime p dividing $a_1, \ldots, a_n$ but such that p^2 does not divide a_n.

4. Show that $\Phi_p = X^{p-1} + X^{p-2} + \cdots + X + 1$ (where p is a prime number) is irreducible over $\mathbb{Q}$.
   Hint: Apply Eisenstein's irreducibility criterion to the polynomial $\Phi_p(X + 1)$.

5. Show that if p is a prime number then

$$\Phi_{p^r} = X^{p^{r-1}(p-1)} + X^{p^{r-1}(p-2)} + \cdots + X^{p^{r-1}} + 1$$

is irreducible over $\mathbb{Q}$.

6. Discuss whether the following polynomials are irreducible over $\mathbb{Q}$: $X^3 + X + 1$; $X^4 + X^2 + 1$; $X^3 + X^2 + X + 1$; $X^6 + X^4 + 1$; $X^4 - 4X^2 + 8X - 4$.

7. Let $K \subseteq K' \subseteq K''$ be fields. Show that $[K'':K] = [K'':K'] \cdot [K':K]$.

8. Let $K \subseteq K' \subseteq K''$ be fields. Show that $K'' \mid K$ is an algebraic extension if and only if $K' \mid K$ and $K'' \mid K'$ are algebraic extensions.

9. Prove that $L \mid K$ has finite degree if and only if L is a finitely generated algebraic extension of K.

10. Find the degree over $\mathbb{Q}$ of the following fields: $\mathbb{Q}(i,\sqrt{3})$, where $i = \sqrt{-1}$; $\mathbb{Q}((1 + i)/2)$; $\mathbb{Q}(\sqrt{2},\sqrt{5},\sqrt{10})$; $\mathbb{Q}(\sqrt{2 + \sqrt{2 + \sqrt{2}}}, i)$.

11. Show that the following numbers are algebraic over $\mathbb{Q}$, determine the minimal polynomial and the conjugates over $\mathbb{Q}$:

$$1 + \sqrt{2}; \quad \sqrt{2 + \sqrt{2}}; \quad \sqrt{2 + \sqrt{2 + \sqrt{2}}}; \quad 1 - 2\sqrt{-1}; \quad \sqrt{2} + \sqrt{3} + \sqrt{5}.$$

12. Let $L \mid K$ be an algebraic extension. Prove:
(a) if K is finite then L is countable.
(b) if K is infinite then $\#(L) = \#(K)$.

13. Show that there exist uncountably many transcendental complex numbers.
Hint: Use the previous exercise.

14. Show that an algebraically closed field cannot be finite.

15. Let $X_1, X_2, \ldots, X_n$ be indeterminates and consider the symmetric polynomials

$$s_1 = X_1 + X_2 + \cdots + X_n$$
$$s_2 = X_1 X_2 + X_1 X_3 + \cdots + X_2 X_3 + \cdots + X_{n-1} X_n$$
$$s_3 = X_1 X_2 X_3 + X_1 X_2 X_4 + \cdots + X_{n-2} X_{n-1} X_n$$
$$\cdots$$
$$s_k = \sum_{i_1 < i_2 < \cdots < i_k} X_{i_1} X_{i_2} \cdots X_{i_k} \quad (\text{where } k \leqslant n)$$
$$\cdots$$
$$s_n = X_1 X_2 \cdots X_n.$$

Show that if Y is any indeterminate then

$$Y^n - s_1 Y^{n-1} + s_2 Y^{n-2} - \cdots + (-1)^k s_k Y^{n-k} + \cdots + (-1)^n s_n = \prod_{k=1}^{n} (Y - X_k).$$

16. Let R be a domain. A polynomial $f \in R[X_1, \ldots, X_n]$ is said to be *symmetric* if for every permutation σ of $\{1, 2, \ldots, n\}$ we have

$$f(X_{\sigma(1)}, X_{\sigma(2)}, \ldots, X_{\sigma(n)}) = f(X_1, X_2, \ldots, X_n).$$

Show that if f is a symmetric polynomial, there exists a symmetric polynomial $g \in R[X_1, \ldots, X_n]$ such that $f(X_1, \ldots, X_n) = g(s_1, s_2, \ldots, s_n)$.
Hint: Define the weight of a monomial $a X_1^{e_1} X_2^{e_2}, \ldots, X_n^{e_n}$ as being $e_1 + 2e_2 + \cdots + ne_n$; define the weight of a polynomial as being the maximum of the weights of its monomials; the proof is done by double induction on n and on the degree d of f; consider $f(X_1, \ldots, X_{n-1}, 0)$, express it as $g_1(s_1^0, s_2^0, \ldots, s_{n-1}^0)$ where $s_1^0, s_2^0, \ldots, s_{n-1}^0$ are the elementary symmetric polynomials on the indeterminates $X_1, \ldots, X_{n-1}$; observe that

$$g_1(s_1, s_2, \ldots, s_{n-1})$$

has degree at most d in $X_1, \ldots, X_n$; then $f_1(X_1, \ldots, X_n) = f(X_1, \ldots, X_n) - g_1(s_1, \ldots, s_{n-1})$ has degree at most d and it is symmetric; also $f_1(X_1, \ldots, X_{n-1}, 0) = 0$ so f_1 is a multiple of X_n and by symmetry of $X_1 X_2 \cdots X_n$; define f_2 by $f_1 = s_n f_2$ hence its degree is less than d; continue by induction.

17. Let $p_0 = n$ and $p_k = X_1^k + X_2^k + \cdots + X_n^k$ where $k \geqslant 1$ and $X_1, X_2, \ldots, X_n$ are indeterminates. Prove *Newton formulas*:
   (a) If $k \leqslant n$ then

$$p_k - p_{k-1}s_1 + p_{k-2}s_2 - \cdots + (-1)^{k-1}p_1 s_{k-1} + (-1)^k k s_k = 0.$$

   (b) If $k > n$ then

$$p_k - p_{k-1}s_1 + \cdots + (-1)^n p_{k-n}s_n = 0.$$

Hint: Let $f(T) = \Pi_{i=1}^n (T - X_i)$, where T is a new indeterminate; write the quotient $f'(T)/f(T)$ as a rational fraction in $T, X_1, \ldots, X_n$ (where f' denotes the derivative of f with respect to T); then develop in formal power series and after multiplying both sides by $f(T)/T^n$, equate the coefficients of equal powers of T.

18. Let $f = X^3 + X^2 - X + 1$ and let x_1, x_2, x_3 be the roots of f in $\mathbb{C}$. Determine $x_1^3 + x_2^3 + x_3^3$.

19. Let K be a field of characteristic p. Show that the mapping $\theta : K \to K$ defined by $\theta(x) = x^p$ for every $x \in K$, is an isomorphism from K into K. Moreover, if K is finite it is an automorphism.

20. Show that if K is a field of characteristic p then $K(X)$ is an inseparable extension of $K(X^p)$.

21. Show that if K is a field of characteristic 0 and $f \in K[X]$ is an irreducible polynomial then its roots are all distinct.

22. Let K be a field of characteristic p, $f \in K[X]$ an irreducible polynomial such that there exists an integer $e \geqslant 1$ and a polynomial $g \in K[X]$ for which $f(X) = g(X^{p^e})$. Then every root of f has multiplicity at least p^e.

23. Show that if K is any infinite field and $f \in K[X_1, \ldots, X_n]$, there exists infinitely many n-tuples $x = (x_1, \ldots, x_n) \in K^n$ such that $f(x_1, \ldots, x_n) \neq 0$.
   Hint: Proceed by recurrence on n.

24. Let V be a vector space of dimension n over K; let $W_1, \ldots, W_m$ be subspaces of V, distinct from V. Show that if K is an infinite field then $W_1 \cup \cdots \cup W_m \neq V$.
   Hint: Use the previous exercise.

25. Prove the theorem of the primitive element: If L is a separable extension of finite degree over a field K, then there exists an element $t \in L$ such that $L = K(t)$.

Hint: Consider first the case when K is finite; then, letting K be infinite, consider the sets $\{x \in L \mid \sigma_i(x) \neq \sigma_j(x)\}$ where σ_i, σ_j are distinct K-isomorphisms from L into an algebraic closure of K; conclude, using the previous exercise.

26. Find a primitive element over $\mathbb{Q}$ for each of the following fields: $\mathbb{Q}(\sqrt{2}, i)$; $\mathbb{Q}(\sqrt{2}, \sqrt{3})$; $\mathbb{Q}(\sqrt{2} + \sqrt{3}, \sqrt{2} - \sqrt{5})$.

27. Determine the smallest normal extension K of $\mathbb{Q}$ containing $\sqrt[3]{2}$; what is the degree of K over $\mathbb{Q}$? Find a primitive element of K over $\mathbb{Q}$.

28. Give an example of an extension of degree 4 of $\mathbb{Q}$ which is not normal.

29. Determine the $\mathbb{Q}$-isomorphisms of the following fields: $\mathbb{Q}(\sqrt[3]{2})$; $\mathbb{Q}((1 + i)/2)$; $\mathbb{Q}(\sqrt{2}, \sqrt{3})$; $\mathbb{Q}(\sqrt{2} + \sqrt{3})$; $\mathbb{Q}(\sqrt{2 + \sqrt{2}})$.

30. Determine the Galois group over $\mathbb{Q}$ of the following polynomials: $X^3 + X + 1$; $X^4 - X^2 + 1$; $X^3 - 2$; $X^3 - 1$; $X^3 + 1$; $X^4 - 5$.

31. For each of the above polynomials, determine the subgroups of the Galois group and the corresponding fields of invariants.

32. Determine the Galois group of $(X^2 - p_1)(X^2 - p_2) \ldots (X^2 - p_n)$ over $\mathbb{Q}$, where $p_1, p_2, \ldots, p_n$ are distinct primes.

33. If L is a finite separable extension of K, prove that there exist only finitely many fields K' such that $K \subseteq K' \subseteq L$.

34. Determine which roots of unity belong to the following fields: $\mathbb{R}$; $\mathbb{Q}(i)$; $\mathbb{Q}(\sqrt{-2})$; $\mathbb{Q}(\sqrt{-3})$; $\mathbb{Q}(\sqrt{-5})$.

35. Let K be a finite extension of $\mathbb{Q}$. Show that K contains only finitely many roots of unity.

36. Prove that if K is an algebraically closed field of characteristic 0 or p, if n is not a multiple of p, then the group of nth roots of unity in K is cyclic.

Hint: Let $n = p_1^{e_1} p_2^{e_2} \cdots p_r^{e_r}$ be the decomposition of n into prime factors; let $n_i = n/p_i$; show that there exists $x_i \in K$ such that $x_i^n = 1$ but $x_i^{n_i} \neq 1$; next note that if $m_i = n/p_i^{e_i}$ then $x_i^{m_i}$ has order $p_i^{e_i}$ (for $i = 1, \ldots, r$); conclude with Lemma 1 of Chapter 3.

37. Compute Φ_6, Φ_{12}, Φ_{18}.

38. Prove that $\Phi_{n,\mathbb{Q}}$ is a monic polynomial with coefficients in $\mathbb{Z}$.

Hint: Use the expression for $X^n - 1$ as a product of cyclotomic polynomials.

39. Prove that $\Phi_{n,\mathbb{Q}}$ is irreducible over $\mathbb{Q}$.

Hint: Let f be the minimal polynomial of ζ over $\mathbb{Q}$; by Gauss' lemma $f \in \mathbb{Z}[X]$ and $\Phi_n = f \cdot g$, with $g \in \mathbb{Z}[X]$ monic. Show that for every prime p, not dividing n, it must be $f(\zeta^p) = 0$; if this is not true, consider $h = g(X^p)$, deduce that f divides h, and conclude that Φ_n, reduced modulo p, would have a double root. Show that this is not the case.

40. Let ζ be a primitive nth root of unity. Show that $\mathbb{Q}(\zeta)$ is a Galois extension of degree $\varphi(n)$ over $\mathbb{Q}$ and that $\mathrm{Gal}(\mathbb{Q}(\zeta) \mid \mathbb{Q})$ is isomorphic to the group $P(n)$ of residue classes $\bar{a}$ modulo n, where a is an integer relatively prime to n (see Chapter 3).

Hint: Use the previous exercise; then consider the mapping θ: $\mathrm{Gal}(\mathbb{Q}(\zeta) \mid \mathbb{Q}) \to P(n)$, defined as follows: if $\sigma(\zeta) = \zeta^a$ then $\theta(\sigma) = \bar{a}$.

41. Let p be a prime not dividing n. Show that $\Phi_{n,\mathbb{F}_p}$ is irreducible over $\mathbb{F}_p$ if and only if the class of p modulo n has order $\varphi(n)$ in the multiplicative group $P(n)$ of prime residue classes modulo n.

Hint: $\Phi_{n,\mathbb{F}_p}$ is irreducible if and only if $[\mathbb{F}_p(\zeta): \mathbb{F}_p] = \varphi(n)$, and using a property of finite fields, this means that $\varphi(n)$ is the smallest integer such that n divides $p^{\varphi(n)} - 1$.

42. Show that $\Phi_{8,\mathbb{F}_7}$ is reducible over $\mathbb{F}_7$ and find its decomposition.

43. Show that Φ_{12} is reducible over the fields $\mathbb{F}_5$, $\mathbb{F}_7$, $\mathbb{F}_{11}$, $\mathbb{F}_{13}$ and find in each case its decomposition into irreducible polynomials.

44. Determine the minimal polynomial over $\mathbb{F}_3$ for a primitive 8th root of unity.

45. Determine the minimal polynomial over $\mathbb{F}_2$ for a primitive 7th root of unity.

46. Let K be a finite field. Show that every element of K is the sum of two squares of elements of K.

47. Let $L \mid K$ be a separable extension of degree n, and assume that the characteristic of K does not divide n. Show that there exists $t \in L$ such that $L = K(t)$ and $\mathrm{Tr}_{L\mid K}(t) = 0$.

Hint: If $L = K(t')$ and $X^n + a_1 X^{n-1} + \cdots + a_n$ is the minimal polynomial of t' over K, show that $t = t' + a_1/n$ satisfies the required property.

48. Determine the discriminant of the following polynomials: $aX^2 + bX + c$; $aX^3 + bX^2 + cX + d$; $aX^4 + bX^3 + cX^2 + dX + e$.

Hint: Use Newton formulas.

49. Prove that if $f \in K[X]$ is an irreducible polynomial of degree n, if t is any root of f and f' is the derivative of f, then

$$\mathrm{discr}(f) = (-1)^{n(n-1)/2} \cdot N_{K(t)|K}(f'(t)).$$

Hint: Let $t = t_1, \ldots, t_n$ be the conjugates of t over K, compute explicitly $f'(t_i)$ and compare its values with $\mathrm{discr}(f)$.

50. Determine the discriminant of $\Phi_{n,\mathbb{Q}}$.

Hint: Use the previous exercise.

51. Let K be a field and $\overline{K}$ an algebraic closure of K.

Prove that the subfield generated by the union of all Abelian extensions of K is an Abelian extension K' of K; it is called the *Abelian closure* of K.

52. Prove the following theorem of Warning and Chevalley:

Let K be a finite field of characteristic p, let $f \in K[X_1, \ldots, X_n]$ be a polynomial without constant term, of degree $d < n$. Then the number of zeros of f in K is a multiple of p; in particular, there exists $x_1, \ldots, x_n \in K$, not all x_i being equal to zero, and such that $f(x_1, \ldots, x_n) = 0$.

Hint: (a) Let $g = X_1^{a_1} \ldots X_n^{a_n}$ with $0 < \Sigma_{i=1}^n a_i < (q-1)n$, where $q = p^r = \#(K)$; show that $\Sigma_{x \in K^n} g(x) = 0$ by computing $\Sigma_{y \in K} y^a$ for different values of the integer $a \geqslant 0$.

(b) Note that the number N of zeros of f in K is given by

$$N = \sum_{x \in K^n} [1 - f(x)^{q-1}] \equiv -\sum_{x \in K^n} f(x)^{q-1}$$

and then apply (a) for the monomials of f^{q-1}.

Part One

CHAPTER 3

Residue Classes

Definition 1. Let m be a positive integer. We say that the integers a,b are *congruent modulo m* if they leave the same remainder when divided by m. We write $a \equiv b \pmod{m}$ to express this fact. It is equivalent to say that m divides $a - b$. The relation of "congruence modulo m" is reflexive, symmetric, and transitive; that is, it is an equivalence relation.

The equivalence class of a modulo m is the set $\bar{a} = \{a + km \mid k \in \mathbb{Z}\}$. $\bar{a}$ is called the *residue class of a modulo m*.

If $a \equiv a' \pmod{m}$, $b \equiv b' \pmod{m}$, then $a + b \equiv a' + b' \pmod{m}$, $a \cdot b \equiv a' \cdot b' \pmod{m}$. Hence we may define the addition and the multiplication of equivalence classes: $\bar{a} + \bar{b} = \overline{a + b}$, $\bar{a} \cdot \bar{b} = \overline{a \cdot b}$. With these operations we obtain a ring, which is denoted by $\mathbb{Z}/m$. Its unit element is $\bar{1} = \{a \in \mathbb{Z} \mid a \equiv 1 \pmod{m}\}$.

It follows that the mapping $a \rightarrow \bar{a}$ is a ring-homomorphism from $\mathbb{Z}$ onto $\mathbb{Z}/m$. The kernel of this homomorphism is $(m) = \mathbb{Z}m$ (the ideal of multiples of m). So $\mathbb{Z}/m$ is the quotient ring of $\mathbb{Z}$ by the ideal (m), and we may also use the notation $\mathbb{Z}/(m)$.

The additive group of $\mathbb{Z}/m$ is cyclic with m elements, generated by $\bar{1}$. In fact $m \cdot \bar{1} = \bar{0}$ ($m \cdot \bar{1}$ denotes $\bar{1} + \bar{1} + \cdots + \bar{1}$, m times) and if $n \cdot \bar{1} = \bar{0}$ then n is a multiple of m.

Conversely, we have the following result:

A. *Every cyclic group with m elements is isomorphic to $\mathbb{Z}/m$.*
Proof: Let G be a cyclic group (written additively) with m elements. Let g be a generator of G and consider the mapping $\theta : \mathbb{Z} \rightarrow G$ defined as follows: $\theta(n) = ng$, for every $n \in \mathbb{Z}$. Then θ is a group-homomorphism onto G; we have $ng = 0$ (in G) if and only if n is a multiple of m, because g has order m. Thus, the kernel of θ is (m) and so $G \cong \mathbb{Z}/(m)$. ∎

Now, we shall determine all the generators of the cyclic group $\mathbb{Z}/m$.

B. *Let a be a positive integer; then $\bar{a}$ is a generator of $\mathbb{Z}/m$ if and only if a,m are relatively prime.*

Proof: Let $\bar{a}$ be a generator of $\mathbb{Z}/m$, let $d = gcd(a,m)$ (the greatest common divisor of a,m). Then

$$\frac{m}{d} \cdot a = m \cdot \frac{a}{d} \equiv 0 \pmod{m}.$$

Since $\bar{a}$ has order m, then m divides m/d, that is, $d = 1$.

Conversely, if $gcd(a,m) = 1$ and m' is the order of $\bar{a}$, then m divides $m'a$ (because $m'a \equiv 0 \pmod{m}$) and therefore m divides m'. On the other hand, $m\bar{a}=\bar{0}$, so the order m' of $\bar{a}$ divides m. Thus $m' = m$.   ∎

We are interested in counting the generators of the additive group $\mathbb{Z}/m$; this number depends on m, and we may introduce the following definition:

Definition 2.   For every $m \geqslant 1$, let $\varphi(m)$ denote the number of integers a, $1 \leqslant a \leqslant m$, such that $gcd(a,m) = 1$.

φ is called the *Euler function* or *totient function*. Thus, $\varphi(m)$ is the number of generators of the additive group $\mathbb{Z}/m$. For example, $\varphi(1) = 1$, $\varphi(2) = 1$, $\varphi(3) = 2$ and more generally, for every prime p, $\varphi(p) = p - 1$. In order to determine $\varphi(m)$, for an arbitrary $m > 1$, we shall prove some interesting results about residue classes.

Theorem 1.   *Let $m = \Pi_{i=1}^{r} p_i^{e_i}$ be the decomposition of $m > 1$ into prime powers (with $e_i > 0$ for $i = 1, \ldots, r$). Then there exists a ring-isomorphism*

$$\bar{\theta} \colon \mathbb{Z}/m \xrightarrow{\sim} \Pi_{i=1}^{r} \mathbb{Z}/p_i^{e_i}$$

(Cartesian product of the rings $\mathbb{Z}/p_i^{e_i}$, for $i = 1, \ldots, r$).
Proof: Let $\theta \colon \mathbb{Z} \to \Pi_{i=1}^{r} \mathbb{Z}/p_i^{e_i}$ be the mapping defined as follows:
$\theta(n) = (\nu_1, \ldots, \nu_r)$ where ν_i denotes the residue class of n modulo $p_i^{e_i}$.

θ is a ring-homomorphism whose kernel is equal to the ideal (m) of multiples of m, because n is a multiple of m if and only if n is a multiple of each $p_i^{e_i}$ ($i = 1, \ldots, r$). Thus θ induces a one-to-one homomorphism $\bar{\theta}$ from $\mathbb{Z}/m$ to $\Pi_{i=1}^{r} \mathbb{Z}/p_i^{e_i}$.

Finally, the number of elements in $\Pi_{i=1}^{r} \mathbb{Z}/p_i^{e_i}$ is $\Pi_{i=1}^{r} p_i^{e_i} = m$. Thus θ maps $\mathbb{Z}/m$ onto $\Pi_{i=1}^{r} \mathbb{Z}/p_i^{e_i}$.   ∎

As a corollary we have the so-called "*Chinese remainder theorem*" below.

C.   *Given $r \geqslant 1$ distinct prime numbers $p_1, \ldots, p_r$, given integers $e_1 \geqslant 1$, $\ldots, e_r \geqslant 1$, and $a_1, \ldots, a_r \in \mathbb{Z}$, there exists $n \in \mathbb{Z}$ satisfying all the congruences*

$$n \equiv a_i \pmod{p_i^{e_i}} \quad for \quad i = 1, \ldots, r.$$

Moreover, $n' \in \mathbb{Z}$ satisfies the above congruences if and only if

$$n' \equiv n \pmod{\Pi_{i=1}^{r} p_i^{e_i}}.$$

Proof: Let v_i be the residue class of a_i modulo $p_i{}^{e_i}$, let $v \in \mathbb{Z}/m$ be the unique residue class modulo m such that $\bar{\theta}(v) = (v_1, \ldots, v_r)$. Then n satisfies the congruences

$$n \equiv a_i \pmod{p_i{}^{e_i}} \quad \text{for} \quad i = 1, \ldots, r$$

if and only if v is the class of n modulo m. The last assertion is obvious. ∎

In order to determine $\varphi(m)$, we consider the set $P(m)$ of all non-zero residue classes $\bar{a}$ modulo m, where $gcd(a,m) = 1$. These are called the *prime residue classes modulo m*, and $P(m)$ has therefore $\varphi(m)$ elements. We remark that if $\bar{a} \in P(m)$ then every element $a + km$ in the class $\bar{a}$ is relatively prime to m.

D. *$P(m)$ is the multiplicative group of all invertible elements of the ring $\mathbb{Z}/m$.*
Proof: Let $a \in \mathbb{Z}$ be such that $gcd(a,m) = 1$. Then there exist integers s,t such that $s . a + t . m = 1$. Hence $\bar{s} . \bar{a} = \bar{1}$ and $\bar{a}$ is invertible in the ring $\mathbb{Z}/m$.

Conversely, if $\bar{a} \in \mathbb{Z}/m$ is invertible in the ring $\mathbb{Z}/m$, then $\bar{s} . \bar{a} = \bar{1}$ (for some $s \in \mathbb{Z}$), so $sa \equiv 1 \pmod{m}$ and $gcd(a,m) = 1$, that is $\bar{a} \in P(m)$.

It is easy to see that in any ring the set of all invertible elements constitutes a multiplicative group. Thus, $P(m)$ is a multiplicative group. ∎

Let us note in particular that $\mathbb{Z}/p$ is a field (when p is prime), since every non-zero residue class modulo p is invertible. Actually $\mathbb{Z}/p$ is the prime field of characteristic p, and as such we denote it also by $\mathbb{F}_p$.

As a corollary, we obtain the following congruence (from Fermat):

E. *If $gcd(a,m) = 1$ then*

$$a^{\varphi(m)} \equiv 1 \pmod{m}$$

Proof: Since $\bar{a} \in P(m)$, a multiplicative group of order $\varphi(m)$, then by Lagrange's theorem on finite groups $\bar{a}^{\varphi(m)} = \bar{1}$, that is $a^{\varphi(m)} \equiv 1 \pmod{m}$. ∎

In particular, for any prime number p we have

$$a^{p-1} \equiv 1 \pmod{p}.$$

F. *If $m = p_1^{e_1}, \ldots, p_r^{e_r}$ is the prime-power decomposition of m, then*

$$P(m) \cong \prod_{i=1}^{r} P(p_i^{e_i}),$$

and

$$\varphi(m) = \prod_{i=1}^{r} \varphi(p_i^{e_i}) = m \prod_{i=1}^{r} \left(1 - \frac{1}{p_i}\right).$$

Proof: We consider again the ring-isomorphism $\bar{\theta}: \mathbb{Z}/m \xrightarrow{\sim} \Pi_{i=1}^r \mathbb{Z}/p_i^{ei}$ of Theorem 1. A residue class $\bar{a}$ modulo m is invertible if and only if its image $\bar{\theta}(\bar{a})$ is invertible in the Cartesian product of rings $\mathbb{Z}/p_i^{ei}$; in other words, each component of $\bar{\theta}(\bar{a})$ is invertible in $\mathbb{Z}/p_i^{ei}$. Thus $\bar{\theta}(P(m)) = \Pi_{i=1}^r P(p_i^{ei})$, proving the isomorphism of multiplicative groups $P(m) \cong \Pi_{i=1}^r P(p_i^{ei})$.

By counting the number of elements in these groups it follows that $\varphi(m) = \Pi_{i=i}^r \varphi(p_i^{ei})$.

To evaluate $\varphi(p^e)$ (for a prime p, and $e \geqslant 1$), we just note that if $1 \leqslant a \leqslant p^e$ $gcd(a, p^e) = 1$ then a is not a multiple of p, and conversely. Thus $P(p^e)$ has $p^e - p^{e-1}$ elements; that is $\varphi(p^e) = p^e[1 - (1/p)]$. Hence $\varphi(m) = \Pi_{i=1}^r p_i^{ei} \cdot \Pi_{i=1}^r [1 - (1/p_i)] = m \Pi_{i=1}^r [1 - (1/p_i)]$. ∎

Incidentally, if $gcd(m, n) = 1$ then $\varphi(mn) = \varphi(m) \cdot \varphi(n)$, as follows immediately from the above. We express this fact by saying that φ is a *multiplicative function*.

The Euler function also possesses the following interesting property:

G. *If* $n \geqslant 1$ *then* $n = \Sigma_{d|n} \varphi(d)$ *(the sum being taken over all divisors* $d \geqslant 1$ *of* n*).*
Proof: Let $S = \{1, 2, \ldots, n\}$ and for each divisor $d \geqslant 1$ of n let $C(d) = \{s \in S \mid gcd(n, s) = d\}$. This gives rise to a partition of S into pairwise disjoint subsets. Therefore $n = \#S = \Sigma_{d|n} \#C(d)$. Now, if $n = dd'$, then $s \in C(d')$ if and only if $s/d' \leqslant n/d' = d$ and $gcd(s/d', d) = 1$. Thus, the number of elements in $C(d')$ is equal to $\varphi(d)$; therefore $n = \Sigma_{d|n} \varphi(d)$. ∎

Up to now, we have studied the additive group of residue classes modulo m, which is cyclic, and whose generators are the elements of the multiplicative group $P(m)$.

In the sequel, we shall study the structure of the group $P(m)$. First, we point out another way of describing $P(m)$:

H. *The group* $\text{Aut}(\mathbb{Z}/m)$ *of automorphisms of the additive group* $\mathbb{Z}/m$ *is isomorphic to* $P(m)$.
Proof: We will show that the ring of endomorphisms $\text{End}(\mathbb{Z}/m)$ of the additive group $\mathbb{Z}/m$ is isomorphic to the ring $\mathbb{Z}/m$. Since $\text{Aut}(\mathbb{Z}/m)$ is just the group of invertible elements of $\text{End}(\mathbb{Z}/m)$ then the conclusion follows from (**D**).

Let $\theta: \text{End}(\mathbb{Z}/m) \to \mathbb{Z}/m$ be defined by $\theta(f) = f(\bar{1})$. Evidently $\theta(f + g) = (f + g)(\bar{1}) = f(\bar{1}) + g(\bar{1}) = \theta(f) + \theta(g)$ so θ is a homomorphism of additive groups. The kernel of θ is 0, for if $f(\bar{1}) = \bar{0}$ and $n > 0$ then $f(\bar{n}) = f(n \cdot \bar{1}) = n \cdot f(\bar{1}) = 0$, so $f = 0$. Thus θ is a one-to-one mapping. Moreover, given $\bar{a} \in \mathbb{Z}/m$ if we define $h: \mathbb{Z}/m \to \mathbb{Z}/m$ by $h(\bar{b}) = \bar{a} \cdot \bar{b}$ then $h \in \text{End}(\mathbb{Z}/m)$ and $\theta(h) = h(\bar{1}) = \bar{a}$. Thus θ is an isomorphism of additive groups

It only remains to show that θ preserves multiplication. Accordingly, suppose that $\theta(f) = \bar{a}$, $\theta(g) = \bar{b}$ with $a,b > 0$; then $\theta(f \cdot g) = f(g(\bar{1})) = f(\bar{b}) = f(b \cdot \bar{1}) = b \cdot f(\bar{1}) = b \cdot \bar{a} = \bar{b} \cdot \bar{a}$ as desired. ∎

In view of **(f)**, it suffices to study the groups $P(p^e)$, where p is a prime number, $e \geqslant 1$. For this purpose, we require two lemmas from the theory of groups.

Lemma 1. *Let G be a multiplicative group, let $x,y \in G$ be elements such that $x \cdot y = y \cdot x$, and the orders h of x, and k of y are relatively prime. Then the order of $x \cdot y$ is hk.*
Proof: From $x \cdot y = y \cdot x$ we deduce $(x \cdot y)^{hk} = (x^h)^k \cdot (y^k)^h = e$ (identity element of G). Thus $x \cdot y$ has finite order l dividing hk.

Since $(x \cdot y)^l = x^l \cdot y^l = e$ then $x^l = y^{-l}$. The order of this last element divides both h and k, hence it must be 1 (because h,k are relatively prime). Thus $x^l = y^{-l} = e$, so h divides l, k divides l and therefore hk divides l. This shows that $x \cdot y$ has order $l = hk$. ∎

The least upper bound of the orders of the elements of a group G is called the *exponent* of G. If the exponent is finite, it is a positive integer, equal to the maximum of the orders of the elements of G. By Lagrange's theorem, if G is finite the exponent of G divides the order of G.

Lemma 2. *Let G be a multiplicative Abelian group with finite exponent g. Then the order of every element of G divides g.*
Proof: Let x be an element of order g, let $y \in G$ be an element of order k not dividing g. Then there exists a prime p and an integer $n \geqslant 1$ such that p^n divides k but not g. If $k = p^n k'$ and $y' = y^{k'}$ then y' has order p^n. Hence $x \cdot y'$ has order $gp^n > g$ (by Lemma 1). This contradicts the definition of g. ∎

I. *$P(p)$ is a cyclic group.*
Proof: We know that $P(p)$ has $\varphi(p) = p - 1$ elements. It is enough to show that the exponent h of $P(p)$ is equal to $p - 1$, so there will exist an element of order $p - 1$.

As we said before, h divides $p - 1$. By Lemma 2, the order of every element of $P(p) = \mathbb{F}_p^*$ divides h. That is, $\bar{x}^h = \bar{1}$ for every $\bar{x} \in P(p)$. Therefore, every element of $P(p)$ is a root of the polynomial $X^h - 1 \in \mathbb{F}_p[X]$. This polynomial has at most h roots; thus $p - 1 \leqslant h$. This shows that $h = p - 1$. ∎

From the proof, it follows that every element of $P(p)$ may be viewed as a $(p - 1)$th root of unity.

Since $P(p)$ is cyclic with $p - 1$ elements, it is isomorphic to the additive group $\mathbb{Z}/(p - 1)$ (by (**A**)), hence it has $\varphi(p - 1)$ generators (by (**B**)). Each generator of $P(p)$ is called a *primitive root modulo p*. It is also customary to say that the integer x, $1 \leqslant x \leqslant p - 1$, is a primitive root modulo p, when its class modulo p generates $P(p)$. If $\bar{x}$ is a primitive root modulo p then $\bar{x}^h$ is a primitive root modulo p if and only if $\gcd(h, p - 1) = 1$.

Let us remark that no quick procedure is known for the determination of the smallest integer a, $1 \leqslant a \leqslant p - 1$, such that $\bar{a} \in P(p)$ is a primitive root modulo p.

Next, we consider the groups $P(p^e)$, where $e \geqslant 2$. First, we treat the case where $p \neq 2$.

J. *If $p \neq 2$, $e \geqslant 1$, then $P(p^e)$ is a cyclic group and*

$$P(p^e) \cong \mathbb{Z}/(p - 1) \times \mathbb{Z}/p^{e-1}.$$

Proof: We may assume that $e \geqslant 2$.

Let $\bar{a}$ denote the residue class of a modulo p^e, and $\bar{\bar{a}}$ the residue class of a modulo p. Let $f : P(p^e) \to P(p)$ be defined by $f(\bar{a}) = \bar{\bar{a}}$. It is obviously a well-defined group homomorphism onto $P(p)$. Its kernel is

$$C = \{\bar{a} \in P(p^e) \mid a \equiv 1 \ (\text{mod } p)\}.$$

C is a subgroup of $P(p^e)$ having order $\varphi(p^e)/\varphi(p) = p^{e-1}$ because

$$P(p^e)/C \cong P(p).$$

C is a cyclic group with generator $\overline{1 + p}$. It is enough to show that $\overline{(1 + p)^{p^{e-2}}} \neq \bar{1}$, that is, $(1 + p)^{p^{e-2}} \not\equiv 1 \ (\text{mod } p^e)$. This is true for $e = 2$ and let us assume it true for $e - 1$, that is $(1 + p)^{p^{e-3}} \not\equiv 1 \ (\text{mod } p^{e-1})$. Therefore $(1 + p)^{p^{e-3}} = 1 + rp^{e-2}$, where $r \not\equiv 0 \ (\text{mod } p)$. Raising to the pth power, we have

$$(1 + p)^{p^{e-2}} = 1 + \binom{p}{1} rp^{e-2} + \binom{p}{2} r^2 p^{2(e-2)} + \cdots + r^p p^{p(e-2)}$$

$$= 1 + rp^{e-1} + sp^e.^*$$

Hence $(1 + p)^{p^{e-2}} \not\equiv 1 \ (\text{mod } p^e)$.

Now, let $B = \{\bar{a} \in P(p^e) \mid \bar{a}^{p-1} = \bar{1}\}$. B is obviously a subgroup of $P(p^e)$ and $B \cap C = \{\bar{1}\}$ since B has no element (except $\bar{1}$) of order a power of p. Thus $P(p^e)$ contains the subgroup $BC \cong B \times C$. B then has order at most $\varphi(p^e)/p^{e-1} = p - 1$.

---

* This is false if and only if $p = 2$ and $e = 3$.

We have $\bar{a}^{p^{e-1}} \in B$ for every $\bar{a} \in P(p^e)$. Since $f(\bar{a}^{p^{e-1}}) = \bar{\bar{a}}^{p^{e-1}} = \bar{\bar{a}}$ then B contains the distinct elements $\bar{1}^{p^{e-1}}, \bar{2}^{p^{e-1}}, \ldots, \overline{(p-1)}^{p^{e-1}}$ (since they have distinct images by f). Thus B has $p - 1$ elements and $P(p^e) = BC \cong B \times C$.

Finally, B is a cyclic group. Let b be a primitive root modulo p, then $\bar{b}^{p^{e-1}}$ has order d dividing $p - 1$. From $b^{p^{e-1}} \equiv b \pmod{p}$ it follows that $b^d \equiv (b^d)^{p^{e-1}} \equiv 1 \pmod{p}$ so $p - 1$ divides d, thus $d = p - 1$.

By Lemma 1, $P(p^e)$ has an element of order $(p - 1)p^{e-1} = \varphi(p^e)$, so it is a cyclic group and it is isomorphic to $\mathbb{Z}/(p - 1) \times \mathbb{Z}/p^{e-1}$. ∎

We need a special treatment for the case $p = 2$.

K. $P(4) = \{\bar{1}, \bar{3}\}$ *is a cyclic group generated by* $\bar{3}$. *If* $e \geqslant 3$ *then* $P(2^e) \cong \mathbb{Z}/2 \times \mathbb{Z}/2^{e-2}$ *and it is not a cyclic group.*
Proof: The first assertion is obvious. Let us assume $e \geqslant 3$.

Let $\bar{a}$ denote the residue class of a modulo 2^e and $\bar{\bar{a}}$ the residue class of a modulo 4. Let $f: P(2^e) \to P(4)$ be defined by $f(\bar{a}) = \bar{\bar{a}}$. It is obviously a well-defined group-homomorphism onto $P(4)$. Its kernel is

$$C = \{\bar{a} \in P(2^e) \mid a \equiv 1 \pmod{4}\}.$$

C is a subgroup of $P(2^e)$ having order $\varphi(2^e)/\varphi(4) = 2^{e-2}$ because $P(2^e)/C \cong P(4)$.

C is a cyclic group generated by $\bar{5}$. It is enough to show that $\bar{5}^{2^{e-3}} \neq \bar{1}$. Indeed $5^{2^{e-3}} \equiv (1 + 2^2)^{2^{e-3}} \equiv 1 + 2^{e-3} \cdot 2 \not\equiv 1 \pmod{2^e}$.

We show now that $P(2^e)$ is isomorphic to the Cartesian product

$$\{1, -1\} \times C.$$

Let $\theta: P(2^e) \to \{1, -1\} \times C$ be the mapping so defined: $\theta(\bar{a}) = ((-1)^r, a^*)$ where

$$a^* = \begin{cases} a & \text{when} \quad a \equiv 1 \pmod{4} \\ -a & \text{when} \quad a \equiv -1 \pmod{4} \end{cases}$$

(we remark that if $\bar{a} \in P(2^e)$ then a is odd) and

$$r = \begin{cases} 0 & \text{when} \quad a \equiv 1 \pmod{4} \\ 1 & \text{when} \quad a \equiv -1 \pmod{4}. \end{cases}$$

It is obvious that θ is a group-homomorphism and from $a = (-1)^r a^*$ we conclude that θ is one-to-one. Since $P(2^e)$ and $\{1, -1\} \times C$ have 2^{e-1} elements it follows that θ is an isomorphism.

Since $\{1, -1\}$ and C are isomorphic to the additive groups $\mathbb{Z}/2$ and $\mathbb{Z}/2^{e-2}$ respectively, then $P(2^e) \cong \mathbb{Z}/2 \times \mathbb{Z}/2^{e-2}$.

To see that $P(2^e)$ is not cyclic, we just observe that the order of every element of $P(2^e)$ divides 2^{e-2}. ∎

As a consequence, we indicate the values of m for which $P(m)$ is a cyclic group. In such case, each generator of $P(m)$, or each integer a of this residue class, $1 \leqslant a \leqslant m - 1$, is called a *primitive root modulo m*.

L. *$P(m)$ is a cyclic group if and only if $m = 2, 4, p^e, 2p^e$, where $e \geqslant 1$ and p is an odd prime.*
Proof: By **(I)**, **(J)**, **(K)**, $P(m)$ is cyclic for each of the given values of m, noting that $P(2p^e) \cong P(2) \times P(p^e) = P(p^e)$.

To prove the converse, we note that if p is any prime then $P(p^e)$ has even order, except when $p = 2$, $e = 1$. By **(F)** it suffices to show that if G has order $2r$ and H has order $2s$ then $G \times H$ is not cyclic. Indeed, for every $x \in G$, $y \in H$ we have $(x . y)^{2rs} = (1,1)$. Therefore, no element of $G \times H$ has order $(2r) . (2s)$. ∎

We want to conclude this chapter by proving the structure theorem of finite Abelian groups. It is a theoretical result analogous to the theorems of structure of $P(m)$. We shall not require this fact until Chapter 8.

Theorem 2. *Let G be a finite Abelian group (written multiplicatively). Then G is isomorphic to a Cartesian product of cyclic groups.*
Proof (Artin):   Let k be an integer with the following properties:

(1) there exist elements $x_1, \ldots, x_k$ in G such that every element of G is a product of powers of $x_1, \ldots, x_k$;

(2) k is the smallest integer satisfying (1).

If $k = 1$ then G consists of all the powers of x_1, so G is a cyclic group.

Let us assume the theorem true for all groups having a system of generators with less than k elements.

Since G is a finite group, there exist integers $e_1, \ldots, e_k$, not all equal to zero, such that

$$\prod_{i=1}^{k} x_i^{e_i} = 1 \tag{1}$$

Let b be the minimum of the absolute values of all non-zero exponents which appear in all possible relations of type (1); thus $b > 0$ and by renumbering and taking inverses, if necessary, we may assume that $b = e_1$, for some relation (1).

$$\text{Let} \quad \prod_{i=1}^{k} x_i^{f_i} = 1 \tag{2}$$

be any other relation, with $f_i \in \mathbb{Z}$, not all equal to 0. Then $b = e_1$ divides f_1. In fact, if $f_1 = qe_1 + r$ with $0 < r < e_1 = b$, dividing (2) by the qth power of (1) we obtain

$$x_1{}^r \prod_{i=2}^{k} x_i^{f_i - qe_i} = 1,$$

which is against the definition of b.

Similarly, $b = e_1$ divides all exponents e_i. In fact, if $e_i = q_i b + r_i$ with $0 < r_i < b$, we consider the system of generators $\{x_1 x_i{}^{q_i}, x_2, \ldots, x_k\}$ of G; it satisfies the relation

$$(x_1 x_i{}^{q_i})^{e_i} x_2{}^{e_2} \cdots x_i{}^{r_i} \cdots x_k{}^{e_k} = 1 \tag{3}$$

with $0 < r_i < b$, which is a contradiction.

Now, let $e_i = q_i b$ and $y_1 = x_1 x_2{}^{q_2} \cdots x_k{}^{q_k}$, so $\{y_1, x_2, \ldots, x_k\}$ is also a system of generators of G and the element y_1 has order $b = e_1$. In fact, if $y_1{}^f = 1$ then $x_1{}^f x_2{}^{fq_2} \cdots x_k{}^{fp_k} = 1$ and therefore, by what we proved above, b divides f; on the other hand, $y_1{}^b = x_1{}^{e_1} x_2{}^{e_1 q_2} \cdots x_k{}^{e_1 q_k} = 1$.

The group G', generated by $\{x_2, \ldots, x_k\}$ is isomorphic to a Cartesian product of cyclic groups (by induction on k).

Next, we show that $G \cong G_1 \times G'$, where G_1 is the group generated by y_1. Indeed, if $x \in G$, we may write $x = y_1{}^{c_1} y'$ with $c_1 \in \mathbb{Z}$, $y' \in G'$. The elements $y_1{}^{c_1}$, y' are uniquely determined by x, for if $y_1{}^{c_1} y' = y_1{}^{d_1} z'$ then $y_1{}^{c_1 - d_1} \cdot (y'z'^{-1}) = 1$; this is a relation of type $x_1{}^{c_1 - d_1} \cdot x_2{}^{f_2} \cdots x_k{}^{f_k} = 1$, thus by the above proof b divides $c_1 - d_1$, so $y_1{}^{c_1 - d_1} = 1$, and $y_1{}^{c_1} = y_1{}^{d_1}$ so also $y' = z'$.

Therefore, the mapping $x \to (y_1{}^{c_1}, y')$ is an isomorphism between G and $G_1 \times G'$, so G is a Cartesian product of cyclic groups. ∎

The preceding theorem contains no uniqueness assertion. For example, Theorem 1 states that every cyclic group of order m is isomorphic to the Cartesian product of cyclic groups of prime-power order. So, any uniqueness statement can at most hold for decompositions into cyclic groups of prime-power order.

We shall prove that this is indeed true, and for this purpose we shall first consider the Abelian groups of prime-power order.

The first basic fact to note holds also for non-Abelian groups, being a consequence of the first Sylow theorem. As we only require this fact for Abelian groups, we shall indicate here a simpler direct proof:

M. *If G is a finite Abelian group of order m and p is a prime dividing m, then G has an element of order p.*

Proof: By Theorem 2, $G \cong G_1 \times \cdots \times G_k$ where G_i is a cyclic group (for $i = 1, \ldots, k$), with order m_i. Since p divides $m = m_1 m_2 \cdots m_1$ then p

divides m_i (for some index i, $1 \leqslant i \leqslant k$). If x_i is a generator of G_i and $m_i = pm_i'$ if $y_i \in G$ corresponds to

$$(1, \ldots, x_i^{m_i'}, \ldots, 1) \in G_1 \times \cdots \times G_i \times \cdots \times G_k,$$

then y_i has order p. ∎

Another proof independent of Theorem 2 is the following:

Let n be the exponent of the group G. We shall prove by induction that the order of G, $\#(G)$ *divides a power of* n. We may assume that $\#(G) > 1$.

Let $x \in G$, $x \neq e$ (identity element of G), then $x^n = e$. If H is the subgroup of G generated by x, then $\#(H)$ divides n. Since the order of every element of the quotient group G/H divides n, the exponent n' of G/H divides n. By induction, $\#(G/H)$ divides a power of n', hence of n. Thus

$$\#(G) = \#(G/H) \cdot \# (H)$$

divides a power of n.

Now, if p is a prime dividing $m = \#(G)$ then p divides a power of the exponent n, so p divides n; that is, $n = pn'$. If $x \in G$ has order n then $x^{n'}$ has order p. ∎

If p is a prime number, a finite group G is said to be a *p-group* when its order is a power of p. In view of (**M**) we deduce:

N.   *A finite Abelian group G is a p-group if and only if the order of every element of G is a power of p.*[*]

Proof: If G is a p-group, by Lagrange's theorem the order of every element of G must be a power of p. Conversely, if q is a prime different from p and dividing the order of G, by (**M**) there exists an element in G having order q. ∎

In order to derive the uniqueness theorem, we first prove the uniqueness of decomposition into Cartesian product of p-groups (for different primes p):

O.   *Let G be a finite Abelian group of order $m = \Pi_{i=1}^r p_i^{e_i}$ (with $r > 0$, $e_i \geqslant 1$, and $p_1, \ldots, p_r$ distinct primes). Then G is isomorphic to the Cartesian product of the p_i-groups $G_{(p_i)} = \{x \in G \mid$ order of x is a power of $p_i\}$:*

$$G \cong G_{(p_1)} \times \cdots \times G_{(p_r)}.$$

This decomposition is unique in the following strong sense: if θ is an isomorphism $\theta: G \to G_1 \times \cdots \times G_s$, where each G_i is a q_i-group ($q_1, \ldots, q_s$ being distinct primes), then $s = r$, there exists a permutation π of $\{1, \ldots, r\}$ such that $q_{\pi(i)} = p_i$ and if $H_i = \theta^{-1}(\{1\} \times \cdots \times G_i \times \cdots \times \{1\})$ then $H_{\pi(i)} = G_{(p_i)}$ (for $i = 1, \ldots, r$).

[*] This result holds also for non-Abelian groups.

Proof: If $r = 1$ then $G = G_{(p_1)}$, as follows from Lagrange's theorem. So, we may assume that $r \geq 2$.

For every prime p let $G_{(p)} = \{x \in G \mid \text{order of } x \text{ is a power of } p\}$. By (M) and Lagrange's theorem $G_{(p)} \neq \{1\}$ if and only if p divides m; that is, $p = p_i$ for some i, $1 \leq i \leq r$. By (N), each $G_{(p_i)}$ is a p_i-group.

We shall show that G is the internal direct product of the subgroups $G_{(p_1)}, \ldots, G_{(p_r)}$.

For every i let $m_i = m/p_i{}^{e_i}$, hence the integers $m_1, \ldots, m_r$ are relatively prime and there exist integers $h_1, \ldots, h_r$ such that $\sum_{i=1}^{r} h_i m_i = 1$.

If $x \in G$ we may write $x = x^{\sum_{i=1}^{r} h_i m_i} = \prod_{i=1}^{r} x^{h_i m_i}$, and $(x^{h_i m_i})^{p_i{}^{e_i}} = x^{h_i m} = 1$ (by Lagrange's theorem). Thus, $x^{h_i m_i}$ has order power of p_i, so $x^{h_i m_i} \in G_{(p_i)}$. From $x = \prod_{i=1}^{r} x^{h_i m_i}$ it follows that the subgroup of G generated by $G_{(p_1)}, \ldots, G_{(p_r)}$ is equal to G itself. Moreover, for every i we have $G_{(p_i)} \cap H_i = \{1\}$, where H_i denotes the subgroup generated by all $G_{(p_j)}$ with $j \neq i$. In fact, if $y \in G_{(p_i)} \cap H_i$ then the order of y is $p_i{}^{f_i}$ (for some $f_i \geq 0$). Since $y \in H_i$ then $y = \prod_{j \neq i} y_j$ where $y_j \in G_{(p_j)}$ for every $j \neq i$. Let $p_j{}^{f_j}$ be the order of y_j, therefore by Lemma 1, $k = \prod_{j \neq i} p_j{}^{f_j}$, is the order of y. Therefore $k = p_i{}^{f_i}$. We conclude that the order of y is necessarily 1, so $y = 1$. Thus, we have shown that G is the internal direct product of the subgroups $G_{(p_1)}, \ldots, G_{(p_r)}$: $G \cong G_{(p_1)} \times \cdots \times G_{(p_r)}$. In particular, since the order of $G_{(p_i)}$ is a power of p_i and $m = \prod_{i=1}^{r} p_i{}^{e_i}$, by the uniqueness of the decomposition of m into prime factors, it follows that $\#(G_{(p_i)}) = p_i{}^{e_i}$ for every $i = 1, \ldots, r$.

Let us assume now that $\theta : G \to G_1 \times \cdots \times G_s$ is an isomorphism, where G_i is a q_i-group, and $q_1, \ldots, q_s$ are distinct primes. Since $m = \prod_{i=1}^{r} p_i{}^{e_i}$ and also $m = \#(G) = \prod_{j=1}^{s} \#(G_i) = \prod_{j=1}^{s} q_j{}^{f_j}$ then necessarily $s = r$, there is a permutation π of $\{1, \ldots, s\}$ such that $q_{\pi(i)} = p_i$, $G_{\pi(i)}$ is a p_i-group, and $\#(G_{\pi(i)}) = p_i{}^{e_i}$. Let $H_{\pi(i)}$ be the subgroup of G which corresponds by the isomorphism θ to the subgroup $\{1\} \times \cdots \times G_{\pi(i)} \times \cdots \times \{1\}$; then $H_{\pi(i)}$ is a p_i-group contained in G, hence $H_{\pi(i)} \subseteq G_{(p_i)}$ and since $\#(H_{\pi(i)}) = \#(G_{(p_i)}) = p_i{}^{e_i}$ we have $H_{\pi(i)} = G_{(p_i)}$. ∎

Now, we are ready to prove the main uniqueness theorem:

Theorem 3. *Every finite Abelian group is isomorphic to a Cartesian product of cyclic groups with prime-power orders. Moreover, if $G \cong G_1 \times \cdots \times G_r \cong G_1' \times \cdots \times G_s'$ where each G_i, G_i' is a cyclic group of prime-power order, then $r = s$ and there exists a permutation π of $\{1, \ldots, r\}$ such that $G'_{\pi(i)} \cong G_i$.*
Proof: By Theorem 2, G is isomorphic to a Cartesian product of cyclic groups; by Theorem 1 G is isomorphic to a Cartesian product of cyclic groups of prime-power orders.

In view of (O), in order to prove the uniqueness of the decomposition, there is no loss of generality in assuming that the group itself is a p-group.

In fact, the p-groups appearing in any decomposition of G correspond to the primes dividing $\#(G) = m$, and for every such prime p the product of p-groups in the decomposition must be isomorphic to the subgroup $G_{(p)}$.

Thus, let G be a p-group,

$$G \cong G_1 \times \cdots \times G_r \cong G_1' \times \cdots \times G_s',$$

where G_i, G_i' are cyclic groups, with generators x_i, x_i' having orders $\#(G_i) = p^{e_i}$, $\#(G_i') = p^{f_i}$, respectively. There is no loss of generality if we assume that

$$e_1 \geqslant e_2 \geqslant \cdots \geqslant e_r, \qquad f_1 \geqslant f_2 \geqslant \cdots \geqslant f_s.$$

Let $G^* = \{x \in G \mid x^p = 1\}$, so G^* is a subgroup of G. From

$$G \cong G_1 \times \cdots \times G_r$$

it follows that $G^* \cong G_1^* \times \cdots \times G_r^*$ where G_i^* is the subgroup of G_i generated by $x_i^{p^{e_i-1}}$. Thus $\#(G^*) = \prod_{i=1}^r \#(G_i^*) = p^r$ since each G_i^* has order p. Similarly, from $G \cong G_1' \times \cdots \times G_s'$ we deduce that $\#(G^*) = p^s$; therefore $r = s$.

Now, let us prove that $e_1 = f_1, \ldots, e_r = f_r$. It is enough to show that $e_i \geqslant f_i$ for every $i = 1, \ldots, r$, since $\prod_{i=1}^r p^{e_i} = \#(G) = \prod_{i=1}^r p^f$.

If j is the smallest index such that $e_j < f_j$, let $G^{**} = \{x \in G \mid \text{there exists } y \in G \text{ such that } y^{p^{e_j}} = x\}$, then G^{**} is a subgroup of G. From $G \cong G_1 \times \cdots \times G_r$, it follows that $G^{**} \cong G_1^{**} \times \cdots \times G_{j-1}^{**}$ where each G_i^{**} is the cyclic group generated by $x_i^{p^{e_j}}$ $(i = 1, \ldots, j - 1)$. On the other hand, from $G \cong G_1' \times \cdots \times G_r'$ it follows that

$$G^{**} \cong G_1'^{**} \times \cdots \times G_{j-1}'^{**} \times G_j'^{**} \times \cdots$$

where each $G_i'^{**}$ is the cyclic group generated by

$$x_i'^{p^{e_j}} \ (i = 1, \ldots, j - 1, j, \ldots)$$

and certainly $G_j'^{**}$ is not trivial. By what we have just proved, the number of cyclic p-groups in any decomposition of G^{**} must be invariant, and so we have arrived at a contradiction. Thus $e_i \geqslant f_i$ for every $i = 1, \ldots, r$ showing the actual equality. ∎

EXERCISES

1. Show that for every natural number m there exists only finitely many numbers n such that $\varphi(n) = m$. In particular, find all integers n such that $\varphi(n) = 2$, $\varphi(n) = 3$, $\varphi(n) = 4$, $\varphi(n) = 6$.

2. Determine the positive integers such that
   (a) $\varphi(n) = (1/2)n$.
   (b) $\varphi(n) = n - 1$.
   (c) $\varphi(n) = \varphi(2n)$.
   (d) $\varphi(3n) = \varphi(4n) = \varphi(6n)$.
   (e) $\varphi(n) = 12$.
   (f) $\varphi(n)$ divides n.
   (g) $2\varphi(n)$ divides n.
   (h) $\varphi(n) \equiv 2$ (mod 4).

3. Show that if $d = gcd(m,n)$ then $\varphi(mn) = (d\varphi(m)\varphi(n))/\varphi(d)$.

4. Determine the sum of all integers m such that $1 \leqslant m < n$ and

$$gcd(m,n) = 1.$$

5. Prove that if d divides n then $\varphi(d)$ divides $\varphi(n)$.

6. Let p, q be distinct prime numbers. Prove that

$$p^{q-1} + q^{p-1} \equiv 1 \ (\text{mod } pq).$$

7. Let m, n be relatively prime positive integers. Prove that

$$m^{\varphi(n)} + n^{\varphi(m)} \equiv 1 \quad (\text{mod } mn).$$

8. Determine explicitly the multiplication table of the integers modulo 12; verify which residue classes are invertible and find their order in the multiplicative group $P(12)$.

9. Solve the following numerical congruences:
   (a) $3x \equiv 12$ (mod 17).
   (b) $4x \equiv 16$ (mod 57).
   (c) $5x \equiv 12$ (mod 18).
   (d) $20x \equiv 60$ (mod 80).

10. Solve the following system of congruences:

$$\begin{cases} x \equiv 1 \ (\text{mod } 2) \\ x \equiv 2 \ (\text{mod } 3) \\ x \equiv 3 \ (\text{mod } 5). \end{cases}$$

11. Solve the following system of congruences:

$$\begin{cases} 2x \equiv 5 \ (\text{mod } 7) \\ 4x \equiv 4 \ (\text{mod } 9) \\ 2x \equiv 6 \ (\text{mod } 25). \end{cases}$$

12. Solve the following system of congruences:

$$\begin{cases} x \equiv 1 \ (\text{mod } 2) \\ x \equiv 2 \ (\text{mod } 3) \\ x \equiv 3 \ (\text{mod } 4) \\ x \equiv 4 \ (\text{mod } 5). \end{cases}$$

13. Let $\{x_0, x_1, \ldots, x_{n-1}\}$ be a complete set of residues modulo $n > 1$, let a be an integer relatively prime to n and b any integer. Show that

$$\{ax_0 + b, ax_1 + b, \ldots, ax_{n-1} + b\}$$

is a complete set of residues modulo n.

14. Let m, n be relatively prime positive integers. Let $\{x_0, x_1, \ldots, x_{m-1}\}$, $\{y_0, y_1, \ldots, y_{n-1}\}$ be complete sets of residues modulo m and n respectively. Show that $\{nx_i + my_j \,|\, i = 0, 1, \ldots, m - 1; j = 0, 1, \ldots, n - 1\}$ is a complete set of residues modulo mn.

15. Let p be a prime number. Show that

$$1^n + 2^n + \cdots + (p-1)^n \equiv \begin{cases} -1 \ (\text{mod } p) & \text{when } p - 1 \text{ divides } n \\ 0 \ (\text{mod } p) & \text{when } p - 1 \text{ does not divide } n. \end{cases}$$

16. Let p be a prime number and let n be a natural number dividing $p - 1$. Show that the congruence $x^n \equiv 1 \ (\text{mod } p)$ has exactly n roots.

17. Prove that any integer

$$a = a_0 + a_1 \cdot 10 + a_2 \cdot 10^2 + \cdots + a_n \cdot 10^n$$

is congruent modulo 9 to the sum of its digits: $a \equiv a_0 + a_1 + \cdots + a_n$ (mod 9). Establish this as a particular case of a more general fact.

18. Prove *Wilson's theorem*: $(m - 1)! \equiv -1 \ (\text{mod } m)$ if and only if m is a prime number.

19. Show that if p is an odd prime number then

$$\left\{\left(\frac{p-1}{2}\right)!\right\}^2 \equiv (-1)^{(p+1)/2} \ (\text{mod } p).$$

20. Let $f \in \mathbb{Z}[X]$ be a polynomial of degree greater than 0. Show that there exist infinitely many integers n such that $f(n)$ is not a prime number.

21. Let $f \in \mathbb{Z}[X]$ have degree n. Assume that there exists an integer m such that the prime number p divides $f(m)$, $f(m + 1)$, $f(m + 2)$, $\ldots$, $f(m + n)$. Show that p divides $f(x)$ for every integer x.

22. Determine all the generators of the cylic groups $P(17)$, $P(31)$, $P(27)$.

23. Let q be equal to 2, 4, p^e or $2p^e$ (where p is an odd prime), and let a be an integer relatively prime to q. Show that the congruence $x^n \equiv a \pmod{q}$ has a solution if and only if $a^{\varphi(q)/d} \equiv 1 \pmod{q}$, where $d = gcd(n, \varphi(q))$.

24. Let p be a prime number, and a a positive integer such that p does not divide a. Show that if $a^p \equiv a \pmod{p^2}$ then $(a+p)^p \not\equiv a+p \pmod{p^2}$.

25. Show that there exist primitive roots r modulo p such that

$$r^{p-1} \not\equiv 1 \pmod{p^2}.$$

26. Let $p \neq 2$ be a prime number, and let r be a primitive root modulo p such that $r^{p-1} \not\equiv 1 \pmod{p^2}$. Show that $r \bmod p^m$ generates $P(p^m)$ for every $m > 1$.

27. Let $a, b \geqslant 3$ be relatively prime integers and b odd. Show that $P(ab)$ is not a cyclic group.

28. Let p be a prime number and e a positive integer dividing $p - 1$. Show that there are exactly $\varphi(e)$ residue classes modulo p having order e in the multiplicative group $P(p)$.

29. Let p be a prime number and n, r any natural numbers. Show that

$$\binom{p^r n}{p^r} \equiv n \pmod{pn}$$

$\left(\binom{a}{b} \text{ denotes the number of combinations of } a \text{ letters in groups of } b \text{ letters} \right)$.

Hint: Show that

$$\binom{p^r n - 1}{p^r - 1} = \prod_{k=1}^{p^r-1} \left(\frac{p^r n}{k} - 1 \right) \equiv (-1)^{p^r-1} \pmod{p}.$$

30. Let r be a given primitive root modulo p. If m is an integer, $1 \leqslant m \leqslant p - 1$, and $m \equiv r^a \pmod{p}$, where $0 \leqslant a \leqslant p - 1$, we say that a is the *index* of m with respect to r (modulo p), and we write $a = ind_r(m)$. Show that $ind_r(mn) \equiv ind_r(m) + ind_r(n) \pmod{p-1}$.

31. Let r be the smallest primitive root modulo 29; determine the indices of the prime residue classes modulo 29; with respect to r.

32. Using the table of indices, compute the least positive residue modulo 17 of the following integers:
(a) $a = 432 \times 8328$.
(b) $b = 38^{919}$.
(c) $c = ((3^3)^3)^3$.

33. Using the table of indices, compute the least positive residue modulo 29 of the following integers:
  (a) $a = 583 \times 1875$.
  (b) $b = 1051^{875}$.
  (c) $c = (5^5)^5$.

34. Using the table of indices, solve the congruence:

$$25x \equiv 15 \ (\mathrm{mod}\ 29).$$

35. Let $a_1, \ldots, a_r$ be pairwise relatively prime positive integers, $a = \prod_{i=1}^r a_i$, $t = lcm\{\varphi\,|a_1), \ldots, \varphi(a_r)\}$. Show that if $b \in P(a)$ then $b^t \equiv 1 \ (\mathrm{mod}\ a)$.

36. Compute the 4 last digits of:
  (a) $((9^9)^9)^9$.
  (b) $9^{(9^{(9^9)})}$.
  Hint: For (b) use the previous exercise.

37. Let p be a prime number, $r \geqslant 2$ an integer, $f \in \mathbb{Z}[X]$ and assume that $f(a) \equiv 0 \ (\mathrm{mod}\ p^{r-1})$, where $0 \leqslant a \leqslant p^{r-1} - 1$. Let $f' \in \mathbb{Z}[X]$ be the derivative of f. Show that:
  (a) if $f'(a) \not\equiv 0 \ (\mathrm{mod}\ p)$ there exists a unique integer b such that $f(b) \equiv 0$ $(\mathrm{mod}\ p^r)$ and $0 \leqslant b \leqslant p^r - 1$; moreover, $b \equiv a \ (\mathrm{mod}\ p^r)$.
  (b) if $f'(a) \equiv 0 \ (\mathrm{mod}\ p)$ then there are p integers $b_0, \ldots, b_{p-1}$ such that $f(b_i) \equiv 0 \ (\mathrm{mod}\ p^r)$, $0 \leqslant b_i \leqslant p^r - 1$, when $f(a) \equiv 0 \ (\mathrm{mod}\ p^r)$ and these integers satisfy $b_i \equiv a \ (\mathrm{mod}\ p^r)$; on the other hand, if $f(a) \not\equiv 0$ $(\mathrm{mod}\ p^r)$ there exists no integer b such that $f(b) \equiv 0 \ (\mathrm{mod}\ p^r)$.

38. Solve the congruences
  (a) $x^3 + 5x - 8 \equiv 0 \ (\mathrm{mod}\ 5^2)$.
  (b) $x^3 - 3x + 2 \equiv 0 \ (\mathrm{mod}\ 245)$.

39. Find the decomposition into irreducible factors of the following polynomials:
  (a) $X^4 + 2X - 3 \in \mathbb{F}_5[X]$.
  (b) $X^4 + 3X^3 - 2X^2 - 9X - 3 \in \mathbb{F}_{11}[X]$.
  (c) $X^4 - 3X^3 - 4X^2 + X - 4 \in \mathbb{F}_7[X]$.

40. Any mapping f from the set of positive integers with values in a domain R is called an *arithmetical function*. Moreover, if $f(mn) = f(m) \cdot f(n)$ when m, n are relatively prime, then f is said to be a *multiplicative function*. Show that if f is multiplicative, if g is defined by $g(n) = \Sigma_{d|n} f(d)$, then g is also multiplicative.

41. Let $\sigma(n)$ denote the sum of positive divisors of $n \geqslant 1$, and let $\tau(n)$ denote the number of positive divisors of n. Show that σ, τ are multiplicative functions.
  Hint: Use the preceding exercise.

42. If $n = \prod_{i=1}^{r} p_i^{e_i}$ is the prime decomposition of n, show that

$$\sigma(n) = \prod_{i=1}^{r} \frac{p_i^{e_i+1} - 1}{p_i - 1},$$

$$\tau(n) = \prod_{i=1}^{r} (e_i + 1).$$

43. Show that $\sum_{d|n} [\tau(d)]^3 = [\sum_{d|n} \tau(d)]^2$.

44. Let R be a domain, let A be the set of all arithmetical functions with values in R; that is, the set of sequences $s = (s_1, s_2, \ldots, s_n, \ldots)$ of elements of R. On the set A we define the following operations:

$$(s_n) + (s_n') = (s_n + s_n'),$$

and

$$(s_n) * (s_n') = (t_n) \quad \text{where} \quad t_n = \sum_{dd'=n} s_d s_{d'}'.$$

Besides, if $r \in R$, $(s_n) \in A$ we define a scalar multiplication as

$$r(s_n) = (rs_n).$$

Show:

(a) A is a commutative ring and also an R-module; the zero element is $0 = (0,0, \ldots)$, the unit element is $e = (1,0,0, \ldots)$.

(b) For every $s = (s_n)$, $s \neq 0$, let $\pi(s)$ be the smallest integer n such that $s_n \neq 0$; let $\pi(0) = \infty$, where $\infty \cdot n = \infty \cdot \infty = \infty$ for every integer $n > 0$. Show that $\pi(s * s') = \pi(s) \cdot \pi(s')$, hence A is a domain.

(c) $s \in A$ is invertible in A if and only if $\pi(s) = 1$ and s_1 is invertible in the ring R.

45. Let $u = (1,1,1, \ldots) \in A$ and let $\mu \in A$ be its inverse; that is, $u * \mu = e$. The arithmetical function μ is called the *Möbius function*. Prove that $\mu(1) = 1$, $\mu(n) = 0$ when some square divides n, and $\mu(n) = (-1)^r$ when

$$n = p_1 p_2 \cdot \ldots \cdot p_r$$

(product of r distinct primes); deduce that μ is a multiplicative function.

46. Show Möbius inversion formula: If $s, s' \in A$ then $u * s = s'$ if and only if $s = \mu * s'$. As an application, prove the following relations:

(a) $\sum_{d|n} \mu(d) = \begin{cases} 1 & \text{when } n = 1 \\ 0 & \text{when } n \neq 1. \end{cases}$

(b) $\varphi(n) = \sum_{d|n} \mu\left(\frac{n}{d}\right) \cdot d$.

(c) $n = \sum_{d|n} \sigma(d)\mu\left(\frac{n}{d}\right)$.

(d) $1 = \sum_{d|n} \tau(d)\mu\left(\frac{n}{d}\right)$.

47. Prove:
   (a) $\tau(n)$ is odd if and only if n is a square.
   (b) $\tau(2^n - 1) \geqslant \tau(n)$ when $n \geqslant 1$.
   (c) $\tau(2^n + 1)$ is at least equal to the number of odd positive divisors of n.

48. Show that if $d = gcd(m,n)$ then

$$\sigma(m)\sigma(n) = \sum_{t|d} t\sigma\left(\frac{mn}{t^2}\right).$$

Hint: First consider the case where m, n are powers of the same prime number.

49. Prove that if $m > 0$, $n > 1$ then

$$\frac{\sigma(m)}{m} < \frac{\sigma(mn)}{mn} \leqslant \frac{\sigma(m)\sigma(n)}{mn}.$$

50. Prove that n is equal to the product of its proper positive divisors if and only if $n = p^3$ or $n = p_1 p_2$, $p_1 \neq p_2$ (where p, p_1, p_2 are prime numbers).

51. Let Λ be the *Mangoldt arithmetical function* defined as follows:
$\begin{cases} \Lambda(n) = \log p \text{ when } n \text{ is a power of some number } p. \\ \Lambda(n) = 0, \text{ otherwise.} \end{cases}$
   Prove:

   (a) $\log n = \sum_{d|n} \Lambda(d)$.

   (b) $\Lambda(n) = \sum_{d|n} \mu(d)\log\left(\frac{n}{d}\right) = -\sum_{d|n} \mu(d)\log d$.

52. Let λ be the *Liouville arithmetical function*, defined as follows (for $n = \prod_{i=1}^{r} p_i^{e_i}$, each p_i prime number, $e_i \geqslant 1$):

$$\lambda(n) = \begin{cases} 1 & \text{when } \sum_{i=1}^{r} e_i \text{ is even} \\ -1 & \text{when } \sum_{i=1}^{r} e_i \text{ is odd.} \end{cases}$$

Prove:

(a) $\lambda(mn) = \lambda(m)\lambda(n)$ for any positive integers m, n.

(b) $\displaystyle\sum_{d|n} \lambda(d) = \begin{cases} 1 & \text{when } n \text{ is a square} \\ 0 & \text{otherwise.} \end{cases}$

53. For every real number x let $[x]$ denote the unique integer such that

$$[x] \leqslant x < [x] + 1.$$

Prove:
(a) $[x] + [y] \leqslant [x + y]$.
(b) $[x/n] = [[x]/n]$ for every positive integer n.
(c) the number of multiples of n which do not exceed x is $[x/n]$.
(d) $[x] + [y] + [x + y] \leqslant [2x] + [2y]$.

54. Let f, g be arithmetical functions such that $f(n) = \sum_{d|n} g(d)$. Show that

$$\sum_{m=1}^{n} f(m) = \sum_{m=1}^{n} [n/m]g(m).$$

55. Prove that

$$\sum_{m=1}^{n} \left[\frac{n}{m}\right] \varphi(m) = n(n + 1)/2.$$

56. Prove that if $n \geqslant 1$ then $\sum_{m=1}^{n} \tau(m) = \sum_{m=1}^{n} [n/m]$.

57. Let $n \geqslant 1$ and $k = [\sqrt{n}]$. Show that

$$\sum_{m=1}^{n} \tau(m) = 2\left(\sum_{m=1}^{k} [n/m]\right) - k^2.$$

58. The numbers $F_n = 2^{2^n} + 1$ ($n \geqslant 0$ integer) are called *Fermat numbers*. Prove:
(a) if $r > 0$ then F_n divides $F_{n+r} - 2$.
(b) any two Fermat numbers are relatively prime.
(c) if $a \geqslant 2$ and $a^r + 1$ is prime then a is even and n is a power of 2.
(d) 641 divides F_5 (Euler).

We note in this respect that the only Fermat numbers which are known to be prime are F_0, F_1, F_2, F_3, F_4. On the other hand, it has been shown that F_n is not prime when $5 \leqslant n \leqslant 16$, and $n = 18, 19, 21, 23, 25, 26, 27, 30, 32, 36, 38, 39, 42, 55, 58, 63, 73, 77, 81, 117, 125, 144, 150, 207, 226, 228, 250, 267, 268, 284, 316, 452, 1945$.* It is an open question whether the number of Fermat primes is finite.

* Wrathall, C. P., *Math. of Comp.*, **18**, 1964, 324–325.

59. A natural number n is said to be *perfect* when n is equal to the sum of its proper divisors. Prove: n is an even perfect number if and only if

$$n = 2^{p-1}(2^p - 1),$$

where p, $2^p - 1$ are primes. Give examples of even perfect numbers.

Note: It is not known whether there exists any odd perfect number. Any such number must have at least 6 distinct prime factors and be greater than 10^{36}.[*] Any prime q of the type $q = 2^p - 1$ (where p is prime) is called a *Mersenne prime*. The known Mersenne primes for $p < 21000$ are those for which p is equal to 2, 3, 5, 7, 13, 17, 19, 31, 61, 89, 107, 127, 521, 607, 1279, 2203, 2281, 3217, 4253, 4423, 9689, 9941, 11213, 19937.[‡] It is not yet known whether there are infinitely many Mersenne primes; equivalently, it is not known whether there are infinitely many even perfect numbers.

60. Show that if n is odd and has at most 2 distinct prime factors, then $\sigma(n) < 2n$, hence n is not a perfect number.

61. Let n be an odd perfect number. Prove that $n = p^r m^2$, where p is a prime number not dividing m and $p \equiv 1 \pmod 4$, $r \equiv 1 \pmod 4$. On the other hand, given m, there is at most one odd perfect number of the type $p^r m^2$, with p prime not dividing m.

62. Prove that the nth cyclotomic polynomial is expressible directly by means of the Möbius function as follows:

$$\Phi_n = \prod_{d \mid n} (X^d - 1)^{\mu(n/d)}.$$

63. Let f, g, h be arithmetical functions, $f * g = h$ and assume that h is multiplicative. Prove that f is multiplicative if and only if g is multiplicative. As an application deduce anew that if $g(n) = \Sigma_{d \mid n} f(d)$ and g is multiplicative then so is f.

64. If R is a unique factorization domain, show that every arithmetical function with values in R is the product of a finite number of "prime" arithmetical functions.

65. Show that the ring of arithmetical functions with values in the domain R is isomorphic to the ring of *unrestricted formal power series*
$S = R[[[X_1, \ldots, X_n, \ldots]]]$. Explicitly, S consists of all countable infinite formal sums of monomials in the variables X_i with coefficients in R. The addition in S is componentwise, while the multiplication follows the same pattern as for polynomials (note that for every monomial $m \in S$ there exist only finitely many monomials m', $m'' \in S$ such that $m'm'' = m$).

---

[*] Tuckerman, B., *I.B.M. Research Reports*, 1971.
[‡] Gillies, D. B., *Math. of Comp.*, **18**, 1964, 93–98.

Note: a recent result of Cashwell and Everett* establishes that if R is a field then the ring of arithmetical functions with values in this field is a unique factorization domain.

66. Let G be an Abelian group of order m and assume that for every prime p dividing m, G has exactly $p - 1$ elements of order p. Show that G is a cyclic group.

67. Show that a finite Abelian group G is cyclic if and only if

$$G \cong \mathbb{Z}/p_1^{e_1} \times \cdots \times \mathbb{Z}/p_r^{e_r}$$

where $p_1, \ldots, p_r$ are distinct prime numbers.

68. We say that a finite Abelian group G is *indecomposable* if it is not possible to write $G \cong G_1 \times G_2$, where G_1, G_2 are Abelian groups of strictly smaller order. Show that G is indecomposable if and only if it is cyclic of prime-power order.

69. Show that the number of pairwise non-isomorphic Abelian groups of order $m = \prod_{i=1}^{r} p_i^{e_i}$ is $\prod_{i=1}^{r} \pi(e_i)$, where we define $\pi(e)$ as follows: it is the number of non-increasing sequences of integers $n_1 \geqslant n_2 \geqslant \cdots \geqslant n_j > 0$ such that $\sum_{i=1}^{j} n_i = e$.

70. Determine the number of pairwise non-isomorphic Abelian groups of order 8, 16200.

71. Let $G \cong \mathbb{Z}/p^{e_1} \times \mathbb{Z}/p^{e_2} \times \cdots \times \mathbb{Z}/p^{e_k}$ where $e_1 \geqslant e_2 \geqslant \cdots \geqslant e_k > 0$. Let $G_r = \{x \in G \mid p^r x = 0\}$. Show that G_r is a subgroup of G having order

$$p^{ri + e_{i+1} + \cdots + e_k}$$

where i is the unique index such that $e_i \geqslant r > e_{i+1}$ (with the convention that $e_0 = \sum_{j=1}^{k} e_j$, $e_{k+1} = 0$).

72. Let G be as in the previous exercise. Show that the number of elements of order p^r in G is equal to

$$p^{ri + e_{i+1} + \cdots + e_k} - p^{(r-1)j + e_{j+1} + \cdots + e_k},$$

where i, j are such that $e_i \geqslant r > e_{i+1}$, $e_j \geqslant r - 1 > e_{j+1}$ (with the same convention as in the previous exercise). As a corollary, show that G has $p^k - 1$ elements of order p; thus G is cyclic if and only if it has $p - 1$ elements of order p.

* "The Ring of Number Theoretic Functions," *Pacific J. Math.*, **9**, 1959, 975–985.

73. Let G be an Abelian group such that every non-zero element has order p. Show that G is a vector space over the field $\mathbb{F}_p$. If G is finite, show that $G \cong \mathbb{Z}/p \times \cdots \times \mathbb{Z}/p$ (for a finite number of copies of $\mathbb{Z}/p$). Such Abelian groups are called *elementary Abelian p-groups*.

74. Show that if G is an elementary Abelian p-group of order p^n and $1 \leqslant r \leqslant n$ then the number of distinct subgroups of order p^r of G is

$$\frac{(p^n - 1)(p^n - p)(p^n - p^2), \ldots, (p^n - p^{r-1})}{(p^r - 1)(p^r - p)(p^r - p^2), \ldots, (p^r - p^{r-1})}.$$

In particular, there are $1 + p + p^2 + \cdots + p^{n-1}$ subgroups of order p^{n-1} or also of order p. Moreover, the number of subgroups of order p^r is the same as the number of subgroups of order p^{n-r}.

CHAPTER 4

Quadratic Residues

In this section we investigate the following question. Let $m > 1$ and a be an integer relatively prime to m. When is the residue class $\bar{a}$ a square in the multiplicative group $P(m)$? In other words, when does there exist an integer x such that $x^2 = a \pmod{m}$?

Definition 1. If $m > 1$ and a are integers, and $gcd(a,m) = 1$, we say that a is *a quadratic residue modulo m* when $\bar{a}$ is a square in $P(m)$. Otherwise, we say that a is *a non-quadratic residue modulo m*.

The first results will reduce the problem to that of finding the quadratic residues modulo an odd prime p or 4 or 8.

A. *If $m = p_1^{e_1} \ldots p_r^{e_r}$ is the prime-power decomposition of m, if a is an integer relatively prime to m, then a is a quadratic residue modulo m if and only if it is a quadratic residue modulo $p_i^{e_i}$ (for all $i = 1, \ldots, r$).*

Proof: By Chapter 3, **(F)**, $P(m) \cong \prod_{i=1}^{r} P(p_i^{e_i})$.

An element of a Cartesian product of groups is a square if and only if its components are squares. If $\bar{a} \in P(m)$, its component in the group $P(p_i^{e_i})$, by the above isomorphism, is the residue class of a modulo $p_i^{e_i}$. Thus a is a quadratic residue modulo m if and only if it is a quadratic residue modulo $p_i^{e_i}$ (for all $i = 1, \ldots, r$). ∎

The study of squares in $P(p^e)$ will now be reduced to that of squares in $P(p)$, when $p \neq 2$:

B. *Let p be an odd prime, $e > 1$, let w_0 be a primitive root modulo p. If a is an integer prime to p, the following conditions are equivalent:*
(1) *a is a quadratic residue modulo p^e.*
(2) *a is a quadratic residue modulo p.*
(3) *$a \equiv w_0^t \pmod{p}$ where t is even.*

Proof: $1 \to 2$. Let x be an integer such that $a \equiv x^2 \pmod{p^e}$. Then $a \equiv x^2 \pmod{p}$.

$2 \to 3$ From $a \equiv x^2 \pmod{p}$ and $a \equiv w_0^t \pmod{p}$, $x \equiv w_0^u \pmod{p}$ it follows that $t \equiv 2u \pmod{p-1}$ because the group $P(p)$ has order $p-1$. Thus t is even.

$3 \rightarrow 1$ By Chapter 3, (**J**) $P(p^e)$ is the cyclic group generated by $\bar{w}(\overline{1+p})$ where $w = w_0^{p^{e-1}}$. Hence $a \equiv w^s(1 + p)^s \pmod{p^e}$. Since $w_0^{p^{e-1}} \equiv w_0 \pmod{p}$ and $(1 + p)^s \equiv 1 \pmod{p}$ then $a \equiv w_0{}^s \pmod{p}$. By hypothesis $q \equiv w_0{}^t \pmod{p}$, hence $s \equiv t \pmod{p - 1}$ and from t even we deduce that s is even, say, $s = 2u$. We conclude that $a \equiv [w^u(1 + p)^u]^2 \pmod{p^e}$. ∎

For $p = 2$, we have:

C. *Let a be an odd integer. Then:*
  (1) *a is a quadratic residue modulo 4 if and only if $a \equiv 1 \pmod 4$.*
  (2) *a is a quadratic residue modulo 8 if and only if $a \equiv 1 \pmod 8$.*
  (3) *a is a quadratic residue modulo 2^e (where $e > 3$) if and only if a is a quadratic residue modulo 8.*
Proof: Since $P(4) = \{\bar{1}, \bar{3}\}$, $P(8) = \{\bar{1}, \bar{3}, \bar{5}, \bar{7}\}$ then the only square in $P(4)$, $P(8)$ is the residue class of 1.

Let $e > 3$. If a is an odd integer, by Chapter 3, (**K**) we may write

$$a \equiv (-1)^{e'} 5^{e''} \pmod{2^e}$$

where $e' \in \{0,1\}$, $0 \leqslant e'' < 2^{e-2}$.

If x is an integer such that $x^2 \equiv a \pmod{2^e}$, letting $x \equiv (-1)^{f'} 5^{f''} \pmod{2^e}$, where $f' \in \{0,1\}$, $0 \leqslant f'' < 2^{e-2}$, it follows that $2f' \equiv e' \pmod 2$ and $2f'' \equiv e'' \pmod{2^{2-e}}$. These congruences have a solution if and only if e', e'' are even, that is $a \equiv 5^{e''} \pmod{2^e}$, where e'' is even. This is equivalent to $a \equiv 1 \pmod 8$. ∎

Let us note here that in $P(2^e)$ there are exactly 2^{e-3} squares, hence $2^{e-1} - 2^{e-3} = 3 . 2^{e-3}$ non-squares.

Putting together these results, we have:

D. *Let $m > 1$, a be relatively prime integers. Then a is a quadratic residue modulo m if and only if: (1) a is a quadratic residue modulo p, for every odd prime p dividing m;*
  (2) *$a \equiv 1 \pmod 4$ if 4 divides m but 8 does not divide m;*
  (3) *$a \equiv 1 \pmod 8$ if 8 divides m.*
Proof: This results immediately from (**A**), (**B**), and (**C**). ∎

In order to determine the quadratic residues modulo p, we introduce the following terminology:

Definition 2. Let p be an odd prime, let a be a non-zero integer not a multiple of p. We define the *Legendre symbol* (a/p) of a, relative to p, as follows:

$$(a/p) = \begin{cases} 1 \text{ when } a \text{ is a quadratic residue modulo } p \\ -1 \text{ when } a \text{ is a non-quadratic residue modulo } p \end{cases}$$

E. *The Legendre symbol has the following properties:*
(1) *if $a \equiv b \pmod{p}$ then $(a/p) = (b/p)$.*
(2) $(ab/p) = (a/p)(b/p)$.

Proof: The first assertion is immediate.

Let $\bar{w}$ be a primitive root modulo p. If a,b are integers, not multiples of p, we may write $a \equiv w^r \pmod{p}$, $b \equiv w^s \pmod{p}$ where $0 \leqslant r,\, s < p - 1$. By **(B)**, we have $(a/p) = 1$ if and only if r is even, and similarly $(b/p) = 1$ when s is even. Since $ab \equiv w^t \pmod{p}$, with $t \equiv r + s \pmod{p - 1}$ then t is even if and only if r,s have the same parity. This proves the second assertion. ∎

F. *For every odd prime p, there are as many square residue classes as there are non-square residue classes modulo p^e, where $e \geqslant 1$.*

Proof: First we assume $e = 1$. Consider the mapping $\sigma: P(p) \to P(p)^2$, defined by $\sigma(\bar{x}) = \bar{x}^2$. Then $\sigma(\bar{x}) = \sigma(\bar{y})$ if and only if $\bar{x} = \bar{y}$ or $\bar{x} = -\bar{y}$, because $P(p) = \mathbb{F}_p^{*}$. Since $p \neq 2$ this shows that $P(p)$ has twice as many elements as $P(p)^2$, so there are as many square as non-square residue classes modulo p.

If $e > 1$ we consider the group-homomorphism $f: P(p^e) \to P(p)$, defined by $f(\bar{a}) = \bar{\bar{a}}$, where $\bar{a}$ is the residue class of a modulo p^e and $\bar{\bar{a}}$ is the residue class of a modulo p. By **(B)** $\bar{a}$ is a square if and only if $\bar{\bar{a}}$ is a square. Therefore $P(p^e)$ has as many squares as non-squares. ∎

Let us note that the above result does not hold for $p = 2$ (as we have already remarked) as well as for products of different primes (for example, when $m = 15$ there are only 2 square residue classes and 6 non-square residue classes modulo 15).

We may find whether an integer is a quadratic residue modulo p by explicit determination of the multiplication in the group $P(p)$ or by first determining a primitive root modulo p. For large primes this may be rather involved. We shall be interested in simpler methods.

G. (Euler's criterion). *Let p be an odd prime, let a be an integer not a multiple of p. Then*

$$(a/p) \equiv a^{(p-1)/2} \pmod{p}.$$

Proof: Let $a \equiv w^t \pmod{p}$, where w is a primitive root modulo p and $0 \leqslant t < p - 1$. Since $\bar{w}$ is not a square in $P(p)$ (as follows from **(B)**) and $w^{(p-1)/2} \equiv -1 \pmod{p}$ we have

$$\left(\frac{a}{p}\right) = \left(\frac{w^t}{p}\right) = \left(\frac{w}{p}\right)^t = (-1)^t \equiv (w^{(p-1)/2})^t = (w^t)^{(p-1)/2}$$

$$\equiv a^{(p-1)/2} \pmod{p}. \quad \blacksquare$$

As a corollary, we deduce:

H. -1 *is a square modulo p if and only if* $p \equiv 1$ (mod 4).
Proof: $(-1/p) \equiv (-1)^{(p-1)/2}$ (mod p)   implies   the   equality $(-1/p) = (-1)^{(p-1)/2}$ (since these integers are either 1 or -1). So $(-1/p) = 1$ exactly when $p \equiv 1$ (mod 4). ∎

If p is large Euler's criterion is not convenient, since it gives rise to lengthy computations.

A better criterion is due to Gauss. If p does not divide a there exists a unique integer s, $1 \leqslant s \leqslant (p-1)/2$ such that $a \equiv s$ (mod p) or $a \equiv -s$ (mod p).

I.   (Gauss' criterion). *Let p be an odd prime, let a be an integer, not a multiple of p, and let v be the number of elements ka in the set*

$$\left\{ a, 2a, \ldots, \frac{p-1}{2} a \right\}$$

such that $ka \equiv -s$ (mod p), *where* $0 < s \leqslant (p-1)/2$. *Then* $(a/p) = (-1)^v$.
Proof: If $1 \leqslant k < k' \leqslant (p-1)/2$ then $ka \not\equiv k'a$ (mod p), otherwise p would divide $k - k'$. Also, $ka \not\equiv -k'a$ (mod p), because p does not divide $k + k'$.

Thus, all integers $s = 1, \ldots, (p-1)/2$ are such that s or $-s$ is congruent modulo p to some multiple $ka(1 \leqslant k \leqslant (p-1)/2)$. Taking into account the definition of v we have:

$$a. \, 2a. \, \ldots . \frac{p-1}{2} a \equiv (-1)^v . \, 1. \, 2. \, \ldots . \frac{p-1}{2} \text{ (mod } p).$$

We deduce that $a^{(p-1)/2} \equiv (-1)^v$ (mod p). By **(G)**, we conclude that $(a/p) = (-1)^v$. ∎

We can use this criterion to determine when 2 is a quadratic residue:

J.   2 *is a quadratic residue modulo p if and only if* $p \equiv \pm 1$ (mod 8); *explicitly* $(2/p) = (-1)^{(p^2-1)/8}$.
Proof: We apply Gauss' criterion. Among the integers $2, 4, \ldots, p-1$, those satisfying $p/2 < 2k \leqslant p-1$ are the ones such that $2k \equiv -s$ (mod p), with $0 < s \leqslant (p-1)/2$. Their number v is equal to the number of integers k, such that $p/4 < k \leqslant (p-1)/2$. If $p \equiv 1$ (mod 4) there are $(p-1)/2 - (p+3)/4 + 1 = (p-1)/4$ such integers. If $p \equiv -1$ (mod 4), there are $(p-1)/2 - (p+1)/4 + 1 = (p+1)/4$ such integers.

Now, if $p \equiv 1$ (mod 8) then $(p-1)/4$ is even, if $p \equiv 5$ (mod 8) then $(p-1)/4$ is odd, if $p \equiv 3$ (mod 8), then $(p+1)/4$ is odd and if $p \equiv 7$ (mod 8) then $(p+1)/4$ is even. In conclusion, $(2/p) = 1$ if and only if $p \equiv \pm 1$ (mod 8).

The last expression is obvious, if we note that $(p^2 - 1)/8$ is even exactly when $p \equiv \pm 1 \pmod 8$. ∎

The preceding criterion leads also to a large number of divisions, whenever p is a large prime.

We shall indicate now a relationship between Legendre symbols relative to different primes; this will be used later as the basis for a very satisfactory method of computation of the Legendre symbol.

Theorem 1. (Gauss quadratic reciprocity law). *If p, q are distinct odd primes, then*

$$(p/q)(q/p) = (-1)^{\frac{p-1}{2} \cdot \frac{q-1}{2}}$$

Proof: By **(I)**, we have $(q/p) = (-1)^\nu$ where ν is the number of integers x, $1 \leqslant x \leqslant (p-1)/2$, such that $qx = py + r$ where $-p/2 < r < 0$, and y is an integer.

We must have $1 \leqslant y \leqslant (q-1)/2$, because y is not negative and

$$py = xq - r < \frac{p-1}{2}q + \frac{p}{2} < \frac{p}{2}(q+1)$$

hence $y < (q+1)/2$, so $y \leqslant (q-1)/2$.

Similarly $(p/q) = (-1)^\mu$, where μ is the number of integers y, $1 \leqslant y \leqslant (q-1)/2$, such that $py = qx + s$ where $-q/2 < s < 0$ and x is an integer; again, we have $1 \leqslant x \leqslant (p-1)/2$.

Therefore $(q/p)(p/q) = (-1)^{\nu+\mu}$.

We observe that $\nu + \mu$ is the number of pairs of integers (x,y) such that $1 \leqslant x \leqslant (p-1)/2$, $1 \leqslant y \leqslant (q-1)/2$ and $-p/2 < qx - py < q/2$.

Let us consider the following sets of pairs of integers:

$$S = \{(x,y) \mid 1 \leqslant x \leqslant (p-1)/2, 1 \leqslant y \leqslant (q-1)/2\}.$$
$$S_1 = \{(x,y) \in S \mid qx - py \leqslant -p/2\}.$$
$$S_0 = \{(x,y) \in S \mid -p/2 < qx - py < q/2\}.$$
$$S_1' = \{(x,y) \in S \mid q/2 \leqslant qx - py\}.$$

The mapping $\theta : S \to S$, defined by $\theta(x,y) = (x',y')$ where $x' = (p+1)/2 - x$, $y' = (q+1)/2 - y$ has the following properties (which are easy to verify): θ is a one-to-one mapping from S onto S, θ^2 is the identity mapping, $\theta(S_1) = S_1'$, $\theta(S_1') = S_1$, so that $\theta(S_0) = S_0$.

Therefore, $\#(S) = \#(S_1) + \#(S_0) + \#(S_1') \equiv \#(S_0) \pmod 2$ so

$$\frac{p-1}{2} \cdot \frac{q-1}{2} \equiv \nu + \mu \pmod 2.$$

Thus,

$$(p/q) \cdot (q/p) = (-1)^{\frac{p-1}{2} \cdot \frac{q-1}{2}}. \quad ∎$$

We may also rewrite the above relation as follows:

$$(p/q) = (q/p)(-1)^{\frac{p-1}{2} \cdot \frac{q-1}{2}}.$$

This form is obtained by noting that $(q/p)^2 = 1$.

The following corollary is immediate: *if p,q are distinct odd primes and p or q is congruent to 1 modulo 4 then* $(p/q) = (q/p)$. *Otherwise,* $(p/q) = -(q/p)$.

We have now a very effective method of computation of the Legendre symbol (a/p), where $p \neq 2$. Indeed, if a is a non-zero integer not multiple of p, $a = (-1)^d 2^e \prod_{i=1}^{r} q_i^{e_i}$ where $d \in \{0,1\}$, $e \geqslant 0$, $e_i \geqslant 1$, and each q_i is an odd prime distinct from p, then by (E) we have

$$(a/p) = (-1/p)^d (2/p)^e \prod_{i=1}^{r} (q_i/p)^{e_i}.$$

Using (H), (J), we have only to compute (q_i/p); this may be done by application of (E) and successive reductions using Gauss' reciprocity law.

We illustrate the method by means of a numerical example. Let $p = 2311$, $a = 1965 = 3 \times 5 \times 131$. Then

$$\left(\frac{1965}{2311}\right) = \left(\frac{3}{2311}\right)\left(\frac{5}{2311}\right)\left(\frac{131}{2311}\right).$$

$$\left(\frac{3}{2311}\right) = \left(\frac{2311}{3}\right)(-1)^{1 \times 1155} = -\left(\frac{2311}{3}\right) = -\left(\frac{1}{3}\right) = -1$$

where we used Gauss' reciprocity law and the congruence $2311 \equiv 1 \pmod 3$.

$$\left(\frac{5}{2311}\right) = \left(\frac{2311}{5}\right)(-1)^{2 \times 1155} = \left(\frac{2311}{5}\right) = \left(\frac{1}{5}\right) = 1$$

by using the reciprocity law and the congruence $2311 \equiv 1 \pmod 5$.

$$\left(\frac{131}{2311}\right) = \left(\frac{2311}{131}\right)(-1)^{65 \times 1155} = -\left(\frac{2311}{131}\right)$$

$$= -\left(\frac{84}{131}\right) = -\left(\frac{2}{131}\right)^2 \cdot \left(\frac{3}{131}\right) \cdot \left(\frac{7}{131}\right)$$

$$= -\left(\frac{3}{131}\right) \cdot \left(\frac{7}{131}\right) = -\left(\frac{131}{3}\right)\left(\frac{131}{7}\right)(-1)^{65 \times 1}(-1)^{65 \times 3}$$

$$= -\left(\frac{131}{3}\right)\left(\frac{131}{7}\right) = -\left(\frac{2}{3}\right)\left(\frac{5}{7}\right)$$

$$= -(-1)\left(\frac{7}{5}\right)(-1)^{2 \times 3} = \left(\frac{7}{5}\right) = \left(\frac{2}{5}\right) = -1$$

using several times the reciprocity law, the congruences $2311 \equiv 84 \pmod{131}$, $131 \equiv 2 \pmod 3$, $131 \equiv 5 \pmod 7$ $7 \equiv 2 \pmod 5$, and the fact that 3, 5 are not congruent to ± 1 modulo 8. Therefore $(1965/2311) = 1$, that is, 1965 is a square modulo 2311.

Recapitulating our results, we arrive at the following interesting observation. The fact that -1 is a quadratic residue modulo p depends only on the residue class of p modulo 4. For 2, it depends on the residue class of p modulo 8. Finally, for an odd prime q, $q \neq p$, it depends only on the residue class of p modulo $4q$.

This is easily seen, for if p' is a prime of the form $p' = p + 4kq$ (k integer) then

$$\left(\frac{q}{p'}\right) = \left(\frac{p'}{q}\right)(-1)^{\frac{q-1}{2} \cdot \frac{p+4kq-1}{2}}$$

$$= \left(\frac{p}{q}\right)(-1)^{\frac{q-1}{2} \cdot \frac{p-1}{2}}(-1)^{\frac{q-1}{2} \cdot 2kq}$$

$$= \left(\frac{p}{q}\right)(-1)^{\frac{q-1}{2} \cdot \frac{p-1}{2}} = \left(\frac{q}{p}\right).$$

We may now consider the following inverse problem. Given -1, or 2, or an odd prime q, how many odd primes p does there exist such that -1, 2 or q is a quadratic residue modulo p?

We have:

$(-1/p) = 1$ if and only if p is a prime in the arithmetic progression $\{1, 5, 9, \ldots, 4n + 1, \ldots\}$.

$(2/p) = 1$ if and only if p is a prime in one of the arithmetic progressions $\{1, 9, 17, \ldots, 8n + 1, \ldots\}$ or $\{7, 15, \ldots, 8n - 1, \ldots\}$.

Similarly, let p_0 be an odd prime dividing $q - 1$ or $q - 4$ (if $q - 1$ is a power of 2) then $(q/p_0) = (1/p_0) = 1$, or $(q/p_0) = (4/p_0) = 1$; therefore there exists at least one prime p_0 such that q is a quadratic residue modulo p_0; then for every prime p in the arithmetical progression $\{p_0, p_0 + 4q, p_0 + 8q, \ldots, p_0 + 4nq, \ldots\}$ we have $(q/p) = 1$.

Dirichlet's theorem on arithmetic progressions states:

In any arithmetical progression $\{k, k + m, k + 2m, \ldots, k + nm, \ldots\}$ where $0 < k \leqslant m$, and k, m are relatively prime, there exist infinitely many prime numbers.

We shall omit the proof of this remarkable theorem (see [3], [9]).

Applying this theorem, we obtain at once the answer to the problem considered above:

K. *For each of the numbers -1, 2 or any prime $q \neq 2$, there exist infinitely many primes p such that -1, 2 or q is a quadratic residue modulo p.*

For the case of an odd prime q, we shall compute explicitly (q/p), where p is any odd prime.

L. *Let q be an odd prime; q is a quadratic residue modulo the odd prime $p \neq q$ if and only if p is congruent modulo $4q$ to one of the following integers:* $\pm 1^2, \pm 3^2, \pm 5^2, \ldots, \pm(q-2)^2.$

Proof: If $p \equiv (2a + 1)^2 \pmod{4q}$ then $p \equiv 1 \pmod 4$ and by Gauss' reciprocity law

$$\left(\frac{q}{p}\right) = \left(\frac{p}{q}\right)(-1)^{\frac{p-1}{2}\cdot\frac{q-1}{2}} = (-1)^{\frac{p-1}{2}\cdot\frac{q-1}{2}} = 1.$$

If $p \equiv -(2a + 1)^2 \pmod{4q}$, then $p \equiv -1 \pmod 4$ and again

$$\left(\frac{q}{p}\right) = \left(\frac{p}{q}\right)(-1)^{\frac{p-1}{2}\cdot\frac{q-1}{2}} = \left(\frac{-1}{q}\right)(-1)^{\frac{p-1}{2}\cdot\frac{q-1}{2}}$$

$$= (-1)^{\frac{q-1}{2}\left(1+\frac{p-1}{2}\right)} = 1$$

(by Euler's criterion).

Conversely, let us assume that $(q/p) = 1$. By Gauss' reciprocity law and Euler's criterion we deduce that

$$\left(\frac{p}{q}\right) = (-1)^{\frac{p-1}{2}\cdot\frac{q-1}{2}} = \left(\frac{(-1)^{\frac{p-1}{2}}}{q}\right)$$

hence $\left(\dfrac{p(-1)^{\frac{p-1}{2}}}{q}\right) = 1$. Thus there exists x such that

$$p(-1)^{(p-1)/2} \equiv x^2 \pmod q.$$

Since $x^2 \equiv (q - x)^2 \pmod q$ and x or $q - x$ is odd, we may assume for example that x is odd, hence $x^2 \equiv 1 \pmod 4$. If $p \equiv 1 \pmod 4$ then $p \equiv x^2 \pmod q$; from $x^2 \equiv 1 \pmod 4$ we deduce that $p \equiv x^2 \pmod{4q}$. If $p \equiv -1 \pmod 4$ then $p \equiv -x^2 \pmod q$; from $p \equiv -x^2 \pmod 4$ we conclude that $p \equiv -x^2 \pmod{4q}$. ∎

We may illustrate this result with some numerical examples.
If $q = 3$, then

$$\left(\frac{3}{p}\right) = \begin{cases} 1 & \text{when } p \equiv \pm 1 \pmod{12} \\ -1 & \text{when } p \equiv \pm 5 \pmod{12} \end{cases}$$

Indeed, $a = 1$ is the only number such that $0 < a < 12$, $a \equiv 1 \pmod 4$ and $(a/3) = 1$. Thus, $(3/p) = 1$ exactly when $p \equiv \pm 1 \pmod{12}$.

Similarly, if $q = 11$, we consider the squares $1^2, 3^2, 5^2, 7^2, 9^2$, whose residue classes modulo 44 are $1, 9, 25, 5, 37$. By our result, $(11/p) = 1$ when p is congruent modulo 44 to any one of the integers $\pm 1, \pm 5, \pm 9, \pm 25, \pm 37$.

The result which follows may be called the "*global property of quadratic residues*." Its interest lies in the fact that a property is deduced for an integer whenever a similar property relative to residue classes modulo p, holds for all primes p.

Theorem 2. *An integer is a square if and only if it is a quadratic residue modulo p, for every prime p.*

Proof: If $a = b^2$ then $a \equiv b^2 \pmod{p}$ for every prime p. Conversely, let us assume that a is not a square, so it is of the form $a < 0$ or $a = m^2 p_1 \ldots p_r$ where $p_1, \ldots, p_r$ are distinct primes, $r \geqslant 1$ and if $i < r$ then $p_i \neq 2$.

Case 1: $a > 0$

We shall show that $(a/p) = -1$ for some prime p.

First we prove that if q is an odd prime there exists an integer u such that $u \equiv 1 \pmod{4}$, q does not divide u and $(u/q) = -1$. Indeed, we exclude from the set of q integers $\{1, 5, 9, \ldots, 4q - 3\}$ those which are the least positive residues of $1^2, 3^2, \ldots, (q-2)^2$ modulo $4q$, since by (L) these numbers are quadratic residues modulo q. We also exclude q when $q \equiv 1 \pmod{4}$ or $3q$ when $q \equiv -1 \pmod{4}$. It remains a set with $q - (q-1)/2 - 1 = (q-1)/2 \geqslant 1$ elements. If u belongs to this set we have $(u/q) = -1$. We apply this fact for $q = p_r$ when p_r is odd. If $p_r = 2$ we take $u = 5$.

By the Chinese remainder theorem there exists an integer x satisfying the following congruences:

$$\begin{cases} x \equiv 1 \pmod{p_1} \\ \ldots \\ x \equiv 1 \pmod{p_{r-1}} \\ x \equiv u \pmod{4p_r}. \end{cases}$$

By Dirichlet's theorem there exists a prime p such that

$$p \equiv x \pmod{4 p_1 \ldots p_{r-1} p_r}.$$

Then

$$\left(\frac{a}{p}\right) = \left(\frac{p_1}{p}\right) \cdots \left(\frac{p_{r-1}}{p}\right)\left(\frac{p_r}{p}\right)$$

$$= \left(\frac{p}{p_1}\right)(-1)^{\frac{p_1-1}{2} \cdot \frac{p-1}{2}} \cdots \left(\frac{p}{p_{r-1}}\right)(-1)^{\frac{p_{r-1}-1}{2} \cdot \frac{p-1}{2}} \cdot \left(\frac{p_r}{p}\right) = -1,$$

since $p \equiv x \equiv 1 \pmod{p_i}$ for all $i = 1, \ldots, r-1$; $p \equiv x \equiv 1 \pmod{4}$; $(2/p) = -1$, since $p \equiv 5 \pmod{8}$ when $p_r = 2$:

$$(p_r/p) = (p/p_r)(-1)^{\frac{p-1}{2} \cdot \frac{p-1}{2}} (u/p_r) = -1$$

because $p \equiv u \pmod{p_r}$.

Case 2: $a < 0$.

If $a = -m^2$, let p be a prime such that $p \equiv -1 \pmod 4$ (for example $p = 3$); then $(a/p) = (-1/p) = -1$.

If $a = -m^2 p_1 \ldots p_r$ where $p_1, \ldots, p_r$ are distinct primes and $r \geqslant 1$, we consider a prime p such that $p \equiv 1 \pmod 4$ and $(-a/p) = -1$, which exists by the first case. Then $(a/p) = (-1/p)(-a/p) = -1$. ∎

We want to present now another more penetrating proof of Gauss' reciprocity law examining one consequence of this method. This proof will illustrate the possibility of deriving properties of integers by considerations in algebraic extensions of the field $\mathbb{Q}$ of rational numbers. This is just one instance of a very fruitful method, and we shall later encounter more applications of this idea.

We assume that the reader has a familiarity with the basic concepts of the theory of commutative fields as found in the Introduction and in several textbooks.

Let p be an odd prime, let $K_0 = \mathbb{Q}$ or $K_0 = \mathbb{F}_q$ (where q is prime distinct from p). Let ζ be a primitive pth root of unity (in an algebraic closure of K_0). Thus $1, \zeta, \zeta^2, \ldots, \zeta^{p-1}$ are all the pth roots of 1 in the algebraic closure, and $\zeta^p = 1$. We agree to write $\zeta^{\bar{a}} = \zeta^a$, where $\bar{a}$ denotes the residue class of a modulo p. From $0 = \zeta^p - 1 = (\zeta - 1)(\zeta^{p-1} + \zeta^{p-2} + \cdots + \zeta + 1)$, $\zeta \neq 1$ it follows that $\zeta^{p-1} + \zeta^{p-2} + \cdots + \zeta + 1 = 0$.

For every $\bar{a} \in P(p)$ we shall consider the sum

$$\tau(\bar{a}) = \sum_{\bar{x} \in P(p)} \left(\frac{x}{p}\right) \zeta^{ax},$$

which is an element of the field $K = K_0(\zeta)$. It is called the *Gaussian sum over* K_0 *belonging to* $\bar{a}$. The *principal Gaussian sum* is

$$\tau(\bar{1}) = \sum_{\bar{x} \in (Pp)} \left(\frac{x}{p}\right) \zeta^x.$$

M.   *For every* $\bar{a} \in P(p)$ *we have* $\tau(\bar{a}) = (a/p)\tau(\bar{1})$.
Proof: Let $\bar{a} . \bar{x} = \bar{y}$ for every $\bar{x} \in P(p)$. Since $P(p)$ is a finite multiplicative group then

$$\left(\frac{a}{p}\right)\tau(\bar{a}) = \left(\frac{a}{p}\right) \sum_{\bar{x} \in P(p)} \left(\frac{x}{p}\right) \zeta^{ax} = \sum_{\bar{x} \in P(p)} \left(\frac{ax}{p}\right) \zeta^{ax} = \sum_{\bar{y} \in P(p)} \left(\frac{y}{p}\right) \zeta^y = \tau(\bar{1}).$$

We deduce, by multiplication with (a/p) that $\tau(\bar{a}) = (a/p) . \tau(\bar{1})$. ∎

Now, we compute the square of the principal Gaussian sum over the field K_0. It is convenient to denote by $\tilde{1}$ the unit element of K_0.

N. $\tau(\bar{1})^2 = (-1)^{(p-1)/2} p \cdot \tilde{1}$, *or explicitly*

$$\tau(\bar{1})^2 = \begin{cases} p \cdot \tilde{1} \ when \ p \equiv 1 \ (\text{mod } 4) \\ -p \cdot \tilde{1} \ when \ p \equiv 3 \ (\text{mod } 4) \end{cases}$$

In particular, if K_0 has characteristic different from p then $\tau(\bar{1}) \neq 0$.
Proof: The statement is proved by a straightforward computation.

$$\tau(\bar{1})^2 = \left[\sum_{\bar{x} \in P(p)} \left(\frac{x}{p} \right) \zeta^x \right] \cdot \left[\sum_{\bar{y} \in P(p)} \left(\frac{y}{p} \right) \zeta^y \right]$$

$$= \sum_{\bar{x}, \bar{y} \in P(p)} \left(\frac{xy}{p} \right) \zeta^{x+y}.$$

Let us write $\bar{y} = \bar{x} \cdot \bar{t}$ (which is possible, since $P(p)$ is a multiplicative group); hence

$$\left(\frac{xy}{p} \right) = \left(\frac{x^2 t}{p} \right) = \left(\frac{t}{p} \right),$$

so

$$\tau(\bar{1})^2 = \sum_{\bar{x}, \bar{t} \in P(p)} \left(\frac{t}{p} \right) \zeta^{x(1+t)}$$

$$= \sum_{\bar{t} \in P(p)} \left(\frac{t}{p} \right) \left[\sum_{\bar{x} \in P(p)} \zeta^{x(1+t)} \right].$$

If $\bar{1} + \bar{t} \neq \bar{0}$ then $\{ \bar{x}(\bar{1} + \bar{t}) \mid \bar{x} \in P(p) \} = P(p)$, thus,

$$\sum_{\bar{x} \in P(p)} \zeta^{x(1+t)} = \zeta + \zeta^2 + \cdots + \zeta^{p-1} = -\tilde{1}.$$

If $\bar{1} + \bar{t} = \bar{0}$ then $\sum_{\bar{x} \in P(p)} \zeta^{\bar{x}(1+\bar{t})} = (p-1) \cdot \tilde{1}$. Therefore,

$$\tau(\bar{1})^2 = \left(\frac{-1}{p} \right)(p-1) \cdot \tilde{1} + \sum_{-\bar{1} \neq \bar{t} \in P(p)} \left(\frac{t}{p} \right)(-\tilde{1})$$

$$= \left(\frac{-1}{p} \right) p \cdot \tilde{1} - \sum_{\bar{t} \in P(p)} \left(\frac{t}{p} \right) \cdot \tilde{1} = \left(\frac{-1}{p} \right) p \cdot \tilde{1} = (-1)^{(p-1)/2} p \cdot \tilde{1}$$

because there are as many quadratic as non-quadratic residues modulo p (by (**F**), (**H**)). In particular, if K_0 as characteristic different from p then $\tau(\bar{1}) \neq 0$. ∎

The above expression of $\tau(\bar{1})^2$ will soon be used in an important instance.

O. *Let q be an odd prime different from p. For the principal Gaussian sum (for the prime p) over $\mathbb{F}_q$ we have:*

$$\tau(\tilde{1})^{q-1} = \left(\frac{q}{p}\right) \cdot \tilde{1}.$$

Proof:

$$\tau(\tilde{1})^q = \left[\sum_{\bar{x}\in P(p)} \left(\frac{x}{p}\right)\zeta^x\right]^q = \sum_{\bar{x}\in P(p)} \left(\frac{x}{p}\right)\zeta^{qx}$$

$$= \left(\frac{q}{p}\right)\left[\sum_{\bar{x}\in P(p)} \left(\frac{qx}{p}\right)\zeta^{qx}\right] = \left(\frac{q}{g}\right)\tau(\tilde{1}),$$

since $\mathbb{F}_q(\zeta)$ has characteristic q. We conclude that $\tau(\tilde{1})^{q-1} = (q/p) \cdot \tilde{1}$ because $\tau(\tilde{1}) \neq 0$. ∎

We are now ready to indicate a *new proof of the quadratic reciprocity law.*

Let p, q be distinct odd primes. We compute in the algebraic closure of $\mathbb{F}_q$ the value of the Legendre symbol (q/p). Let $p^* = (-1)^{(p-1)/2}p$ and let $\tilde{1}$ be the unit element of $\mathbb{F}_q$. By (**O**)

$$\left(\frac{q}{p}\right) \cdot \tilde{1} = \tau(\tilde{1})^{q-1} = [\tau(\tilde{1})^2]^{(q-1)/2} = (p^*)^{(q-1)/2} \cdot \tilde{1}.$$

By Euler's criterion we have the equality in the field $\mathbb{F}_q$:

$$\left(\frac{q}{p}\right) = \left(\frac{(-1)^{(p-1)/2}p}{q}\right) = \left(\frac{-1}{q}\right)^{(p-1)/2}\left(\frac{p}{q}\right) = (-1)^{\frac{p-1}{2}\cdot\frac{q-1}{2}}\left(\frac{p}{q}\right),$$

so that $(p/q)(q/p) = (-1)^{\frac{p-1}{2}\cdot\frac{q-1}{2}}$. ∎

This method of proof implies the following interesting result:

P. *If $K \mid \mathbb{Q}$ is an algebraic extension of degree 2 then there exists a root of unity ζ such that $K \subseteq \mathbb{Q}(\zeta)$.*

Proof: We may assume that $K = \mathbb{Q}(\sqrt{d})$ where d is an integer with no square factors. Indeed, by the theorem of the primitive element (see Chapter 2, (6)) there exists an element t such that $K = \mathbb{Q}(t)$. The minimal polynomial of t is of degree 2, $X^2 + aX + b$. Replacing t by $t' = t + (a/2)$, it follows that $K = \mathbb{Q}(t')$ and t' is a root of $X^2 - (d_1/d_2)$, where $(d_1/d_2) = b - (a^2/4)$, with $d_1, d_2 \in \mathbb{Z}$, $d_2 \neq 0$. So $K = \mathbb{Q}(\sqrt{d_1/d_2}) = \mathbb{Q}(\sqrt{d_1 d_2})$, hence $K = \mathbb{Q}(\sqrt{d})$, where d is an integer with no square factors. Thus $d = \pm 2^e p_1 \cdot \ldots \cdot p_r$, where $e = 0$ or 1, $r \geqslant 0$ and each p_i is an odd prime. It follows that $K = \mathbb{Q}(\sqrt{d}) \subseteq \mathbb{Q}(\sqrt{-1}, \sqrt{2}, \sqrt{p_1}, \ldots, \sqrt{p_r})$.

$\sqrt{-1}$ is a primitive 4th root of unity ζ_4. $\sqrt{2}$ is expressible in terms of a primitive 8th root of unity ζ_8, because

$$(\zeta_8 + \zeta_8^{-1})^2 = \zeta_8^2 + \zeta_8^{-2} + 2 = \zeta_4 + \zeta_4^{-1} + 2 = 2,$$

thus $\mathbb{Q}(\sqrt{2}) \subseteq \mathbb{Q}(\zeta_8)$.

Finally, if p is any prime, $p \neq 2$, then $\pm p = [\tau(\bar{1})]^2$ (by (N)), therefore, $\sqrt{p} = \tau(\bar{1}) \in \mathbb{Q}(\zeta_p)$ or $\sqrt{p} = \sqrt{-1}\tau(\bar{1}) \in \mathbb{Q}(\zeta_4, \zeta_p)$, where ζ_p is a primitive pth root of unity.

Combining these facts, $K = \mathbb{Q}(\sqrt{d}) \subseteq \mathbb{Q}(\zeta_8, \zeta_{p_1}, \ldots, \zeta_{p_r}) \subseteq \mathbb{Q}(\zeta_m)$, where $m = 8p_1, \ldots, p_r$ and ζ_m is a primitive mth root of unity. ∎

The preceding result has a far-reaching generalization, which is the classical Kronecker and Weber theorem:

If $K \mid \mathbb{Q}$ is an Abelian extension (that is, a Galois extension with Abelian Galois group) of finite degree, then there exists a root of unity ζ, such that $K \subseteq \mathbb{Q}(\zeta)$.

We postpone the proof of this theorem until Chapter 12, since it requires deep considerations of arithmetical nature.

We conclude this section by indicating a generalization of Legendre's symbol, which is useful in the study of quadratic number fields.

Let a be a non-zero integer, and let b be an odd integer relatively prime to a, $|b| = \prod_{p \mid b} p^{\beta_p}$ (p odd prime, $\beta_p \geqslant 1$).

We define the *Jacobi symbol* $[a/b]$ by

$$\left[\frac{a}{b}\right] = \left[\frac{a}{-b}\right] = \prod_{p \mid b} \left(\frac{a}{p}\right)^{\beta_p}.$$

In particular $[a/1] = [a/-1] = 1$. Let us note that since $\gcd(a,b) = 1$, if $\beta_p \geqslant 1$, then p does not divide a, so that the Legendre symbol (a/p) has a meaning.

The Jacobi symbol has value 1 or -1. If $b = p$ is an odd prime number then $[a/p] = (a/p)$. So, we shall write (a/b) instead of $[a/b]$, without ambiguity.

Below we list some of the properties of the Jacobi symbol (under above assumptions about the numerator and the denominator):

Q. (1) *If $a \equiv a' \pmod{b}$ then $(a/b) = (a'/b)$.*

(2) $\left(\dfrac{aa'}{b}\right) = \left(\dfrac{a}{b}\right)\left(\dfrac{a'}{b}\right).$

(3) $\left(\dfrac{a}{bb'}\right) = \left(\dfrac{a}{b}\right)\left(\dfrac{a}{b'}\right).$

(4) *If the class of a modulo $b > 1$ is a square in $P(b)$ then $(a/b) = 1$.*
Proof: These properties follow easily from the corresponding properties for the Legendre symbol.

To show (4) we note that if $a \equiv x^2$ (mod b) and if p is a prime dividing b then $a \equiv x^2$ (mod p), so $(a/p) = 1$. Thus $(a/b) = 1$.   ∎

Let us observe, however, that it may happen that a modulo b is not a square in $P(b)$ and yet $(a/b) = 1$. For example $(2/9) = (2/3)^2 = 1$, but 2 is not a square modulo 9.

To deduce other properties, let us first observe the following simple facts. If $b = \Pi_{p|b}\, p^{\beta_p}$ is an odd integer then $b \equiv 1$ (mod 4) if and only if there is an even number of primes p such that β_p is odd and $p \equiv 3$ (mod 4). Hence, $\Sigma_{p|b}\,\beta_p \cdot (p - 1)/2 \equiv (b - 1)/2$ (mod 2). Similarly, $b \equiv \pm 1$ (mod 8) if and only if there is an even number of primes p such that β_p is odd and $p \equiv \pm 3$ (mod 8). Thus $\Sigma_{p|b}\,\beta_p\,(p^2 - 1)/8 \equiv (b^2 - 1)/8$ (mod 2).

For odd integers $b > 1$ we have:

R.
$$\left(\frac{-1}{b}\right) = (-1)^{(b-1)/2} = \begin{cases} 1 & \text{when } b \equiv 1 \ (\text{mod } 4) \\ -1 & \text{when } b \equiv 3 \ (\text{mod } 4) \end{cases}$$

$$\left(\frac{2}{b}\right) = (-1)^{(b^2-1)/8} = \begin{cases} 1 & \text{when } b \equiv \pm 1 \ (\text{mod } 8) \\ -1 & \text{when } b \equiv \pm 3 \ (\text{mod } 8). \end{cases}$$

Proof: By definition and (**H**) we have:

$$\left(\frac{-1}{b}\right) = \prod_{p|b}\left(\frac{-1}{p}\right)^{\beta_p} = (-1)\exp \sum \beta_p \cdot \frac{p-1}{2} = (-1)^{(b-1)/2}.$$

In   the   same   way $(2/b) = \Pi_{p|b}\,(2/p)^{\beta_p} = (-1)\exp \sum \beta_p \cdot (p^2 - 1)/8 = (-1)^{(b^2-1)/8}$.

The *reciprocity law for the Jacobi symbol* is the following:

S.
$$\left(\frac{a}{b}\right) = (-1)^{\frac{a-1}{2}\cdot\frac{b-1}{2}}\left(\frac{b}{a}\right)$$

when a, b are relatively prime odd integers and $b \geqslant 3$.
Proof: First let us assume that $a > 0$ and let $a = \Pi_{q|a}\, q^{\alpha_q}$, $b = \Pi_{p|b}\, p^{\beta_p}$ be the prime decompositions of a, b. The primes dividing b are different from those dividing a. Thus by the quadratic reciprocity law for the Legendre symbol:

$$\left(\frac{a}{b}\right) = \prod_{p|b}\left(\frac{a}{p}\right)^{\beta_p} = \prod_{p|b}\prod_{q|a}\left(\frac{q}{p}\right)^{\alpha_q\beta_p}$$

$$= \prod_{p|b}\prod_{q|a}\left(\frac{p}{q}\right)^{\alpha_q\beta_p}(-1)\exp\sum_{p,q}\alpha_q \cdot \frac{q-1}{2} \cdot \beta_p \cdot \frac{p-1}{2}$$

$$= \prod_{q|a}\left(\frac{b}{q}\right)^{\alpha_q}(-1)\exp\sum_{p,q}\alpha_q \cdot \frac{q-1}{2} \cdot \beta_p \cdot \frac{p-1}{2}.$$

Since

$$\sum_{q|a} \alpha_q \cdot \frac{q-1}{2} \equiv \frac{a-1}{2} \pmod 2,$$

and

$$\sum_{p|b} \beta_p \cdot \frac{p-1}{2} \equiv \frac{b-1}{2} \pmod 2,$$

we have

$$\left(\frac{a}{b}\right) = \prod_{q|a} \left(\frac{b}{q}\right)^{\alpha_q} (-1)^{\frac{a-1}{2} \cdot \frac{b-1}{2}} = \left(\frac{b}{a}\right)(-1)^{\frac{a-1}{2} \cdot \frac{b-1}{2}}.$$

Now, if $a < 0$ then

$$\left(\frac{a}{b}\right) = \left(\frac{-1}{b}\right)\left(\frac{|a|}{b}\right) = (-1)^{(b-1)/2}(-1)^{\frac{b-1}{2} \cdot \frac{|a|-1}{2}} \cdot \left(\frac{b}{|a|}\right)$$

$$= (-1)^{\frac{b-1}{2}\{1 + \frac{|a|-1}{2}\}} \left(\frac{b}{a}\right).$$

Since $1 + (|a| - 1)/2 \equiv (a - 1)/2 \pmod 2$ when $a < 0$ then

$$\left(\frac{a}{b}\right) = (-1)^{\frac{b-1}{2} \cdot \frac{a-1}{2}} \left(\frac{b}{a}\right).$$

EXERCISES

1. Determine the squares in the following groups: $P(7)$, $P(11)$, $P(12)$, $P(16)$, $P(49)$.

2. Compute the following Legendre symbols:

$$\left(\frac{6}{11}\right), \qquad \left(\frac{8}{17}\right), \qquad \left(\frac{18}{23}\right).$$

3. Compute the following Legendre symbols:

$$\left(\frac{205}{307}\right), \qquad \left(\frac{18}{461}\right), \qquad \left(\frac{753}{811}\right), \qquad \left(\frac{48}{1117}\right).$$

4. Determine the primes p modulo 68 such that 17 is a quadratic residue modulo p.

5. Determine the primes p modulo 20 for which 5 is a quadratic residue modulo 20.

6. Determine the odd primes p for which 7 is a quadratic residue modulo p.

7. Determine the primes p modulo 12 such that -3 is a quadratic residue modulo p.

8. Determine the odd primes p for which 10 is a quadratic residue modulo p.

9. Determine the odd primes p for which -2 is a quadratic residue modulo p.

10. Prove that 7 is a primitive root modulo every prime $p = 2^{2^k} + 1$ ($k \geqslant 1$).
   Hint: First show that $2^k \equiv 2$ or $4 \pmod 6$, then $2^{2^k} \equiv 2$ or $4 \pmod 7$; next compute $(7/p)$.

11. If p is a prime number, $p \equiv 1 \pmod 4$, prove that there exists an integer x such that $1 + x^2 = mp$, with $0 < m < p$.

12. Let p be an odd prime. Show that there exist integers x, y and m, $0 < m < p$ such that $1 + x^2 + y^2 = mp$.

13. Prove that for every integer $a > 1$ there exist infinitely many integers n which are not prime and such that $a^{n-1} \equiv 1 \pmod n$.
   Hint: For every odd prime p not dividing $a(a^2 - 1)$ consider $n = (a^{2p} - 1)/(a^2 - 1)$.

14. Prove that $2^{1092} \equiv 1 \pmod{1093^2}$ (Landau). In this respect, let us mention that it seems to be only rarely that a prime number p satisfies $2^{p-1} \equiv 1 \pmod{p^2}$; namely, $p = 1093, p = 3511$ are the only primes less than 10^5 for which this is true.
   Hint: In order to show that $2^{182} \equiv -1 \pmod{p^2}$ establish that $3^{14} \equiv 4p + 1 \pmod{p^2}$, $3^2 \cdot 2^{26} \equiv -469p - 1 \pmod{p^2}$, $3^{14} \cdot 2^{182} = -4p - 1 \pmod{p^2}$.

15. Let p be an odd prime. Show that

$$\sum_{n=1}^{p-2} \left(\frac{n(n+1)}{p} \right) = -1.$$

   Hint: Make use of the fact that n has an inverse modulo p.

16. Let p be an odd prime. Prove that the number of pairs of consecutive integers $n, n + 1$, with $1 \leqslant n \leqslant p - 2$, which are quadratic residues modulo p is equal to $\frac{1}{4}(p - 4 - (-1/p))$.
   Hint: Use the previous exercise after evaluating the product $(1 + (n/p))(1 + ((n + 1)/p))$.

17. Let p be an odd prime. Show that the product of the quadratic residues a modulo p, where $1 \leqslant a \leqslant p - 1$, is congruent to $-(-1/p)$ modulo p.
   Hint: Group the quadratic residues by pairs.

18. Let $a, b, c \in \mathbb{Z}$, and let p be an odd prime not dividing a. Show that there exists $x \in \mathbb{Z}$, such that $ax^2 + bx + c \equiv 0 \pmod p$ if and only if $b^2 - 4ac$ is a quadratic residue modulo p.

19. Prove that there are $\frac{1}{2}p(p-1)$ quadratic residues modulo p^2 and these are the solutions of the congruence

$$x^{\frac{1}{2}p(p-1)} \equiv 1 \;(\mathrm{mod}\; p^2).$$

20. Let p be a prime number, $p \equiv 1 \;(\mathrm{mod}\; 4)$, and let $a_1, a_2, \ldots, a_{(p-1)/2}$ be all the quadratic residues modulo p such that $1 < a_i \leqslant p - 1$. Prove that

$$\sum_{i=1}^{(p-1)/2} a_i = p(p-1)/4.$$

21. Show that if the prime number p divides $839 = 38^2 - 5 \cdot 11^2$, then $(5/p) = 1$. From this deduce that 839 is a prime number.
  Hint: Determine the primes for which 5 is a quadratic residue.

22. Show with the same method that 757 is a prime number.
  Hint: Write $757 = 55^2 - 7 \cdot 18^2$.

23. Let p be a prime number, a an integer not multiple of p and assume that there exist integers x, y such that $p = x^2 + ay^2$. Show that $-a$ is a quadratic residue modulo p.

24. Assume that the congruence $x^n \equiv a \;(\mathrm{mod}\; m)$ is solvable for every integer $m > 1$. Prove that a is the nth power of a natural number.

25. Let p be an odd prime, $1 \leqslant a < p$ and $n = ap + 1$ or $n = ap^2 + 1$. Prove that if $2^a \not\equiv 1 \;(\mathrm{mod}\; n)$ and $2^{n-1} \equiv 1 \;(\mathrm{mod}\; n)$ then n is a prime number.
  Hint: Show that the order d of 2 modulo n is a multiple of p, hence p divides $\varphi(n)$; noting that p does not divide n, deduce that $n = qm$, where q is a prime number, $q \equiv 1 \;(\mathrm{mod}\; p)$; conclude by proving that $m > 1$ leads to a contradiction in both cases $n \equiv 1 \;(\mathrm{mod}\; p)$, $n \equiv 1 \;(\mathrm{mod}\; p^2)$.

26. Let p be an odd prime, $m \geqslant 2$, $1 \leqslant a < 2^m$, $n = 2^m a + 1$ and assume that $(n/p) = -1$. Prove that n is a prime number if and only if

$$p^{(n-1)/2} \equiv -1 \;(\mathrm{mod}\; n).$$

  Hint: If n is prime, use Gauss' reciprocity law and Euler's criterion. Conversely, consider the order d of p modulo q (a prime factor of n); show that 2^m divides d, hence $q \equiv 1 \;(\mathrm{mod}\; 2^m)$ and from this conclude that $n = q$.

27. Apply the previous exercise to show that the Fermat number F_n (see Exercise 58, Chapter 3) is prime if and only if F_n divides

$$3^{(F_n-1)/2} + 1.$$

28. Find the smallest prime p which may be written simultaneously in the forms $p = x_1^2 + y_1^2 = x_2^2 + 2y_2^2 = x_3^2 + 3y_3^2$, where $x_i, y_i \in \mathbb{Z}$.

29. Let p be a prime number, $p > 7$, $p \equiv 3 \pmod 4$. Prove (Euler):
  (a) $2p + 1$ is prime if and only if $2^p \equiv 1 \pmod{2p + 1}$.
  (b) If $2p + 1$ is prime then the Mersenne number M_p is not a prime number.
  (c) Show successively that $23 \mid M_{11}$, $47 \mid M_{23}$, $167 \mid M_{83}$, $263 \mid M_{131}$, $359 \mid M_{179}$, $383 \mid M_{191}$, $479 \mid M_{239}$, $503 \mid M_{251}$.
  Hint: use Exercise 25, this chapter.

30. Show that if n is not a square, there exist infinitely many prime numbers p such that n is not a quadratic residue modulo p.

31. Prove the following particular case of Dirichlet's theorem: There exist infinitely many prime numbers in the arithmetic progression

$$\{12k + 7 \mid k = 0,1,2,\ldots\}.$$

Hint: After computing the Legendre symbol $(-3/p)$, show that $4a^2 + 3$ is divisible by a prime in the given arithmetic progression; conclude considering integers of the form $4(p_1 p_2 \ldots p_n)^2 + 3$.

32. Prove the following particular case of Dirichlet's theorem: In the arithmetic progression $\{1 + 2^n k \mid k = 0,1,2,\ldots\}$, $n \geqslant 1$, there exist infinitely many primes.
  Hint: First establish that if $p \neq 2$ is prime and p divides $2^{2^{n-1}} + 1$ then $p \equiv 1 \pmod{2^n}$.

33. Prove the following particular case of Dirichlets' theorem: In the arithmetic progression $\{1 + q^n k \mid k = 0,1,2,\ldots\}$, $n \geqslant 1$, q prime number, there exist infinitely many prime numbers.
  Hint: First establish that if p is a prime number, $p \neq q$, dividing $1 + x^{q^{n-1}} + x^{2q^{n-1}} + \cdots + x^{(q-1)q^{n-1}}$ then $p \equiv 1 \pmod{q^n}$. For this purpose, write $y = x^{q^{n-1}}$ and note that

$$1 + y + y^2 + \cdots + y^{q-1} = (y - 1)^{q-1} + \binom{q}{1}(y - 1)^{q-2}$$

$$+ \binom{q}{2}(y - 1)^{q-3} + \cdots + \frac{q}{2}(y - 1) + q.$$

34. The *Kronecker symbol* (a/b) is defined for $a \in \mathbb{Z}$, $a \equiv 0 \pmod 4$ or $a \equiv 1 \pmod 4$, a not a square, and $b \geqslant 1$ integer, in the following way: If $b = p_1 p_2 \ldots p_r$ is the decomposition of b as a product of primes, we put $(a/-b) \equiv (a/b) = (a/p_1)(a/p_2), \ldots, (a/p_r)$; next, if p is an odd prime $(a/p) = 0$ when p divides a, while (a/p) is the Legendre symbol (a/p) when p does not divide a, and finally, $(a/2) = 1$ when $a \equiv 1 \pmod 8$, while $(a/2) = -1$ when $a \equiv 5 \pmod 8$. Prove:

(a) if a, b are such that both the Jacobi and the Kronecker symbols are defined, then these symbols coincide.

(b) $(a/2) =$ Jacobi symbol $(2/|a|)$ when a is odd.

(c) if $gcd(a,b) > 1$ then $(a/b) = 0$, if $gcd(a,b) = 1$ then $(a/b) = \pm 1$.

(d) if $b_1, b_2 \geqslant 1$ then $(a/b_1b_2) = (a/b_1)(a/b_2)$.

(e) if $b > 0$, $gcd(a,b) = 1$, a odd then $(a/b) = (b/|a|)$ (Jacobi symbol in the right).

(f) if $b > 0$, $gcd(a,b) = 1$, $a = 2^r a'$ with a' odd, then

$$\left(\frac{a}{b}\right) = \left(\frac{2}{b}\right)^r (-1)^{\frac{a'-1}{2} \cdot \frac{b-1}{2}} \left(\frac{b}{|a'|}\right)$$

(Jacobi symbols in the right).

(g) if $b_1 \equiv b_2 \pmod{|a|}$ then $(a/b_1) = (a/b_2)$.

(h) for every a there exists b such that $(a/b) = -1$.

Hint: Distinguish cases and subcases.

35. Prove the following properties of the Kronecker symbol:

(a) $\left(\dfrac{a}{|a| - 1}\right) = \begin{cases} 1 & \text{when } a > 0 \\ -1 & \text{when } a < 0 \end{cases}$

(b) if $b \equiv -b' \pmod{|a|}$ then

$$\left(\frac{a}{b}\right) = \begin{cases} \left(\dfrac{a}{b'}\right) & \text{when } a > 0 \\[2ex] -\left(\dfrac{a}{b'}\right) & \text{when } a < 0. \end{cases}$$

Part Two

CHAPTER 5

Algebraic Integers

The arithmetic of the field of rational numbers is mainly the study of divisibility properties with respect to the ring of integers.

Similarly, the arithmetic of an algebraic number field K is concerned with divisibility properties of algebraic numbers relatively to some subring of K, which plays the role of the integers. Accordingly, we shall define the concept of an algebraic integer.

More generally, we introduce the following definition:

Definition 1. Let R be a ring,* and A a subring of R. We say that the element $x \in R$ is an *integer over* A when there exist elements $a_1, \ldots, a_n \in A$ such that $x^n + a_1 x^{n-1} + \cdots + a_n = 0$.

For example, if $A = K$ and $R = L$ are fields, then $x \in L$ is integral over K if and only if it is algebraic over K.

The first basic result about integral elements is the following:

A. *Let R be a ring, A a subring of R and $x \in R$. Then the following properties are equivalent:*

(1) *x is integral over A.*

(2) *the ring $A[x]$ is a finitely generated A-module.*

(3) *there exists a subring B of R such that $A[x] \subseteq B$ and B is a finitely generated A-module.*

Proof: (1) $\rightarrow$ (2) Let us assume that $x^n + a_1 x^{n-1} + \cdots + a_n = 0$ with $a_1, \ldots, a_n \in A$. We shall show that $\{1, x, \ldots, x^{n-1}\}$ is a system of generators of the A-module $A[x]$. Indeed, from $x^n = -(a_1 x^{n-1} + \cdots + a_n)$ it follows that $x^{n+1}, x^{n+2}, \ldots$ are expressible as linear combinations of $1, x, \ldots, x^{n-1}$, with coefficients in A.

(2) $\rightarrow$ (3) It is enough to take $B = A[x]$.

(3) $\rightarrow$ (1) Let $B = A y_1 + \cdots + A y_n$.

Since $x, y_i \in B$ then $x y_i \in B$; thus there exist elements a_{ij} $(j = 1, \ldots, n)$ such that $x y_i = \sum_{j=1}^{n} a_{ij} y_j$ (for all $i = 1, \ldots, n$). Therefore, letting $\delta_{ij} = 1$ when $i = j$, $\delta_{ij} = 0$ when $i \neq j$, we may write $\sum_{j=1}^{n} (\delta_{ij} x - a_{ij}) y_j = 0$ (for all $i = 1, \ldots, n$).

---

* We shall only consider commutative rings with unit element and the image of the unit element by all ring-homomorphisms is again the unit element.

In other words, the system of linear equations $\sum_{j=1}^{n} (\delta_{ij}x - a_{ij})Y_j = 0$ (for all $i = 1, \ldots, n$) has the solution $(y_1, \ldots, y_n)$. Let d be the determinant of the matrix $(\delta_{ij}x - a_{ij})_{i,j}$. By Cramer's rule, we must have $dy_j = 0$ (for all $j = 1, \ldots, n$).

Since $1 \in B$, it may be written in the form $1 = \sum_{j=1}^{n} c_j y_j$ (with $c_j \in A$), hence $d = d \cdot 1 = \sum_{j=1}^{n} c_j dy_j = 0$.

Computing d explicitly

$$d = \det \begin{pmatrix} x - a_{11} & -a_{12} & \cdots & -a_{1n} \\ -a_{21} & x - a_{22} & \cdots & -a_{2n} \\ \cdot & \cdot & & \cdot \\ \cdot & \cdot & & \cdot \\ \cdot & \cdot & & \cdot \\ -a_{n1} & -a_{n2} & \cdots & x - a_{nn} \end{pmatrix},$$

we deduce that d is of the form $0 = d = x^n + b_1 x^{n-1} + \cdots + b_n$ where each $b_i \in A$. This shows that x is integral over A. ∎

With this result we are able to deduce readily several properties of integral elements.

Definition 2. Let R be a ring and A a subring. We say that R is *integral over A* when every element of R is integral over A.

The following fact is evident:

Let R be a ring integral over the subring A, let $\theta: R \to R'$ be a homomorphism from R onto the ring R' and $\theta(A) = A'$. Then R' is integral over the subring A'.

Next, we have:

B. *If R is a domain which is integral over the subring A, if J is a non-zero ideal of R, then $J \cap A \neq 0$.*

Proof: Let $x \in J$, $x \neq 0$; by hypothesis, there exists a monic polynomial $f = X^n + a_1 X^{n-1} + \cdots + a_n \in A[X]$ such that $f(x) = 0$. We may take one such polynomial of minimal degree. Then $a_n \neq 0$, otherwise $x^{n-1} + a_1 x^{n-2} + \cdots + a_{n-1} = 0$ (since $x \neq 0$ and R is a domain). Hence $a_n = -(x^{n-1} + a_1 x^{n-2} + \cdots + a_{n-1})x \in J \cap A$. ∎

Definition 3. Let R be a ring and A a subring of R. If every element of R which is integral over A belongs to A, then A is said to be *integrally closed in R*.

If A is a domain, $R = K$ the field of quotients of A, and A is integrally closed in K, we say that A is an *integrally closed domain*.

C. *Let R be a ring, A a subring, and let $x_1, \ldots, x_n \in R$. If x_1 is integral over A, if x_2 is integral over $A[x_1], \ldots,$ if x_n is integral over $A[x_1, \ldots, x_{n-1}]$, then $A[x_1, \ldots, x_n]$ is a finitely generated A-module.*

Proof: By (**A**), $A[x_1]$ is a finitely generated A-module and $A[x_1, x_2]$ is a finitely generated $A[x_1]$-module. Hence, $A[x_1, x_2]$ is a finitely generated A-module.

The remainder of the proof is done similarly. ∎

D. *Let $A \subseteq B \subseteq C$ be rings. If C is integral over B and B is integral over A then C is also integral over A.*

Proof: Let $x \in C$, so there exist elements $b_1, \ldots, b_n \in B$ such that $x^n + b_1 x^{n-1} + \cdots + b_n = 0$.

But this means that x is integral over the subring $A[b_1, \ldots, b_n]$. By (**A**), the ring $A[b_1, \ldots, b_n, x]$ is a finitely generated module over $A[b_1, \ldots, b_n]$. By (**C**), $A[b_1, \ldots, b_n]$ is a finitely generated A-module, hence $A[x] \subseteq A[b_1, \ldots, b_n, x]$, which is a finitely generated A-module. Thus, by (**A**) again, we deduce that x is integral over A, proving our statement. ∎

E. *Let R be a ring, A a subring, and let A' be the set of all elements $x \in R$ which are integral over A. Then A' is a subring of R, which is integrally closed in R, and integral over A.*

Proof: Clearly $A \subseteq A'$, because every element $a \in A$ is a root of the polynomial $X - a$.

If $x, y \in A'$, then $x + y$, $x - y$, xy belong to the ring $A[x,y]$. By (**C**), $A[x,y]$ is a finitely generated A-module, hence by (**A**), $x + y$, $x - y$, xy are integral over A, so belong to A'.

By (**D**), A' is integrally closed in R. ∎

This result justifies the following definition:

Definition 4. Let R be a ring, and A a subring. The ring A' of all elements of R which are integral over A is called the *integral closure of A in R*.

Let us study the case of a field:

F. *Let R be a domain which is integral over the subring A. Then, R is a field if and only if A is a field.*

Proof: If R is a field, if $x \in A$, $x \neq 0$, consider its inverse x^{-1} in R. It is integral over A, hence there exist elements $a_i \in A$ such that $(x^{-1})^n + a_1(x^{-1})^{n-1} + \cdots + a_n = 0$. Multiplying by x^{n-1} we obtain: $x^{-1} + a_1 + a_2 x + \cdots + a_n x^{n-1} = 0$, hence $x^{-1} \in A$.

Conversely, let A be a field, let $x \in R$, and $x \neq 0$. By (**B**) there exists $a \in Rx \cap A$, $a \neq 0$; so $a = bx$, with $b \in R$. Let $a' \in A$ be the inverse of a, so $1 = a'a = (a'b)x$; hence x is invertible in R and R is a field. ∎

As a corollary:

G. *Let R be a ring integral over the subring A and let Q be a prime ideal of R. Then Q is a maximal ideal of R if and only if $Q \cap A$ is a maximal ideal of A.*
Proof: If Q is a maximal ideal of R then R/Q is a field which is integral over the subring $A/(Q \cap A)$; thus $A/(Q \cap A)$ is a field and $Q \cap A$ is a maximal ideal of A.

Conversely, if $Q \cap A$ is a maximal ideal of A then the domain R/Q is integral over the field $A/(Q \cap A)$. So R/Q is a field and Q is a maximal ideal of R. ∎

The important situation for algebraic numbers is the case where A is an integrally closed domain, with field of quotients K and L is an algebraic extension of K.

H. *Let A be an integrally closed domain with field of quotients K and let L be an algebraic extension of K. If $x \in L$ is integral over A, then its minimal polynomial over K has all its coefficients in A; all conjugates of x over K are also integral over A. If B is the integral closure of A in L then $B \cap K = A$.*
Proof: Let $f \in K[X]$ be the minimal polynomial of x over K. Since x is integral over A, there exists a monic polynomial g with coefficients in A, such that $g(x) = 0$. Hence f divides g.

Let L' be the splitting field of f over K, that is, the field generated over K by the roots of f. Let A' be the integral closure of A in L'. Then $A' \cap K$ is integral over A, hence must be equal to A (which is assumed integrally closed).

The conjugates of x are also roots of g, hence integers over A, so they belong to A'.

The coefficients of f are, up to sign, equal to the elementary symmetric polynomials in the conjugates of x, hence f has coefficients in $A' \cap K = A$.

The last assertion follows from the hypothesis that A is integrally closed. ∎

Another useful property follows:

I. *Let A be an integrally closed domain with field of quotients K and let L be an algebraic extension of K. If B denotes the integral closure of A in L, then every element of L is of the form b/d, where $b \in B$, $d \in A$ $(d \neq 0)$.*
Proof: Let $x \in L$, $x \neq 0$, so x is algebraic over K, hence there exist elements $c_i/d_i \in K$ (with $c_i, d_i \in A$, $d_i \neq 0$ for $i = 1, \ldots, n$), such that $x^n + (c_1/d_1)x^{n-1} + \cdots + c_n/d_n = 0$. Let $d = d_1 \ldots d_n \in A$, then $d^n x^n + (d^{n-1} d_1' c_1)x^{n-1} + \cdots + d^{n-1} d_n' c_n = 0$ where $d_i' = d/d_i \in A$ for

$i = 1, \ldots, n$. It follows that $(dx)^n + (d_1'c_1)(dx)^{n-1} + \cdots + d^{n-1}d_n'c_n = 0$, with $d^{i-1}d_i'c_i \in A$ for $i = 1, \ldots, n$. Thus dx is integral over A, so $dx = b \in B$, $b \neq 0$, and $x = b/d$. $\blacksquare$

Let us now note some important types of integrally closed domains.

J. *Every unique factorization domain is an integrally closed domain.*

Proof: Let K be the field of quotients of the unique factorization domain A. Let $x \in K$, $x \neq 0$, so $x = a/b$ with $a, b \in A$, $a, b \neq 0$, and we may assume that $gcd(a,b) = 1$.

If x is integral over A, there exist elements $c_1, \ldots, c_n \in A$ such that $(a/b)^n + c_1(a/b)^{n-1} + \cdots + c_n = 0$, thus $a^n + c_1ba^{n-1} + \cdots + c_nb^n = 0$. It follows that $a^n = -b(c_1a^{n-1} + \cdots + c_nb^{n-1})$ and therefore b divides a^n (relatively to the ring A). So, b divides a (we use the fact that if b divides cc' and $gcd(b,c) = 1$, then b divides c'); this means that $x = a/b \in A$, showing that A is an integrally closed domain. $\blacksquare$

In particular, since every principal ideal domain is a unique factorization domain, then every principal ideal domain is also integrally closed.

We shall now consider explicitly the above definitions and results for the case of algebraic number fields.

Definition 5. An algebraic number which is a root of a monic polynomial with coefficients in the ring $\mathbb{Z}$ of integers is called an *algebraic integer*.

It is also customary to say that the elements of $\mathbb{Z}$ are *rational integers*.

From our results, we see that the conjugates of an algebraic integer are algebraic integers.

If K is any field of algebraic numbers (of arbitrary degree over $\mathbb{Q}$), if $A_K = A$ denotes the ring of all algebraic integers in K, then A is an integrally closed domain, so $A \cap \mathbb{Q} = \mathbb{Z}$. We deduce that the trace (equal to the sum of conjugates) and the norm (equal to the product of conjugates) of an algebraic integer are rational integers (since they belong to $A \cap \mathbb{Q} = \mathbb{Z}$).

We shall be concerned with the arithmetic in an algebraic number field, relative to the ring of all algebraic integers. The natural question to ask is whether such rings must necessarily be unique factorization domains, and if this is not the case, how is it possible to describe their arithmetic.

By imitating the procedure in the case of the rational integers, we may prove the following result:

K. *Let A be a domain, satisfying the following property: if $Aa_1 \subseteq Aa_2 \subseteq \cdots \subseteq Aa_n \subseteq \cdots$ is an increasing chain of principal ideals of A, then there exists an integer n such that $Aa_n = Aa_{n+1} = \cdots$.*

Then every non-zero element of A which is not a unit may be written as a product of prime elements.

Proof: Let $a \in A$, $a \neq 0$. If a is not a unit element, there exists an element $a_1 \in A$, a_1 not associated with a, such that $a_1 \mid a$; hence $Aa \subset Aa_1$. Similarly, if a_1 is not a unit there exists a_2 such that $Aa \subset Aa_1 \subset Aa_2$. In virtue of the hypothesis, there exists n such that a_n is a unit, so that a_{n-1} must be a prime. Thus, we have shown that if $a \in A$, $a \neq 0$, there exists a prime p_1 such that $p_1 \mid a$; hence there exists $b_1 \in A$, $b_1 \neq 0$, such that $a = p_1 b_1$, so $Aa \subset Ab_1$. But again if b_1 is not a unit there exists a prime p_2 such that $b_1 = p_2 b_2$, so $a = p_1 p_2 b_2$, $Aa \subset Ab_1 \subset Ab_2$. By the hypothesis, there exists m such that b_m is a unit, hence $a = p_1 p_2 \cdots (p_m b_m)$, that is, a is a product of prime elements. ∎

We shall soon see that the ring of all algebraic integers of an algebraic number field K satisfies the condition in the statement of (**K**).* However, we have not established the *uniqueness* of the decomposition into primes. This is therefore the crucial question.

Let us note this reformulation of the unique factorization:

L.   *Let A be a domain such that every element is a product of prime elements. Then the following statements are equivalent:*

   (1) *if* $p_1 \ldots p_r = p_1' \ldots p_s'$ *(where* p_i, p_j' *are prime elements of A) then* $s = r$ *and there is a permutation* σ *of* $\{1, \ldots, r\}$ *such that* $p_i \sim p_{\sigma(i)}'$.

   (2) *if p is a prime and p divides xy (where* $x, y \in A$) *then either* $p \mid x$ *or* $p \mid y$.
Proof: The proof of this statement is exactly similar to the case where A is the ring of integers, therefore we leave the details to the reader. ∎

Concerning the units of the ring of algebraic integers, we have the following easy fact:

M.   *An algebraic integer x is a unit if and only if its norm is* $N(x) = \pm 1$.
Proof: If x is a unit, there exists an algebraic integer x' such that $xx' = 1$; taking the norms, we obtain $N(x) \cdot N(x') = 1$, so $N(x)$ is a unit in the ring $\mathbb{Z}$, that is $N(x) = \pm 1$.

Conversely, if $N(x) = \pm 1$, then letting x' be the product of all conjugates of x, distinct from x, we have $x \cdot x' = \pm 1$; but x' is an algebraic integer, so x divides 1 in the ring A of algebraic integers. ∎

At this point, it is advisable to pause and consider numerical examples.

---

* This is only true if K has finite degree over $\mathbb{Q}$, as we are assuming throughout.

1. INTEGERS OF QUADRATIC NUMBER FIELDS

Let K be a quadratic extension of $\mathbb{Q}$, that is $[K:\mathbb{Q}] = 2$.

As we have already indicated in Chapter 4, (**P**), we have $K = \mathbb{Q}(\sqrt{d})$, where d is a square-free integer.

Every element of K is of type $a + b\sqrt{d}$, where $a,b \in K$. The conjugate of $a + b\sqrt{d}$ is $a - b\sqrt{d}$. Let A denote the ring of all algebraic integers of $\mathbb{Q}(\sqrt{d})$.

N. $a + b\sqrt{d} \in A$ *if and only if* $2a = u \in \mathbb{Z}$, $2b = v \in \mathbb{Z}$ *and* $u^2 - dv^2 \equiv 0 \pmod 4$.

Proof: If $x = a + b\sqrt{d} \in A$ then its conjugate $x' = a - b\sqrt{d}$ is also an algebraic integer. So $x + x' = 2a \in A \cap \mathbb{Q} = \mathbb{Z}$, $x \cdot x' = a^2 = b^2 d \in A \cap \mathbb{Q} = \mathbb{Z}$.

It follows that $(2a)^2 - (2b)^2 d \in 4\mathbb{Z}$ and since $(2a)^2 \in \mathbb{Z}$ then $(2b)^2 d \in \mathbb{Z}$; but d is square-free, thus $2b$ has denominator equal to 1, that is $v = 2b \in \mathbb{Z}$.

Conversely, these conditions imply $a^2 - b^2 d \in \mathbb{Z}$, and since x is a root of $X^2 - 2aX + (a^2 - b^2 d)$, then x is an algebraic integer. ∎

The previous result may be reformulated as follows:

O. *Let* $K = \mathbb{Q}(\sqrt{d})$ *where d is a square-free integer; let A be the ring of all algebraic integers of K. If* $d \equiv 2 \pmod 4$ *or* $d \equiv 3 \pmod 4$ *then* $A = \{a + d\sqrt{d} \mid a,b \in \mathbb{Z}\}$. *If* $d \equiv 1 \pmod 4$ *then* $A = \{u/2 + v/2\sqrt{d} \mid u,v \in \mathbb{Z}, \ u \text{ and } v \text{ have the same parity}\}$.

Proof: We examine all the possible cases in succession.

If $d \equiv 2 \pmod 4$:

u	even	even	odd	odd	
v	even	odd	even	odd	
$u^2 - dv^2 \equiv$	0	2	1	3	(mod 4)

If $d \equiv 3 \pmod 4$:

u	even	even	odd	odd	
v	even	odd	even	odd	
$u^2 - dv^2 \equiv$	0	1	1	2	(mod 4)

If $d \equiv 1 \pmod 4$:

u	even	even	odd	odd	
v	even	odd	even	odd	
$u^2 - dv^2 \equiv$	0	3	1	0	(mod 4)

Therefore, by means of (**N**), we deduce statement (**O**). ∎

Let us note incidentally the following fact, which will be generalized later:

P. *A is a free Abelian group. If $d \equiv 2 \pmod 4$ or $d \equiv 3 \pmod 4$ then* $\{1,\sqrt{d}\}$ *is a basis of A.*

If $d \equiv 1 \pmod 4$ then $\{1, (1 + \sqrt{d})/2\}$ is a basis of A.

Proof: The statement is obvious when $d \equiv 2 \pmod 4$ or $d \equiv 3 \pmod 4$. Let us assume now that $d \equiv 1 \pmod 4$ and let us show that every algebraic integer $u/2 + (v/2)\sqrt{d}$ (with u,v integers of same parity) is a linear combination of 1 and $(1 + \sqrt{d})/2$, with coefficients in $\mathbb{Z}$.

If u,v are even, $u = 2a$, $v = 2b$ with $a,b \in \mathbb{Z}$, so $u/2 + (v/2)\sqrt{d} = a + b\sqrt{d} = (a - b)1 + 2b(1 + \sqrt{d})/2$.

If u,v are odd, then $u - 1, v - 1$ are even, so $u/2 + (v/2)\sqrt{d} = [(1 + \sqrt{d})/2] + \{[(u - 1)/2] + [(v - 1)/2]\sqrt{d}\}$ and this last summand is a linear combination of 1, $(1 + \sqrt{d})/2$ with coefficients in $\mathbb{Z}$. ∎

2. ARITHMETIC IN THE FIELD OF GAUSSIAN NUMBERS

An important example of quadratic number fields is $\mathbb{Q}(\sqrt{-1})$, which is called the field of *Gaussian numbers*. Since $-1 \equiv 3 \pmod 4$, the Gaussian integers have the form $a + bi$, where $a,b \in \mathbb{Z}$.

Let us start determining the units of the ring $\mathbb{Z}[i]$ of Gaussian integers. By (**K**), $x = a + bi$ is a unit if and only if $N(x) = a^2 + b^2 = \pm 1$; however, we must have $N(x) > 0$, so $a^2 + b^2 = 1$ with $a,b \in \mathbb{Z}$. The only possibilities are ± 1, $\pm i$ and these are therefore all the units of $\mathbb{Z}[i]$. Actually, they are roots of unity.

Before determining the primes in $\mathbb{Z}[i]$, we shall prove that $\mathbb{Z}[i]$ is a *Euclidean ring* (see Chapter 1, Exercise 26):

Q. *If $x,y \in \mathbb{Z}[i]$, $y \neq 0$, there exist $q,r \in \mathbb{Z}[i]$ such that $x = qy + r$, $|N(r)| < |N(y)|$.*

Proof: Consider the Gaussian number $x/y \in \mathbb{Q}(i)$, so we may write $x/y = a' + b'i$ with $a',b' \in \mathbb{Q}$.

Let $a,b \in \mathbb{Z}$ be the best approximations to a',b', that is, $|a' - a| \leqslant 1/2$, $|b' - b| \leqslant 1/2$. Let $q = a + bi \in \mathbb{Z}[i]$, so $x = qy + [(a' - a) + (b' - b)i]y$

and

$$N([(a' - a) + (b' - b)i]y) = [(a' - a)^2 + (b' - b)^2]N(y) \leqslant (1/2)N(y) < N(y)$$

(because $y \neq 0$ implies $N(y) \neq 0$). ∎

If the ring A of all algebraic integers of an algebraic number field K is a Euclidean ring we say that the *Euclidean algorithm holds in A* and also that K is a *Euclidean field*.

R. *If K is an euclidean field then A is a principal ideal domain.**
Proof: Let J be an ideal of A, $J \neq A$, $J \neq 0$. Then for every non-zero element $y \in J$ we have $N(y) \in \mathbb{Z}$ and $1 < |N(y)|$ (by (**M**)).

Let $y \in J$, $y \neq 0$, be an element such that $|N(y)|$ is minimum. We show that if $x \in J$ then x is a multiple of y. In fact, by the Euclidean algorithm, there exist algebraic integers q, r such that $x = qy + r$, with $|N(r)| < |N(y)|$. Since $r = x - qy \in J$, by the choice of y we must have $N(r) = 0$ so $r = 0$ and $x = qy$, showing that J is a principal ideal of A. ∎

It follows from (**Q**), (**R**) that the ring $\mathbb{Z}[i]$ is a principal ideal domain, hence also a unique factorization domain. By (**L**), it is true that if p is a prime in $\mathbb{Z}[i]$ and $p \mid xy$ then either $p \mid x$ or $p \mid y$.

To complete our description of the arithmetic in $\mathbb{Q}(i)$ we still need to determine the prime elements in $\mathbb{Z}[i]$.

S. *A Gaussian integer is prime if and only if it is associated with one and only one of the following Gaussian integers:*
(1) *any rational prime p, such that $p \equiv 3 \pmod 4$.*
(2) $1 + i$.
(3) $a + bi$ *where $a,b \in \mathbb{Z}$, satisfy: $a > 0$, $b \neq 0$, a is even, $a^2 + b^2 = p$ where p is a rational prime, $p \equiv 1 \pmod 4$.*
Proof: We first show that every Gaussian prime x divides one and only one rational prime.

In fact, $N(x) \in \mathbb{Z}$, so $N(x) = \pm p_1 \ldots p_r$ where each p_i is a rational prime; since x divides $N(x)$ (relatively to $\mathbb{Z}[i]$) then x divides some p_i.

Next, if x divides the distinct rational primes p,p', since there exist rational integers m,m' such that $mp + m'p' = 1$, it follows that x divides 1, that is x is a unit, against our hypothesis.

Now, we shall determine all Gaussian primes $x = a + bi$ dividing a rational prime p.

If $p = 2$, from $x \mid p$ it follows that $a^2 + b^2 = N(x)$ divides $N(p) = p^2 = 4$. The only possibilities are: $a = \pm 1$, $b = \pm 1$, or $a = \pm 2$, $b = 0$, or $a = 0$, $b = \pm 2$. Since ± 1, $\pm i$ are units in $\mathbb{Z}[i]$, then all integers $1 + i$, $1 - i$, $-1 + i$, $-1 - i$ are associated. Finally, $1 + i$ is a Gaussian prime, since its norm is the rational prime 2; so, if $1 + i = yz$, then $N(y)$ or $N(z)$ is ± 1, so y or z is a unit. In the other cases, we get ± 2, or $\pm 2i$, which are associated,

---
* See Chapter 1, Exercise 26.

and from $2 = (1 + i)(1 - i) = -i(1 + i)^2$, it follows that 2 is not a Gaussian prime.

Let $p \equiv 1 \pmod 4$. By Chapter 4, **(H)**, -1 is a square modulo p, that is $\mid n^2 + 1$ (for some $n \in \mathbb{Z}$). But $n^2 + 1 = (n + i)(n - i)$. If x is a Gaussian prime dividing p, then by **(L)** $x \mid n + i$ or $x \mid n - i$.

x and p are not associated in $\mathbb{Z}[i]$, because this would imply that p divides either $n + i$ or $n - i$, hence $(n/p) + (1/p)i$ or $(n/p) - (1/p)i$ would be a Gaussian integer, which is not the case.

It follows that $N(x) \neq N(p)$ and since x divides p then $N(x) = p$. Thus if $x = a + bi$ then $p = N(x) = a^2 + b^2$. From $p \equiv 1 \pmod 4$ it follows that exactly one of a or b is even.

Among such numbers we have $x = a + bi$ with $a^2 + b^2 = p$, $a > 0$, a even, $b \neq 0$.

If b is even, then $\mp i(a + bi) = \pm b \mp ai$, so $a + bi$ is associated with $b - ai$ (or $-b + ai$) with b even, and $b > 0$ (or $-b > 0$).

Moreover, if $x = a + bi$, $y = c + di$ are such that $a^2 + b^2 = p$, $c^2 + d^2 = p$, a,c are even, $a > 0$, $c > 0$, $b \neq 0$, $d \neq 0$, and if $x \sim y$ then $x = y$. Indeed $x = uy$ with $u = 1$ or -1 or i or $-i$. But $u = -1$ would imply $a = -c < 0$, $u = i$ would imply $a = -d$ is odd and $u = -i$ would imply that $a = d$ is odd. This establishes which are the Gaussian primes dividing p, when $p \equiv 1 \pmod 4$:

Finally, let $p \equiv 3 \pmod 4$. If $x = a + bi$ is a Gaussian prime dividing p, then $1 \neq N(x) = a^2 + b^2$ divides p^2, so either $a^2 + b^2 = p$ or $a^2 + b^2 = p^2$. But, from $p \equiv 3 \pmod 4$ it is not possible that $a^2 + b^2 = p$ (since $n^2 \equiv 0$ or $n^2 \equiv 1$ modulo 4, for every $n \in \mathbb{Z}$). Thus $N(x) = N(p)$ hence x/p is a unit, so x is associated with p. ∎

Part (3) of **(S)** may be phrased more explicitly, and constitutes an interesting theorem about rational integers:

T.   *A positive integer $n = p_1^{k_1} \ldots p_s^{k_s}$ is the sum of squares of two integers if and only if k_j is even when $p_j \equiv 3 \pmod 4$.*
Proof: First we note that $2 = 1^2 + 1^2$. If p is a prime number congruent to 1 modulo 4, let $x = a + bi$ be a prime Gaussian integer dividing p. Then $p = N(x) = a^2 + b^2$. Now we observe that if $n_1 = a_1^2 + b_1^2$, $n_2 = a_2^2 + b_2^2$, with $a_j, b_j \in \mathbb{Z}$, then also $n_1 n_2 = (a_1^2 + b_1^2) \cdot (a_2^2 + b_2^2) = (a_1 a_2 - b_1 b_2)^2 + (a_1 b_2 + a_2 b_1)^2$. Altogether, we have shown that if $n = p_1^{k_1} \ldots p_s^{k_s}$, if k_j is even when $p_j \equiv 3 \pmod 4$ then n is the sum of two squares.

Conversely, if $p \equiv 3 \pmod 4$ and $n = p^{2k+1}m$, with $k \geqslant 0$ and m not divisible by p, then n cannot be the sum of two squares. Indeed, let $n = a^2 + b^2$, let $d = gcd(a,b)$, so $a = da_1$, $b = db_1$, and $gcd(a_1,b_1) = 1$. Writing $a_1^2 + b_1^2 = n_1$ then $d^2 n_1 = n$. The exact power of p dividing d has even exponent, hence p divides n_1. Then a_1 and b_1 are not multiples of p, since they are relatively prime. Let $c \in \mathbb{Z}$ be such that $a_1 c \equiv b_1 \pmod{p}$. Then $n_1 = a_1^2 + b_1^2 \equiv a_1^2(1 + c^2) \equiv 0 \pmod{p}$, and therefore $c^2 \equiv -1 \pmod{p}$, that is -1 is a quadratic residue modulo p. Therefore p would be either 2 or $p \equiv 1 \pmod 4$, against the hypothesis. ∎

Before proceeding, let us summarize what we have just learned.

If the Euclidean algorithm holds in the ring of algebraic integers A then A is a principal ideal domain.

It has been shown that there exists only finitely many quadratic Euclidean fields $\mathbb{Q}(\sqrt{d})$, namely those with d equal to one of the following 21 integers:

$$-11, -7, -3, -2, -1, 2, 3, 5, 6, 7, 11, 13, 17, 19, 21, 29, 33, 37,$$
$$41, 57, 73.$$

This rather negative result does not yet exclude the possibility that for all quadratic fields, the ring of algebraic integers be a principal ideal domain. The following classical example shows that this is not the case.

Example: Let us consider the field $K = \mathbb{Q}(\sqrt{-5})$, so the ring of algebraic integers A consists of the numbers of the type $a + b\sqrt{-5}$, where $a, b \in \mathbb{Q}$ (by (**O**)).

We may write $21 = 3 \cdot 7 = (1 + 2\sqrt{-5})(1 - 2\sqrt{-5})$.

Each of the numbers $3, 7, 1 + 2\sqrt{-5}, 1 - 2\sqrt{-5}$ is a prime in $\mathbb{Z}[\sqrt{-5}]$. For example, if $3 = xy$, with x, y not units, then taking norms, $9 = N(x) \cdot N(y)$ therefore $N(x) = N(y) = 3$. If $x = a + b\sqrt{-5}$ then $N(x) = a^2 + 5b^2 = 3$, but this is impossible with $a, b \in \mathbb{Z}$.

Also, the numbers $3, 7, 1 + 2\sqrt{-5}, 1 - 2\sqrt{-5}$ are pairwise non-associated, since $N(3) = 9$, $N(7) = 49$, $N(1 + 2\sqrt{-5}) = N(1 - 2\sqrt{-5}) = 21$, and

$$\frac{1 + 2\sqrt{-5}}{1 - 2\sqrt{-5}} = \frac{-19 + 4\sqrt{-5}}{21} \notin \mathbb{Z}[\sqrt{-5}].$$

This example shows what seems to be even stronger, namely, $\mathbb{Z}[\sqrt{-5}]$ is not a unique factorization domain. (Later, we shall prove that the ring A of all algebraic integers of a field of algebraic numbers is a principal ideal domain if and only if it is a unique factorization domain.)

The hopes being dashed, we may still ask which are the quadratic fields $\mathbb{Q}(\sqrt{d})$ having a principal ideal domain as ring of integers?

Besides the domains with Euclidean algorithms, if $d < 0$ then it must be equal to $d = -19, -43, -67$, or -163 (see Chapter 13, (**7**)). Heilbronn and Linfoot proved in 1934 that there could exist at most one more value of $d < 0$ for which the ring of integers of $\mathbb{Q}(\sqrt{d})$ would be a principal ideal domain. Lehmer has shown, using analytical methods, that $|d| > 5 \times 10^9$, and later Stark proved that $|d|$ should be greater than $\exp(2,2 \times 10^7)$. Finally, in 1967, Stark proved that no such d could exist. We have mentioned these facts, since they serve to illustrate the need of appealing to delicate analytical methods, which is a recurring characteristic in the theory of algebraic numbers.

On the other hand it is still an open question whether there exist infinitely many fields $\mathbb{Q}(\sqrt{d})$, with $d > 0$, whose ring of algebraic integers is a principal ideal domain.

3. INTEGERS OF CYCLOTOMIC FIELDS

Let $K = \mathbb{Q}(\zeta)$, where ζ is a primitive pth root of unity and p is a prime number.

The minimal polynomial of ζ over $\mathbb{Q}$ is $\Phi_p = X^{p-1} + X^{p-2} + \cdots + X + 1$, hence ζ belongs to the ring A of integers of $\mathbb{Q}(\zeta)$. The roots of Φ_p are $\zeta, \zeta^2, \ldots, \zeta^{p-1}$, thus $\Phi_p = \Pi_{i=1}^{p-1}(X - \zeta^i)$; in particular, $p = \Phi_p(1) = \Pi_{i=1}^{p-1}(1 - \zeta^i)$.

Let us note that the elements $1 - \zeta, 1 - \zeta^2, \cdots, 1 - \zeta^{p-1}$ are associated. In fact, if $1 \leqslant i, j \leqslant p - 1$ then there exists an integer k such that $j \equiv ik \pmod{p}$; thus

$$\frac{1 - \zeta^j}{1 - \zeta^i} = \frac{1 - \zeta^{ik}}{1 - \zeta^i} = 1 + \zeta^i + \zeta^{2i} + \cdots + \zeta^{(k-1)i} \in A;$$

similarly, $(1 - \zeta^i)/(1 - \zeta^j) \in A$, so $1 - \zeta^i = u_i(1 - \zeta)$, where u_i is a unit of A. We conclude that $p = u(1 - \zeta)^{p-1}$ where $u = u_1 \ldots u_{p-1}$ is a unit of A.

The element $1 - \zeta$ is not invertible in A, otherwise p would have an inverse which belongs to $A \cap \mathbb{Q} = \mathbb{Z}$. Hence $A(1 - \zeta) \cap \mathbb{Z} = \mathbb{Z}p$, since the ideal $A(1 - \zeta) \cap \mathbb{Z}$ contains p and is not equal to the unit ideal.

U. *A is a free Abelian group with basis $\{1, \zeta, \ldots, \zeta^{p-2}\}$, so $A = \mathbb{Z}[\zeta]$.*
Proof: Obviously, $1, \zeta, \ldots, \zeta^{p-2}$ are linearly independent over $\mathbb{Q}$, otherwise ζ would be a root of a polynomial of degree at most $p - 2$, against the fact that Φ_p is its minimal polynomial.

If $x \in A$ there exist uniquely defined rational numbers $a_0, a_1, \ldots, a_{p-2}$ such that $x = a_0 + a_1\zeta + \cdots + a_{p-2}\zeta^{p-2}$. We shall prove that each $a_i \in \mathbb{Z}$. We have $x\zeta = a_0\zeta + a_1\zeta^2 + \cdots + a_{p-2}\zeta^{p-1}$ and subtracting:

$$x(1 - \zeta) = a_0(1 - \zeta) + a_1(\zeta - \zeta^2) + \cdots + a_{p-2}(\zeta^{p-2} - \zeta^{p-1}).$$

We note that the traces (in $\mathbb{Q}(\zeta) \mid \mathbb{Q}$) of $\zeta, \zeta^2, \ldots, \zeta^{p-1}$ are all equal (since these elements are conjugate). Hence $\mathrm{Tr}(x(1 - \zeta)) = \mathrm{Tr}(a_0(1 - \zeta)) = a_0 \cdot \mathrm{Tr}(1 - \zeta) = a_0[(p - 1) + 1] = a_0 p$.

It will be enough to show that $\mathrm{Tr}(x(1 - \zeta)) \in \mathbb{Z}p$, hence $a_0 \in \mathbb{Z}$. To prove that $a_j \in \mathbb{Z}$, we multiply by ζ^{p-j}, obtaining $x\zeta^{p-j} = a_0 \zeta^{p-j} + a_1^{p-j+1} + \cdots + a_{j-1}\zeta^{p-1} + a_j + a_{j+1}\zeta + \cdots + a_{p-2}\zeta^{p-j-2}$, and expressing ζ^{p-1} in terms of the lower powers of ζ, we may write $x\zeta^{p-j}$ in the form $x\zeta^{p-j} = (a_j - a_{j-1}) + a_1'\zeta + a_2'\zeta^2 + \cdots + a_{p-2}'\zeta^{p-2}$.

By induction $a_{j-1} \in \mathbb{Z}$, so that by the same argument, $a_j - a_{j-1} \in \mathbb{Z}$, thus $a_j \in \mathbb{Z}$.

To compute $\mathrm{Tr}(x(1 - \zeta))$, let $x_1 = x, x_2, \ldots, x_{p-1} \in A$ be the conjugates of x; so

$$\mathrm{Tr}(x(1 - \zeta) = x_1(1 - \zeta) + x_2(1 - \zeta^2) + \cdots + x_{p-1}(1 - \zeta^{p-1})$$
$$= (1 - \zeta)x' \in A(1 - \zeta),$$

since $(1 - \zeta^{i+1})/(1 - \zeta) = 1 + \zeta + \cdots + \zeta^i \in A$. But $\mathrm{Tr}(x(1 - \zeta)) \in A \cap \mathbb{Q} = \mathbb{Z}$, hence $\mathrm{Tr}(x(1 - \zeta)) \in A(1 - \zeta) \cap \mathbb{Z} = \mathbb{Z}p$. ∎

EXERCISES

1. Let x be a root of $X^3 - 2X + 5$. Compute the norm and trace of $2x - 1$ in the extension $\mathbb{Q}(x) \mid \mathbb{Q}$.

2. Let x be an algebraic integer, $x \neq 0$. Let $f \in \mathbb{Q}[X]$ be the minimal polynomial of x. Show that x^{-1} is an algebraic integer if and only if $f(0) = \pm 1$.

3. Let $f \in \mathbb{Z}[X]$ be a monic polynomial, let x be an algebraic number. Show that if $f(x)$ is an algebraic integer then x is an algebraic integer.

4. Let J be a non-zero ideal of the ring A of integers of an algebraic number field K. Show that there exists a positive integer m belonging to J.

5. Give an example of a Gaussian number $x = a + bi$ such that $N(x) = 1$ but x is not an algebraic integer.

6. Find the quotient and the remainder of the following divisions:
   (a) $2 + 3i$ by $1 + i$.
   (b) $3 - 2i$ by $1 + 2i$.
   (c) $4 + 5i$ by $2 - i$.

7. Find the greatest common divisor of the following pairs of Gaussian integers:
   (a) $15 + 12i, 3 - 9i$.
   (b) $6 + 7i, 12 - 3i$.
   (c) $3 + 8i, 12 + i$.

8. Find the decomposition into prime factors of the following Gaussian integers: $12 + i, 6 + 2i, 12, 35 - 12i, 3 + 5i$.

9. Show that if $x = a + bi$ is a Gaussian prime then either $ab = 0$ or $gcd(a,b) = 1$.

10. Determine the prime elements of the ring of algebraic integers of $\mathbb{Q}(\sqrt{2})$.

11. Determine the prime elements of the ring of algebraic integers of $\mathbb{Q}(\sqrt{5})$.

12. Let $K = \mathbb{Q}(\sqrt{d})$, A the ring of integers of K. Show that A is a Euclidean domain if and only if for every $x + y\sqrt{d} \in K$ there exists $a + b\sqrt{d} \in A$ such that $|N((x + y\sqrt{d}) - (a + b\sqrt{d}))| < 1$. Consider the cases where $d \equiv 1 \pmod 4$ and $d \not\equiv 1 \pmod 4$ and derive explicit relations.

13. Prove that $\mathbb{Q}(\sqrt{-1})$, $\mathbb{Q}(\sqrt{-2})$, $\mathbb{Q}(\sqrt{-3})$, $\mathbb{Q}(\sqrt{-7})$, $\mathbb{Q}(\sqrt{-11})$ are the only Euclidean fields $\mathbb{Q}(\sqrt{d})$, with $d < 0$.
   Hint: Use the previous exercise.

14. Prove that if $d = 2, 3, 5, 6, 7, 13, 17, 21$, and 29 then $\mathbb{Q}(\sqrt{d})$ is a Euclidean field.
   Hint: Use exercise 12.

15. Prove that there exist only finitely many integers d such that $d \equiv 2 \pmod 4$ or $d \equiv 3 \pmod 4$ and $\mathbb{Q}(\sqrt{d})$ is a Euclidean field.

16. Prove that $\mathbb{Q}(\sqrt{-19})$ and $\mathbb{Q}(\sqrt{23})$ are not Euclidean fields.

17. Determine the ring of integers of the field $\mathbb{Q}(\sqrt{2},i)$. Prove that this is a Euclidean domain.

18. Determine the ring of integers of the field $\mathbb{Q}(\sqrt{2},\sqrt{3})$ and prove that this is a Euclidean domain.

19. Let ζ be a primitive 5th root of unity. Prove that $\mathbb{Q}(\zeta)$ is a Euclidean field.

20. Determine the ring of integers of the field $\mathbb{Q}(\sqrt[4]{2})$.

21. Let ω be a primitive cubic root of unity. Determine the norm of an arbitrary element $a + b\omega (a,b \in \mathbb{Q})$. Determine the ring of algebraic integers and the units of $\mathbb{Q}(\omega)$.

22. Let ω be a primitive cubic root of unity. Prove that $\mathbb{Q}(\omega)$ is a Euclidean field.

23. Determine the prime elements of the ring of algebraic integers of $\mathbb{Q}(\omega)$, where ω is a primitive cubic root of unity.

24. In the ring $\mathbb{Z}[\sqrt{-5}]$ consider the following ideals: $I = (3, 4 + \sqrt{-5})$, $I' = (3, 4 - \sqrt{-5})$, $J = (7, 4 + \sqrt{-5})$, $J' = (7, 4 - \sqrt{-5})$. Prove that $I \cdot I' = (3)$, $J \cdot J' = (7)$, $I \cdot J = (4 + \sqrt{-5})$, $I' \cdot J' = (4 - \sqrt{-5})$ and show that I, I', J, J' are prime ideals.

25. Find algebraic integers $x, y \in \mathbb{Q}(\sqrt{-5})$ such that x, y have no common factor different from a unit, but $xr + ys \neq 1$ for all algebraic integers $r, s \in \mathbb{Q}(\sqrt{-5})$.

26. Show that the ring of algebraic integers of $(\mathbb{Q}\sqrt{10})$ is not a unique factorization domain.

27. Let K be a Galois extension of degree n over $\mathbb{Q}$. Show that for every non-zero algebraic integer x of K we have $\sum_{i=1}^{n} \sigma_i(x) \cdot \overline{\sigma_i(x)} \geqslant n$, (where $\sigma_1, \ldots, \sigma_n$ are the automorphisms of K and $\bar{a}$ denotes the complex conjugate of a).

Hint: Consider the norm of x and use the fact that the geometric mean of positive numbers does not exceed their arithmetic mean.

28. Let $\zeta_1, \ldots, \zeta_n$ be roots of unity over $\mathbb{Q}$. Show that $|\zeta_1 + \cdots + \zeta_n| \leqslant n$ and the equality holds if and only if $\zeta_1 = \cdots = \zeta_n$. Conclude that if $x = \zeta_1 + \cdots + \zeta_n$ and $|x| < n$ then for every conjugate $\sigma(x)$ of x we have $|\sigma(x)| < n$.

29. Let R be a domain, and let $\mathbf{K}_R$ be the free R-module with basis $\{1, i, j, k\}$.

On $\mathbf{K}_R$ we define an operation of multiplication which is bilinear and has the following multiplication table: 1 is the unit element; $i^2 = j^2 = k^2 = -1$; $ij = -ji = k$; $jk = -kj = i$; $ki = -ik = j$. With this operation $\mathbf{K}_R$ is an R-algebra, called the *algebra of quaternions over* R. We identify R with the subring $R \cdot 1$ of multiples of 1. In $\mathbf{K}_R$ we have the conjugation, defined as follows: if $\alpha = a_0 + a_1 i + a_2 j + a_3 k$ then $\bar{\alpha} = a_0 - a_1 i - a_2 j - a_3 k$. Finally, let $N : \mathbf{K}_R \to R$ be the norm mapping, defined by $N(\alpha) = \alpha\bar{\alpha} = a_0^2 + a_1^2 + a_2^2 + a_3^2 \in R$. Prove:

(a) $\overline{\alpha + \beta} = \bar{\alpha} + \bar{\beta}$, $\overline{\alpha\beta} = \bar{\alpha} \cdot \bar{\beta}$, $\bar{\bar{\alpha}} = \alpha$.

(b) $N(\alpha\beta) = N(\alpha) \cdot N(\beta)$.

(c) α is invertible in $\mathbf{K}_R$ if and only if $N(\alpha)$ is invertible in R.

(d) if R is a field then $N(\alpha) = 0$ if and only if $\alpha = 0$; hence $\mathbf{K}_R$ is a skew-field, that is, a ring which is not commutative and such that every non-zero element is invertible.

(e) the product of two sums of 4 squares in R is a sum of 4 squares in R.

30. Prove *Euler's identity:* In any commutative ring we have
$(x_1^2 + x_2^2 + x_3^2 + x_4^2)(y_1^2 + y_2^2 + y_3^2 + y_4^2) =$
$(x_1 y_1 - x_2 y_2 - x_3 y_3 - x_4 y_4)^2 + (x_1 y_2 + x_2 y_1 + x_3 y_4 - x_4 y_3)^2 +$
$(x_1 y_3 + x_3 y_1 + x_4 y_2 - x_2 y_4)^2 + (x_1 y_4 + x_4 y_1 + x_2 y_3 - x_3 y_2)^2$.

31. A quaternion $\alpha = \frac{1}{2}(a_0 + a_1 i + a_2 j + a_3 k) \in \mathbf{K}_\mathbb{Q}$ is said to be *integral* when all coefficients a_i are integers with the same parity. Let A be the set of all integral quaternions. Show:

(a) A is a subring of $\mathbf{K}_\mathbb{Q}$ containing $\mathbf{K}_\mathbb{Z}$.

(b) A is a free $\mathbb{Z}$-module with basis $\{\frac{1}{2}(1 + i + j + k), i, j, k\}$.

(c) if $\alpha \in A$ then $N(\alpha) \in \mathbb{Z}$.

(d) the only units in the ring A (that is, quaternions α such that α and α^{-1} belong to A) are the following 24 quaternions: $\pm 1, \pm i, \pm j, \pm k,$ $\frac{1}{2}(\pm 1 \pm i \pm j \pm k)$.

32. Let $\alpha, \beta \in \mathbf{K}_\mathbb{Q}$. We say that β is a *right-hand factor* of α when there exists $\gamma \in A$ such that $\alpha = \gamma \beta$. Similarly, we define a left-hand factor. We say that α, β are *associates* when there exists a unit ε of A such that $\alpha = \beta \varepsilon$ or $\alpha = \varepsilon \beta$. Prove:

(a) if $\alpha, \beta \in A$, $\beta \neq 0$, there exists $\gamma, \rho \in A$ such that $\alpha = \beta \gamma + \rho$ and $N(\rho) < N(\beta)$.

(b) let J be a right ideal of A, that is, $J \pm J \subseteq J$, $J \cdot A \subseteq J$. Show that J is principal; that is, there exists $\alpha \in A$ such that $J = \alpha A$.

(c) if $\alpha, \beta \in A$, not both equal to 0, there exists a greatest common right-hand factor δ; δ is unique up to a left-hand unit factor and may be expressed in the form $\delta = \mu \alpha + \nu \beta$, with $\mu, \nu \in A$ (Bézout property).

(d) let $\alpha \in A$, let $b \in \mathbb{Z}$, $b > 0$; show that the greatest common right-hand factor of α, b is equal to 1 if and only if $gcd(N(\alpha), b) = 1$.

(e) if $\alpha \in A$ show that there exists one associate α' of α which belongs to $\mathbf{K}_\mathbb{Z}$.

33. An integral quaternion π is said to be *prime* whenever the only factors of π are units or associates to π. Prove:

(a) if $p \in \mathbb{Z}$ is a prime number then p is not a prime quaternion.

(b) an integral quaternion $\pi \in A$ is prime if and only if $N(\pi)$ is a prime number.

34. Prove the following theorem of Lagrange: every natural number is the sum of the squares of 4 natural numbers.

Hint: Use Exercise 28 to reduce to the case of a prime number; then use Exercises 32, 33 to express p as the norm of a prime quaternion.

CHAPTER 6

Integral Basis, Discriminant

We have seen in the numerical examples of the preceding section that the ring of algebraic integers of a quadratic number field, and also of the cyclotomic field $\mathbb{Q}(\zeta)$ (where ζ is a primitive pth root of unity), are free Abelian groups.

In this chapter, we shall prove this fact in general and establish other interesting properties of the ring of algebraic integers. For this purpose, it will be necessary to develop theories which belong properly to algebra. However, we think that their inclusion in the text will be convenient to certain readers. Others, already well versed in these facts, may just take note of our notation and terminology.

To conclude the chapter, we shall introduce the discriminant, which is a rational integer associated with every algebraic number field. It will be a recurring procedure in the theory to attach numerical invariants to algebraic number fields, and in each case they will serve to measure a certain phenomenon.

To begin, let us recall the notion of rank of a module. Let R be a domain, M a R-module. If n is the maximum number of linearly independent elements of M, then n is called *the rank of M*.

If F is the field of quotients of R, if it is known that the R-module M is contained in a vector space V over F, let FM denote the subspace of V generated by M (it consists of all elements of the form $\Sigma_{i=1}^{r} a_i x_i$, with $a_i \in F$, $x_i \in M \subseteq V$ for $i = 1, \ldots, r$).

Thus for every element $y \in FM$ there exists $a \in R$, $a \neq 0$, such that $ay \in M$. We note that the elements $x_1, \ldots, x_n$ of M are linearly independent over R if and only if they are also linearly independent over F. Hence, M has rank n exactly when the F-vector space FM has dimension n.

Thus, everything is as natural as possible, when M is contained in a vector space V over F. There are several instances in which this occurs, but we just want to mention the very simplest of these cases:

A. *Let R be a domain, F its field of quotients. If M is a free R-module having a basis with n elements, then there exists a F-vector space V containing M; any two bases of M have the same number of elements, equal to the rank of M. Proof:* We take $V = F\xi_1 \oplus \cdots \oplus F\xi_n$ to be the set of all "formal" linear combinations of symbols $\xi_1, \ldots, \xi_n$, with coefficients in F; thus, the elements of V may be written uniquely in the form $\Sigma_{i=1}^n a_i \xi_i$, with $a_i \in F$.

Let $\{x_1, \ldots, x_n\}$ be any basis of the R-module M, so

$$M = Rx_1 \oplus \cdots \oplus Rx_n,$$

The mapping $\theta \colon M \to V$, defined by $\theta(\Sigma_{i=1}^n a_i x_i) = \Sigma_{i=1}^n a_i \xi_i$ is an isomorphism from the R-module M into V. Thus, replacing M by its image, we may consider M as contained in V.

If $y_1, \ldots, y_r$ are elements of $M \subseteq V$, linearly independent over R then by our previous considerations $r \leqslant n$, the dimension of the vector space V over F. In particular, any other basis of the R-module M has at most n elements. By symmetry, any two bases of the R-module M have n elements and n is the rank of M. ∎

Let us prove a weak result, which hints already the main theorem:

B.   *Let R be an integrally closed domain, F its field of quotients, let K be a separable extension of degree n of F, and let A be the integral closure of R in F. Then there exist free R-modules M and M' of rank n, such that $M' \subseteq A \subseteq M$. Explicitly, if $K = F(t)$, with $t \in A$, if d is the discriminant of t in $K \mid F$ then $M' = \mathbb{Z} \oplus \mathbb{Z}t \oplus \cdots \oplus \mathbb{Z}t^{n-1} = \mathbb{Z}[t]$, $M = (1/d)\mathbb{Z}[t]$.*

Proof: Let t' be a primitive element of K over F, so $K = F(t')$. By Chapter 5, **(I)**, we may write $t' = t/b$, with $t \in A$, $b \in R$. Thus $K = F(t)$ and therefore $\{1, t, \ldots, t^{n-1}\}$ is a F-basis of K. Since $t \in A$, then A contains the free R-module M' generated by this basis: $M' = R \oplus Rt \oplus \cdots Rt^{n-1} \subseteq A$; evidently M' is free of rank n.

To prove that A is contained in a free-R-module M, we proceed as follows: Let d be the discriminant of t in $K \mid F$, that is $d = \Pi_{i<j} (t_i - t_j)^2$, where $t_1 = t, t_2, \ldots, t_n$ are the conjugates of t over F; so $d \in F$.

We shall prove that $A \subseteq M = R(1/d) \oplus R(t/d) \oplus \cdots \oplus R[(t^{n-1})/d]$, and M is a free module of rank n. Let $y \in A$, so we may write $y = \Sigma_{j=0}^{n-1} c_j t^j$, where $c_j \in F$ for all $j = 0, 1, \ldots, n-1$. Then $y = \Sigma_{j=0}^{n-1} dc_j(t^j/d)$ and our task is now to show that each element dc_j belongs to R. We have $dc_j \in F$, and since R is integrally closed, it is enough to show that dc_j is integral over R.

Let K' be the smallest normal extension of F containing K, let $y = y_1, \ldots, y_n$ be the conjugates of y over F, so $y_i, t_i \in K'$. From $y = \Sigma_{j=0}^{n-1} c_j t^j$ we deduce $y_i = \Sigma_{j=0}^{n-1} c_j t_i^j$ for all $i = 1, \ldots, n$.

This set of relations indicates that $c_1, \ldots, c_n$ is the solution of the system of linear equations $\Sigma_{j=0}^{n-1} t_i^j X_j = y_i$ $(i = 1, \ldots, n)$ with coefficients $t_i^j \in K'$. To apply Cramer's rule, we note that

$$\delta = \det(t_i^j) = \prod_{<j} (t_i - t_j)$$

(as a Vandermonde determinant) so $\delta^2 = d$ and $\delta c_j = e_j$ where e_j is the determinant of the matrix obtained from $(t_i{}^j)$ be replacing the jth column by the elements $y_1, \ldots, y_n$. Since each y_i, $t_i{}^j$ is integral over R, then δ and e_j are integral over R. Therefore, $dc_j = \delta e_j$ is integral over R, proving the proposition. ∎

The above result holds in particular when $R = \mathbb{Z}$, so we rephrase it now:

B′. *If K is an algebraic number field of degree n, A the ring of algebraic integers, then A contains and is contained in free Abelian groups of rank n.*

This points toward our main theorem of this chapter. However, before tackling this question, we want to establish that the ideals of A are finitely generated; we have seen in Chapter 5, that it is not necessarily true that every ideal of A be principal (that is, with one generator), so the result which follows provides a weak replacement.

Now, ideals of a ring A are nothing else than the submodules of the A-module A. So, we are faced with the question: if R is a ring, if M is a R-module when are all submodules of M finitely generated? We treat this question in a roundabout way, giving a name to the modules with the above property, and finding sufficient condition for modules to belong to the class in question.

Definition 1. Let R be a ring; a R-module M is said to be *Noetherian* whenever every submodule of M is finitely generated.

In particular, M itself is finitely generated.

C. *Let R be a ring, M a R-module. Then the following properties are equivalent:*

(1) *M is a Noetherian R-module.*

(2) *every strictly increasing chain $N_1 \subset N_2 \subset N_3 \subset \cdots$ of submodules of M is finite.*

(3) *every non-empty family of submodules of M has a maximal element (with respect to the inclusion relation).*

Proof:

(1) → (2). Let $N_1 \subseteq N_2 \subseteq N_3 \subseteq \cdots$ be an increasing sequence of submodules of M, and let N be the union of all these submodules. By hypothesis, N is finitely generated, say by the elements $x_1, \ldots, x_n$. For every index $i = 1, \ldots, n$, there exists an index j_i such that $x_i \in N_{j_i}$. If $m \geq j_i$ for all $i = 1, \ldots, n$, then each $x_i \in N_m$ $(i = 1, \ldots, n)$ hence $N = N_m$, so $N_m = N_{m+1} = \cdots$.

(2) → (3). Let $\mathcal{M}$ be a non-empty family of submodules of M. Let $N_1 \in \mathcal{M}$; if N_1 is not a maximal element of $\mathcal{M}$, there exists $N_2 \in \mathcal{M}$,

such that $N_1 \subset N_2$; if N_2 is not a maximal element of $\mathcal{M}$, there exists $N_3 \in \mathcal{M}$, such that $N_1 \subset N_2 \subset N_3$. This procedure must lead to a maximal element of $\mathcal{M}$, otherwise there would exist an infinite strictly increasing chain of submodules of M, against the hypothesis.

(3) $\rightarrow$ (1). Let us assume that there exists a submodule N of M which is not finitely generated. Let $\mathcal{M}$ be the family of all finitely generated submodules of M which are contained in N (for example $0 \in \mathcal{M}$); let N' be a maximal element of $\mathcal{M}$, so $N' \neq N$. If $x \in N$, $x \notin N'$ the module $N' + Rx$ is still finitely generated, and contained in N, hence $N' + Rx \in \mathcal{M}$, with $N' \subset N' + Rx$; this contradicts the maximality of N'. ∎

An important particular case is obtained by considering the R-module $M = R$:

Definition 2. A ring R is said to be *Noetherian* when every ideal of R is finitely generated.

Thus, we may rephrase (**C**):

C'. *Let R be a ring. The following properties are equivalent:*
   (1) *R is a Noetherian ring.*
   (2) *every strictly increasing chain $J_1 \subset J_2 \subset J_3 \subset \cdots$ of ideals of R is finite.*
   (3) *every non-empty family of ideals of R has a maximal element (with respect to the inclusion relation).*

Thus, in particular, every principal ideal domain is a Noetherian ring.
We shall now develop properties of Noetherian modules:

D. *Every submodule and every quotient module of a Noetherian module are Noetherian modules.*
Proof: Let M be a Noetherian R-module and N a submodule. Since every submodule of N is also a submodule of M, the first assertion follows from (**C**), part (1).

Similarly, there is a one-to-one correspondence, preserving inclusion, between the submodules of the quotient module M/N and the submodules of M containing N. Then the second assertion follows also from (**C**), part (2). ∎

In order to be able to use inductive arguments, we establish now the following result:

E. *Let M be an R-module having a submodule N such that N and M/N are Noetherian modules. Then M itself is a Noetherian module.*
Proof: Let $\varphi: M \rightarrow M/N$ be the canonical homomorphism from M onto

M/N. Let M' be any submodule of M. Since M/N is a Noetherian module, there exist finitely many elements $x_1, \ldots, x_n \in M'$ such that their classes modulo N generate the submodule $(M' + N)/N$ of M/N.

If $y \in M'$ then $\varphi(y) \in (M' + N)/N$ so there exist elements $a_1, \ldots, a_n \in R$ such that $\varphi(y) = \Sigma_{i=1}^n a_i \varphi(x_i)$, so $y - \Sigma_{i=1}^n a_i x_i \in \varphi^{-1}(0) = N$; but, on the other hand $y - \Sigma\, a_i x_i \in M'$. Since N is a Noetherian module, the submodule $M' \cap N$ is finitely generated, say by the elements $y_1, \ldots, y_{m'} \in M' \cap N$; hence there exist elements $b_1, \ldots, b_m \in R$ such that $y = \Sigma_{i=1}^n a_i x_i + \Sigma_{j=1}^m b_j y_j$. This shows that $\{x_1, \ldots, x_n, y_1, \ldots, y_m\}$ is a system of generators of M', and so M is a Noetherian module. ∎

As a corollary, we have:

F. *If $M_1, M_2, \ldots, M_n$ are Noetherian R-modules, then the Cartesian product $M_1 \times M_2 \times \cdots \times M_r$ is also Noetherian.*

Proof: It is enough to show the statement for two modules M_1, M_2.

$M_1 \times M_2$ has the Noetherian submodule M_1 such that the quotient module $(M_1 \times M_2)/M_1 \cong M_2$ is also Noetherian. Hence, by (**E**), $M_1 \times M_2$ is a Noetherian module. ∎

G. *If R is a Noetherian ring, if M is a finitely generated R-module, then M is a Noetherian module.*

Proof: Let $x_1, \ldots, x_n$ be generators of the R-module M. Let $R^n = R \times \cdots \times R$ be the Cartesian product of n copies of the R-module R, let $\varphi: R^n \to M$ be the homomorphism from R^n onto M such that

$$\varphi(a_1, \ldots, a_n) = \sum_{i=1}^n a_i\, x_i.$$

Then $M \cong R^n/\mathrm{Ker}(\varphi)$, where

$$\mathrm{Ker}(\varphi) = \left\{ (a_1, \ldots, a_n) \,\middle|\, \sum_{i=0}^n a_i\, x_i = 0 \right\}$$

(kernel of the mapping φ).

By (**F**), R^n is a Noetherian R-module, and by (**D**), M is a Noetherian module. ∎

This result may be applied at once to the ring of algebraic integers:

H. *The ring of algebraic integers of an algebraic number field is a Noetherian ring.*

Proof: Let A be the ring of algebraic integers, so by (**B'**), $A \subseteq G$ where G is a free Abelian group (that is $\mathbb{Z}$-module) of finite rank. Since $\mathbb{Z}$ is a Noetherian ring, by (**G**), G is a Noetherian $\mathbb{Z}$-module, and by (**D**) A is a

Noetherian $\mathbb{Z}$-module. Since $\mathbb{Z} \subseteq A$, every A-submodule of A (that is, ideal) is also a $\mathbb{Z}$-submodule of A (that is, a subgroup). Using (**C**), part (2), A is a Noetherian A-module, that is, a Noetherian ring. $\blacksquare$

Thus, every ideal of the ring of algebraic integers is finitely generated. We shall show later that, even when an ideal is not principal, it may still be generated by 2 elements.

We shall now endeavor to prove that the ring A of algebraic integers of K is a free Abelian group. By (**B′**), it is contained in a free Abelian group of rank $n = [K:\mathbb{Q}]$. It will be enough to prove that every subgroup of a free Abelian group must be free. It is not more difficult to prove the corresponding statement for free modules over a principal ideal domain.

Let R be a domain, K its field of quotients. An R-module M is said to be *torsion-free* when the following property holds: if $a \in R$, $x \in M$ and $ax = 0$, $x \neq 0$ then $a = 0$.

For example:

I. *If R is a domain and M is a free R-module, then M is torsion-free.*
Proof: Let $\{x_1, \ldots, x_n\}$ be a basis of the R-module M. If $x \in M$, $x \neq 0$, it may be written in unique way as a linear combination $x = \sum_{i=1}^{n} a_i x_i$, where $a_i \in A$ and some coefficient a_{i_0} is not 0.

If $a \in A$ and $ax = 0$ then by the uniqueness of expression of 0 as linear combination of the basis elements, $aa_i = 0$ for every $i = 1, \ldots, n$. Since $a_{i_0} \neq 0$ and R is a domain, then $a = 0$, proving that M is torsion-free. $\blacksquare$

The converse holds, when R is a principal ideal domain:

Theorem 1. *Every finitely generated torsion-free module M over a principal ideal domain R is a free module.*
Proof: If $M = 0$ then it is a free module, with an empty basis.

Let $\{x_1, \ldots, x_n\}$ be a set of non-zero generators of M, with $n \geq 1$. If $n = 1$ then $M = Rx_1$ is a free R-module, because if $a_1 x_1 = 0$, then $a_1 = 0$.

We assume now that the theorem holds for modules with less than n generators, where $n > 1$. Let $S = \{y \in M \mid$ there exists $a \in R$, $a \neq 0$, with $ay \in Rx_n\}$. Since R is a domain, S is a submodule of M, containing x_n. The quotient module M/S is torsion-free; indeed, if $a \in R$, $a \neq 0$ and $a\bar{y} = \bar{0}$ in M/S (where $\bar{y}$ denotes the image of y in M/S), then $ay \in S$, so there exists $a' \in R$, $a' \neq 0$, with $a'ay \in Rx_n$; since $a'a \neq 0$ then $y \in S$, that is $\bar{y} = \bar{0}$.

The module M/S is finitely generated by the images $\bar{x}_1, \ldots, \bar{x}_{n-1}$. By induction, M/S is a free R-module with a finite basis. In the following lemma, we shall prove that $M \cong S \oplus (M/S)$.

It will be enough now to show that S is itself a free R-module; then a basis of S, together with a basis of M/S, will constitute a basis of M.

We show that S is isomorphic to a submodule of the field of quotients K of R. Namely, if $y \in S$, $y \neq 0$, let $a,b \in R$, $a \neq 0$, be such that $ay = bx_n$; if $a',b' \in R$, $a' \neq 0$ are also such that $a'y = b'x_n$, then $ba'y = bb'x_n = ab'y$, so $(ba' - ab')y = 0$, thus, $ba' = ab'$, that is, $b/a = b'/a'$ (in K). This allows us to define a mapping $\theta: S \to K$ by putting $\theta(y) = b/a$ (where $y \neq 0$, $ay = bx_n$ with $a,b \in R$, $a \neq 0$) and $\theta(0) = 0$. It is easy to check that θ is an isomorphism of the R-module S into K. But M is a finitely generated R-module and R is a principal ideal domain, hence a Noetherian ring; by **(G)**, M is a Noetherian R-module, so S and $\theta(S)$ are finitely generated R-modules. If $b_1/a_1, \ldots, b_m/a_m$ are generators of $\theta(S)$, then $S \cong \theta(S) \subseteq Rz$, where $z = 1/a_1 \ldots a_m$.

We shall soon prove in a lemma that every submodule of a cyclic R-module is again cyclic (when R is a principal ideal domain). This implies therefore that S is a cyclic module, $S = Ry \supseteq Rx_n \neq 0$, and since S is torsion-free, it must be free.

The proof of the first assertion is completed, except for two lemmas. █

Now, we fill the gaps in the above proof establishing the necessary lemmas.

Lemma 1. *Let R be any domain, let M be a R-module, M' a free R-module and assume that there exists a homomorphism φ from M onto M'. Then there exists an isomorphism ψ from M' into M, such that $M = \text{Ker}(\varphi) \oplus \psi(M')$.*
Proof: Let $\{x_i'\}_{i \in I}$ be a basis of M' and for every $i \in I$ let us choose arbitrarily an element $x_i \in M$ such that $\varphi(x_i) = x_i'$.

We define $\psi: M' \to M$ as follows: if $x' \in M'$, let $x' = \Sigma_{i \in I} a_i x_i'$ be the unique expression of x' as a linear combination of the basis; we put $\psi(x') = \Sigma_{i \in I} a_i x_i$. Then ψ is a homomorphism from M' into M, and $\varphi \circ \psi$ is the identity mapping of M', hence ψ is one-to-one.

Now, we may prove that $M = \text{Ker}(\varphi) \oplus \psi(M')$. If $x \in M$, let $x' = \varphi(x) = \Sigma_{i \in I} a_i x_i'$, then

$$x = \left(x - \sum_{i \in I} a_i x_i\right) + \sum_{i \in I} a_i x_i = (x - \psi(x')) + \psi(x'),$$

with $x - \Sigma_{i \in I} a_i x_i \in \text{Ker}(\varphi)$, $\psi(x') \in \psi(M')$. Also $\text{Ker}(\varphi) \cap \psi(M') = 0$, showing the lemma. █

Lemma 2. *If R is a principal ideal domain, then every submodule of a cyclic R-module is again cyclic; the converse also holds.*
Proof: Let N be a cyclic R-module, so $N = Rx$; if $\theta: R \to N$ is defined by $\theta(a) = ax$, then θ is a homomorphism of R-modules, having a kernel which is an ideal J of R. Since $\theta(R) = N$ then $N \cong R/J$.

If N' is a submodule of N, then its inverse image by θ is an ideal J' of R, $J \subseteq J'$. But R is a principal ideal domain, so $J' = Rb$ and $N' = \theta(J') = R . \theta(b)$, so it is a cyclic R-module.

The converse holds, because every submodule of R, that is, every ideal of R, has to be principal. ∎

We apply this theorem, obtaining the following corollary:

J. *If R is a principal ideal domain then every submodule of a free R-module of rank n is again free, of rank at most n.*
Proof: Let M be free of rank n, so M is torsion-free (by (**I**)). If N is a submodule of M, it is also torsion-free, and finitely generated (by (**G**)), hence by Theorem 1, N is a free R-module.
The assertion concerning the rank of the submodule is trivial. ∎

We return now to the situation of an algebraic number field, and state the following important theorem:

Theorem 2. *If K is an algebraic number field of degree n, then the ring A of algebraic integers is a free Abelian group of rank n.*
Proof: By (**B′**), there exist free Abelian groups M', M of rank n, such that $M' \subseteq A \subseteq M$. By (**J**), A is a free Abelian group of rank necessarily equal to n. ∎

Definition 3. Any basis of the free Abelian group A (ring of algebraic integers) is called an *integral basis* of K.
An integral basis is therefore also a basis of the vector space K over $\mathbb{Q}$, since it has $n = [K:\mathbb{Q}]$ elements.
We may apply (**J**), to obtain:

K. *Let A be the ring of integers of an algebraic number field of degree n. Then every non-zero ideal J of A is a free Abelian group of rank n.*
Proof: By (**J**), J is a free Abelian group of rank at most equal to n. However, if $\{x_1, \ldots, x_n\}$ is an integral basis of A, if $a \in J$, $a \neq 0$, then $\{ax_1, \ldots, ax_n\}$ is a linearly independent set of elements of J. Thus J has rank equal to n. ∎

Actually (**K**) may be considerably improved by a statement which relates two bases of the Abelian groups A and J.

L. *Let R be a principal ideal domain, let M be a free R-module of rank n, and M' a submodule of M of rank m. Then there exists a basis $\{x_1, \ldots, x_n\}$ of M and non-zero elements $f_1, \ldots, f_m \in R$ such that $\{f_1 x_1, \ldots, f_m x_m\}$ is a basis of M'. Moreover, we may choose the elements $f_1, \ldots, f_m$ so that f_i divides f_{i+1} (for $i = 1, \ldots, m-1$).*
Proof: The proof will be done by induction on m. The statement is trivial for $m = 0$.

Let $\{y_1, \ldots, y_n\}$ be any basis of the free R-module M. Every element $y \in M$ may be written in unique way in the form

$$y = \sum_{i=1}^{n} a_i y_i, \quad \text{with } a_i \in R.$$

Let $p_i : M \to R$ be the mapping defined by $p_i(y) = a_i$, so p_i is a linear transformation, namely, the ith projection. Since $M' \neq 0$ there exists an index i such that $p_i(M') \neq 0$.

So, we may consider the non-empty set of linear transformations $u : M \to R$ such that $u(M') \neq 0$. Since R is a principal ideal domain, hence a Noetherian ring, there exists a maximal element in the family of principal ideals $u(M') \neq 0$ so obtained, say $u_1(M') = Rf_1 \neq 0$.

Let $z_1 \in M'$ be such that $u_1(z_1) = f_1$. We shall show that if u is any other linear transformation from M to R we have $f = u(z_1) \in Rf_1$. Indeed, R is a principal ideal domain, therefore the ideal generated by f_1, f is principal, say equal to Rf'; thus there exist elements $r_1, r \in R$ such that $f' = r_1 f_1 + rf = (r_1 u_1 + ru)(z_1)$. For the linear transformation $r_1 u_1 + ru : M \to R$ we have $Rf_1 \subseteq Rf' \subseteq (r_1 u_1 + ru)(M')$. By the maximality of Rf_1 we conclude that $(r_1 u_1 + ru)(M') = Rf_1$ hence $f' \in Rf_1$ and $Rf \subseteq R_1 f$.

In particular, considering the projections p_i associated with the basis $\{y_1, \ldots, y_n\}$ of M, we have $p_i(z_1) \in Rf_1$ (for every $i = 1, \ldots, n$). Therefore $z_1 = \sum_{i=1}^{n} p_i(z_1) \cdot y_i = f_1 x_1$, for some element $x_1 \in M$; we conclude that $f_1 = u_1(z_1) = f_1 \cdot u_1(x_1)$, thus $u_1(x_1) = 1$.

Let N be the kernel of u_1; then $M = Rx_1 \oplus N$ and $M' = Rf_1 x_1 \oplus (N \cap M')$. Indeed, $Rx_1 \cap N = 0$ since $rx_1 \in N$ implies $r = u_1(rx_1) = 0$; in particular $Rf_1 x_1 \cap (N \cap M') = 0$. On the other hand, if $y \in M$ then we may write $y = u_1(y)x_1 + (y - u_1(y)x_1)$ with $u_1(y)x_1 \in Rx_1$ and $u_1(y - u_1(y)x_1) = 0$; this shows that $M = Rx_1 \oplus N$. Also, if $y' \in M'$ then $u_1(y') \in u_1(M') = Rf_1$ hence $u_1(y')x_1 \in Rf_1 x_1 = Rz_1 \subseteq M'$, and $y' - u_1(y')x_1 \in M' \cap N$.

Now $M' \cap N$ is a module of rank $m - 1$ contained in the free module N of rank $n - 1$. By induction, there exists a basis $\{x_2, \ldots, x_n\}$ of N and nonzero elements $f_2, \ldots, f_m \in R$ such that $\{f_2 x_2, \ldots, f_m x_m\}$ is a basis of $M' \cap N$ and f_i divides f_{i+1} (for $i = 2, \ldots, m - 1$). So, $\{x_1, x_2, \ldots, x_n\}$ is a basis of M and $\{f_1 x_1, f_2 x_2, \ldots, f_m x_m\}$ is a basis of M'.

We still have to show that f_1 divides f_2. We prove first that if $u : M \to R$ is any linear transformation, then $u(M' \cap N) \subseteq Rf_1$. Assuming the contrary, we consider the linear transformation $v : M \to R$, which coincides with u on N and coincides with u_1 on Rx_1; explicitly, if $r_1 \in R$, $y \in N$, then $v(r_1 x_1 + y) = r_1 + u(y)$. Then $v(M') = Rf_1 + u(N \cap M')$, and because $u(N \cap M')$ is not contained in Rf_1 we conclude that $v(M')$ contains properly Rf_1, against the maximality of the ideal Rf_1; this is a contradiction.

Now let p_2' be the linear transformation from M to R defined by $p_2'(x_2) = 1$, $p_2'(x_i) = 0$ when $i \neq 2$. Then $Rf_2 = p_2'(M' \cap N) \subseteq Rf_1$ by the preceding considerations, and f_2 divides f_1.   ∎

It is possible to prove that if $\{x_1', \ldots, x_n'\}$ is another basis of M and $f_1', \ldots, f_m' \in R$ are non-zero elements such that f_i' divides f_{i+1}' (for $i = 1, \ldots, m - 1$) and $\{f_1'x_1', \ldots, f_m'x_m'\}$ is a basis of M', then $f_i' = f_i$ (for $i = 1, \ldots, m$).

We shall now introduce a numerical invariant.

M.   *Let K be a field of algebraic numbers, and A the ring of algebraic integers. If $\{x_1, \ldots, x_n\}$, $\{x_1', \ldots, x_n'\}$ are any two integral bases, then*

$$\operatorname{discr}_{K|\mathbb{Q}}(x_1, \ldots, x_n) = \operatorname{discr}_{K|\mathbb{Q}}(x_1', \ldots, x_n').$$

Proof: By definition (see Chapter 2, **(11)**)

$$\operatorname{discr}_{K|\mathbb{Q}}(x_1, \ldots, x_n) = \det(\operatorname{Tr}_{K|\mathbb{Q}}(x_i x_j)) \neq 0.$$

If $x_j' = \Sigma_{i=1}^n a_{ij} x_i$ (with $a_{ij} \in \mathbb{Z}$) for $j = 1, \ldots, n$, then

$$\operatorname{discr}_{K|\mathbb{Q}}(x_1', \ldots, x_n') = [\det(a_{ij})]^2 \operatorname{discr}_{K|\mathbb{Q}}(x_1, \ldots, x_n).$$

Similarly, if $x_j = \Sigma_{i=1}^n b_{ij} x_i'$ (with $b_{ij} \in \mathbb{Z}$) for $j = 1, \ldots, n$, then

$$\operatorname{discr}_{K|\mathbb{Q}}(x_1, \ldots, x_n) = [\det(b_{ij})]^2 \operatorname{discr}_{K|\mathbb{Q}}(x_1', \ldots, x_n');$$

thus, $[\det(a_{ij})]^2 \cdot [\det(b_{ij})]^2 = 1$, so $[\det(a_{ij})]^2 = 1$, and therefore

$$\operatorname{discr}_{K|\mathbb{Q}}(x_1, \ldots, x_n) = \operatorname{discr}_{K|\mathbb{Q}}(x_1', \ldots, x_n').$$

Definition 4.   The discriminant (in $K \mid \mathbb{Q}$) of any integral basis is called the *discriminant of the field K*, and denoted by $\delta_{K|\mathbb{Q}} = \delta_K$. Thus, $\delta_K \in \mathbb{Z}$, $\delta_K \neq 0$.

We shall see later the significance of the discriminant. For the moment, we shall be only interested in computing the discriminant in some special cases.

For this purpose the following remarks will be useful.

N.   *Let $K = \mathbb{Q}(t)$ be an extension of degree n, t an algebraic integer. Then δ_K divides $\operatorname{discr}_{K|\mathbb{Q}}(t)$ and the quotient is the square of an integer.*
Proof: Let $\{x_1, \ldots, x_n\}$ be an integral basis of $K \mid \mathbb{Q}$. We may write

$$t^j = \sum_{i=1}^n a_{ij} x_i \qquad (j = 0, 1, \ldots, n - 1) \quad \text{with } a_{ij} \in \mathbb{Z}.$$

Hence, $\mathrm{discr}_{K|\mathbb{Q}}(t) = \mathrm{discr}_{K|\mathbb{Q}}(1, t, \ldots, t^{n-1}) = (\det(a_{ij}))^2 \cdot \delta_K$ and therefore δ_K divides $\mathrm{discr}_{K|\mathbb{Q}}(t) \in \mathbb{Z}$, and the quotient is the square of an integer. ∎

This limits the search of the discriminant, once a primitive integral element has been found. A further limitation comes from the following result due to Stickelberger:

O. *Let* $K \mid \mathbb{Q}$ *be an extension of degree* n, *and let* $\{y_1, \ldots, y_n\}$ *be any* $\mathbb{Q}$-*basis of* K, *where each* y_i *is an algebraic integer. Then*

$$\mathrm{discr}_{K|\mathbb{Q}}(y_1, \ldots, y_n) \equiv 0 \text{ or } 1 \pmod 4.$$

Proof: By definition, if $\sigma_1, \ldots, \sigma_n$ are the $\mathbb{Q}$-isomorphisms of K then

$$\mathrm{discr}_{K|\mathbb{Q}}(y_1, \ldots, y_n) = [\det(\sigma_i(y_j))]^2 = (P - N)^2 = (P + N)^2 - 4PN$$

where P (respectively N) denotes the sum of the terms with positive sign (respectively negative sign) in the expression of the determinant. Since $P + N$ and PN remain unchanged by the action of each σ_i, then by Galois theory $P + N, PN \in \mathbb{Q}$. On the other hand $P + N$, PN are algebraic integers hence $P + N, PN \in \mathbb{Z}$. Therefore, $\mathrm{discr}_{K|\mathbb{Q}}(y_1, \ldots, y_n)$ is congruent to 0 or to 1 modulo 4. ∎

1. DISCRIMINANT OF QUADRATIC FIELDS

P. *Let* $K = \mathbb{Q}(\sqrt{d})$, *where* d *is a square-free integer. If* $d \equiv 2 \pmod 4$ *or* $d \equiv 3 \pmod 4$ *then the discriminant of* K *is* $\delta_K = 4d$. *If* $d \equiv 1 \pmod 4$ *then* $\delta_K = d$.

Proof: By Chapter 5, (P), if $d \equiv 2 \pmod 4$ or $d \equiv 3 \pmod 4$ then $\{1, \sqrt{d}\}$ is an integral basis. Therefore

$$\delta_K = \begin{vmatrix} \mathrm{Tr}_{K|\mathbb{Q}}(1) & \mathrm{Tr}_{K|\mathbb{Q}}(\sqrt{d}) \\ \mathrm{Tr}_{K|\mathbb{Q}}(\sqrt{d}) & \mathrm{Tr}_{K|\mathbb{Q}}(d) \end{vmatrix} = \begin{vmatrix} 2 & 0 \\ 0 & 2d \end{vmatrix} = 4d.$$

If $d \equiv 1 \pmod 4$, then $\{1, (1 + \sqrt{d})/2\}$ is an integral basis, so

$$\delta_K = \begin{vmatrix} \mathrm{Tr}_{K|\mathbb{Q}}(1) & \mathrm{Tr}_{K|\mathbb{Q}}\left(\dfrac{1 + \sqrt{d}}{2}\right) \\ \mathrm{Tr}_{K|\mathbb{Q}}\left(\dfrac{1 + \sqrt{d}}{2}\right) & \mathrm{Tr}_{K|\mathbb{Q}}\left(\dfrac{1 + \sqrt{d}}{2}\right)^2 \end{vmatrix}$$

$$= \begin{vmatrix} 2 & 1 \\ 1 & \dfrac{1 + d}{2} \end{vmatrix} = (1 + d) - 1 = d.$$

Thus, the possible discriminants of quadratic fields are integers δ such that $\delta \equiv 1 \pmod 4$, or $\delta \equiv 8 \pmod{16}$ or $\delta \equiv 12 \pmod{16}$.

Let us note also in all cases:

Q. *The integers of $\mathbb{Q}(\sqrt{d})$ may be written in the form $(a + b\sqrt{\delta})/2$ where $a, b \in \mathbb{Z}$, $a^2 \equiv \delta b^2 \pmod 4$, and conversely.*

Proof: Indeed, by Chapter 5, **(O)**, this is true when $d \equiv 1 \pmod 4$, since $\delta = d$. If $d \equiv 2 \pmod 4$ or $d \equiv 3 \pmod 4$, then $\delta = 4d$ and if $a + b\sqrt{d}$ is an algebraic integer, $a, b \in \mathbb{Z}$, so

$$a + b\sqrt{d} = (2a + 2b\sqrt{d})/2 = (2a + b\sqrt{\delta})/2$$

with $(2a)^2 \equiv \delta b^2 \pmod 4$; and conversely. ∎

2. DISCRIMINANT OF CYCLOTOMIC FIELDS

R. *Let $K = \mathbb{Q}(\zeta)$, where ζ is a primitive pth root of 1, and p is an odd prime number. Then the discriminant of $K \mid \mathbb{Q}$ is*

$$\delta = (-1)^{(p-1)/2} p^{p-2}.$$

Proof: By Chapter 5, **(U)**, $\{1, \zeta, \zeta^2, \ldots, \zeta^{p-2}\}$ is an integral basis of $\mathbb{Q}(\zeta)$. The minimal polynomial of ζ over $\mathbb{Q}$ is the pth cyclotomic polynomial $\Phi_p = X^{p-1} + X^{p-2} + \cdots + X + 1$. By Chapter 2, **(11)**,

$$\delta = \operatorname{discr}(\Phi_p) = (-1)^{[(p-1)(p-2)]/2} N_{\mathbb{Q}(\zeta)|\mathbb{Q}}(\Phi_p{}'(\zeta))$$

where $\Phi_p{}'$ is the derivative of Φ_p. But $X^p - 1 = (X - 1)\Phi_p$, hence taking derivatives, $pX^{p-1} = \Phi_p + (X - 1)\Phi_p{}'$; thus for every root ζ^j of Φ_p $(j = 1, 2, \ldots, p - 1)$, we have $p(\zeta^j)^{p-1} = (\zeta^j - 1)\Phi_p{}'(\zeta^j)$.

Now, $N_{\mathbb{Q}(\zeta)|\mathbb{Q}}(\Phi_p{}'(\zeta)) = \prod_{j=1}^{p-1} \Phi_p{}'(\zeta^j)$, so we compute:

$$\prod_{j=1}^{p-1} \zeta^j = N_{\mathbb{Q}(\zeta)|\mathbb{Q}}(\zeta) = (-1)^{p-1} \cdot 1 = 1,$$

$$\prod_{j=1}^{p-1} (\zeta^j - 1) = \prod_{j=1}^{p-1} (1 - \zeta^j) = \Phi_p(1) = p;$$

hence,

$$N_{\mathbb{Q}(\zeta)|\mathbb{Q}}(\Phi_p{}'(\zeta)) = \frac{p^{p-1} \cdot 1}{p} = p^{p-2},$$

and therefore

$$\delta = (-1)^{(p-1)(p-2)/2} p^{p-2} = (-1)^{(p-1)/2} p^{p-2}$$

since $[(p - 1)(p - 2)]/2 \equiv (p - 1)/2 \pmod 2$. ∎

EXERCISES

1. Let t be an algebraic integer of degree n over $\mathbb{Q}$, and let $x_1, \ldots, x_n \in \mathbb{Z}[t]$ be such that $\mathrm{discr}(x_1, \ldots, x_n)$ is a square-free rational integer. Prove that $\{x_1, \ldots, x_n\}$ is a basis of the Abelian group $\mathbb{Z}[t]$.

2. Let d be a square-free rational integer, and let δ be the discriminant of the field $\mathbb{Q}(\sqrt{d})$. Show that $\{1, (\delta + \sqrt{\delta})/2\}$ is a basis of the Abelian group $\mathbb{Z}[\sqrt{d}]$.

3. Let J be a non-zero ideal of the ring A of integers of an algebraic number field K. Let $\{x_1, \ldots, x_n\}$ be an integral basis of K. Show that J has a basis $\{y_1, \ldots, y_n\}$ over $\mathbb{Z}$ of the following type:

$$\begin{cases} y_1 = a_{11}x_1 \\ y_2 = a_{21}x_1 + a_{22}x_2 \\ \quad\quad \cdots \\ y_n = a_{n1}x_1 + a_{n2}x_2 + \cdots + a_{nn}x_n \end{cases}$$

with $a_{ij} \in \mathbb{Z}$ satisfying $0 \leqslant a_{ij} < a_{jj} \leqslant a_{11}$ for every $i, j = 1, \ldots, n$.

Hint: 1.   Construct inductively a basis $\{y_1', \ldots, y_n'\}$ for the $\mathbb{Z}$-module J in the following way: using Exercise 4 of Chapter 5, show that there exists the smallest positive integer a_{11}' such that $y_1' = a_{11}'x_1 \in J$; then consider the smallest positive integer a_{22}' for which a linear combination $y_2' = a_{21}'x_2 + a_{22}'x_2$ belongs to J; define $y_3', \ldots, y_n'$ in similar manner; 2. Note that if $\{y_1', \ldots, y_n'\}$ is a basis of the $\mathbb{Z}$-module J and $i \neq j$, $m \in \mathbb{Z}$, then

$$\{y_1', \ldots, y_i' - my_j', y_{i+1}', \ldots, y_n'\}$$

is also a basis of J; using this fact, subtract a suitable multiple of y_{n-1}' from y_n', so that the new coefficient of x_{n-1} is not negative and smaller than $a_{n-1,n-1}'$; repeat this procedure.

4. Let J be the ideal of $\mathbb{Z}[\sqrt{d}]$, which is generated by p and $a + b\sqrt{d}$ (where d is a square-free rational integer, $d \not\equiv 1 \pmod 4$, $a, b \in \mathbb{Z}$, and p is a rational prime number). Show that J has the basis $\{p, a + (b - p[b/p]\sqrt{d}\}$ where $[x]$ denotes the largest integer not exceeding x.

5. Let $K = \mathbb{Q}(\sqrt{-5})$, and A the ring of integers of K. Find a basis over $\mathbb{Z}$ for each of the following ideals of A:
   (a) the ideal generated by 3 and $4 + \sqrt{-5}$.
   (b) the ideal generated by 7 and $4 + \sqrt{-5}$.

6. Let $\{x_1, \ldots, x_n\}$ be a linearly independent set of algebraic integers of the field K of degree n over $\mathbb{Q}$. Prove:
  (a) $\operatorname{discr}_{K|\mathbb{Q}}(x_1, \ldots, x_n)$ is a multiple of the discriminant δ of the field K.
  (b) $\{x_1, \ldots, x_n\}$ is an integral basis if and only if $|\operatorname{discr}_{K|\mathbb{Q}}(x_1, \ldots, x_n)| = |\delta|$.

7. Let $K = \mathbb{Q}(t)$ be an extension of degree n, $t \in A$ (the ring of algebraic integers of K), let d be the discriminant of t in $K \mid \mathbb{Q}$. Among all elements of the form

$$\frac{1}{d}(a_0 + a_1 t + \cdots + a_i t^i) \qquad \text{(with any } a_j \in \mathbb{Z} \text{ and } a_i \neq 0)$$

let x_i be one such that $|a_i|$ is the least possible. Show that $\{x_1, \ldots, x_n\}$ is an integral basis of K.

8. Let $K \mid \mathbb{Q}$ be an extension of degree n, let $\{x_1, \ldots, x_n\}$ be a $\mathbb{Q}$-basis of K composed of algebraic integers. Show that $\{x_1, \ldots, x_n\}$ is an integral basis of $K \mid \mathbb{Q}$ if and only if $|\operatorname{discr}_{K|\mathbb{Q}}(x_1, \ldots, x_n)| \leqslant |\operatorname{discr}_{K|\mathbb{Q}}(y_1, \ldots, y_n)|$ for every $\mathbb{Q}$-basis $\{y_1, \ldots, y_n\}$ of K composed of algebraic integers.

9. Compute the discriminant of the field $\mathbb{Q}(\sqrt{2}, \sqrt{3})$.

10. Compute the discriminant of the field $\mathbb{Q}(\sqrt{2}, i)$.

11. Compute the discriminant of the field $\mathbb{Q}(\sqrt[4]{2})$.

12. Compute the discriminant of the field $\mathbb{Q}(\sqrt[5]{2})$.

13. Let R be a ring, and M a R-module. If $x \in M$, let

$$\operatorname{Ann}(x) = \{a \in R \mid ax = 0\}$$

(the *annihilator* of x) . x is called a *torsion element* when $\operatorname{Ann}(x) \neq 0$ and a *torsion-free element* when $\operatorname{Ann}(x) = 0$. M is called a *torsion module* if all its elements are torsion elements. Show:
  (a) For every $x \in M$ $\operatorname{Ann}(x)$ is an ideal of R.
  (b) If R is a domain, the set T of all torsion elements of M forms a torsion submodule of M, and M/T is a torsion-free R-module.

14. Let R be a commutative ring, let M be a R-module and N a submodule of M. We say that N is a *pure submodule* of M when $N \cap aM = aN$ for every $a \in R$. Prove:
  (a) If $M = N \oplus N'$ then N, N' are pure submodules of M.
  (b) If M is a torsion-free module then the submodule N is pure if and only if $ax \in N$, $a \in R$, $x \in M$ imply $x \in N$.
  (c) Let M be a torsion-free module; show that N is a pure submodule of M if and only if M/N is a torsion-free module.

15. Let R be a principal ideal domain, let M be a free R-module of rank n, and N a submodule of M. Show that N is a pure submodule of M if and only if there exists a submodule N' of M such that $M \cong N \oplus N'$.

16. Let R be a principal ideal domain, let M be a free R-module of rank n, and N a submodule of M (hence N is also a free- R-module). Show that N is a pure submodule of M if and only if every basis of N may be extended to a basis of M.

17. Let $K \subseteq L$ be algebraic number fields, and let $A \subseteq B$ be their rings of algebraic integers. Show that the Abelian group A is a pure subgroup of B. Conclude that every integral basis of A is a part of an integral basis of B.

CHAPTER 7

The Decomposition of Ideals

We have shown that the ring A of algebraic integers of an algebraic number field is Noetherian and integrally closed. However, it is not true in general that A is a principal ideal domain. What can be said about ideals which are not principal?

We shall imitate the theory of divisibility, replacing elements of A by ideals of A, and more generally, elements of K (not in A) by a more general type of ideals.

Definition 1. Let A be any domain, K its field of quotients. An A-module M, contained in K, is said to be a *fractional ideal of A* when there exists an element $a \in A$, $a \neq 0$, such that $a \cdot M \subseteq A$.

Thus, every ideal of A is also a fractional ideal (taking $a = 1$), and if necessary, we shall call it an *integral ideal*.

However, K itself is not a fractional ideal of A (unless $A = K$), otherwise there exists $a \in A$, $a \neq 0$, such that $K = A(1/a)$. Then $1/a^2 = b/a$, with $b \in A$, and so $1/a = b \in A$, showing that $K = A(1/a) = A$.

The set of non-zero fractional ideals of A is endowed with an operation of multiplication: $M \cdot M' = \{\Sigma_{i=1}^{n} x_i x_i' \mid n \geqslant 1, x_i \in M, x_i' \in M'\}$.

It is easy to verify that $M \cdot M'$ is again a fractional ideal. If M, M' are integral ideals, so is $M \cdot M'$.

This operation is commutative, associative and has a unit element, namely the ideal A itself: $M \cdot A = M$. It generalizes the operation defined for integral ideals in Chapter 1, (3).

We say that a non-zero fractional ideal M is invertible when there exists a fractional ideal M' such that $M \cdot M' = A$. We shall return later to this concept. It is enough to observe now the following general fact: M_1, M_2 are invertible fractional ideals if and only if $M_1 \cdot M_2$ is invertible.

Among the fractional ideals, we consider the *principal fractional ideals*, namely those of type Ax, where $x \in K$. Every non-zero principal fractional ideal is invertible.

If M, N are non-zero fractional ideals of A, we say that M *divides* N, and write $M \mid N$, when there exists an integral ideal $J \subseteq A$ such that $M \cdot J = N$.

The following properties are immediate to verify: $M \mid M$ for every non-zero fractional ideal M; if M, M', M'' are fractional ideals, and $M \mid M'$, $M' \mid M''$

then $M \mid M''$; if $M \mid M'$ then $M' \subseteq M$ (because $M' = M \cdot J \subseteq M \cdot A = M$); if $M \mid M'$ and $M' \mid M$ then $M = M'$.

Thus, the relation of divisibility between non-zero fractional ideals is an order relation which implies the reverse inclusion. Actually, we shall see soon that for rings of algebraic integers the divisibility relation is equivalent to the reverse inclusion.

Let us note also that if $a,b \in K^{\cdot}$, then a divides b if and only if the principal fractional ideal Aa divides Ab.

In the theory of divisibility of $\mathbb{Z}$ the prime numbers play a basic role. As we know, $n \in \mathbb{Z}$ is a prime number if and only if $\mathbb{Z}/\mathbb{Z}n$ is a domain different from $\mathbb{Z}$ (and then also a field).

Thus the non-zero prime ideals of $\mathbb{Z}$ are $\mathbb{Z}p$, for all prime numbers. So, the non-zero prime ideals are candidates to play a similar role as prime elements do in the case of the ring $\mathbb{Z}$. As for $\mathbb{Z}$, we have:

A.   *Let A be the ring of integers of an algebraic number field K. Then every non-zero prime ideal P of A is a maximal ideal, and it contains exactly one prime number p. Moreover A/P is a finite field containing $\mathbb{F}_p = \mathbb{Z}/\mathbb{Z}_p$.*
Proof: Let P be a non-zero prime ideal of A. Since A is integral over $\mathbb{Z}$, by Chapter 5, **(B)**, $P \cap \mathbb{Z}$ is a non-zero ideal which is obviously prime. Thus $P \cap \mathbb{Z} = \mathbb{Z}p$ for some prime number p. No other prime $p' \neq p$ is contained in P, otherwise taking m,n integers such that $1 = mp + np'$ we would have $1 \in P$, which is not the case.

But A/P is integral over the field $\mathbb{F}_p = \mathbb{Z}/\mathbb{Z}p$. Hence, by Chapter 5, **(F)** A/P is a field and P is a maximal ideal of A.   ∎

Summarizing, we have shown that A is a Noetherian, integrally closed domain in which every non-zero prime ideal is maximal. These properties are already very strong and have significant implications on the arithmetic of the ring A.

We may state a basic theorem, which includes results of Dedekind, Noether, Krull, and Matusita:

Theorem 1.   *Let A be a domain. Then the following properties are equivalent:*

(1) *A is Noetherian, integrally closed and every non-zero prime ideal of A is maximal.*

(2) *Every non-zero (integral) ideal of A is expressible in a unique way as the product of prime ideals.*

(3) *Every non-zero (integral) ideal of A is the product of prime ideals.*

(4) *The set of non-zero fractional ideals of A is a multiplicative group.*
Proof:

(1) → (2). We begin by showing:

(a) If J is a non-trivial ideal of A, there exist prime ideals $P_1, \ldots, P_r$ of A, such that $P_i \supseteq J$ for all $i = 1, \ldots, r$, but $J \supseteq P_1 P_2 \ldots P_r$.

The assertion holds when J is a prime ideal. If J is not prime, there exist elements $a,a' \notin J$ such that $aa' \in J$. Let $I = J + Aa$, $I' = J + Aa'$, so $I \cdot I' \subseteq J$, but $J \subset I, J \subset I'$; thus $I \neq A$ and $I' \neq A$.

If the statement is already true for the ideals I,I' then it follows automatically for J; namely, there exist prime ideals $P_1, \ldots, P_s$ containing $I \supset J$ and $P_1', \ldots, P_{s'}'$ containing $I' \supset J$, such that $P_1 \ldots P_s \subseteq I, P_1' \ldots P_{s'}' \subseteq J'$, so $P_1 \ldots P_s P_1' \ldots P_{s'}' \subseteq I \cdot I' \subseteq J$.

If however the statement fails for I, for example, then I is not a prime ideal and we may repeat the argument. This process must terminate, otherwise it would give rise to a strictly increasing infinite chain of ideals of A, which cannot exist, because A is a Noetherian ring.*

Now, we claim:

(b) Let P be a non-zero prime ideal of A, let $P^{-1} = \{x \in K \mid xP \subseteq A\}$. Then P^{-1} is a fractional ideal, $A \subset P^{-1}$ and $PP^{-1} = A$.

In fact, P^{-1} is evidently an A-module contained in K, and if $a \in P$, $a \neq 0$, then $aP^{-1} \subseteq PP^{-1} \subseteq A$, so P^{-1} is a fractional ideal. From $1 \in P^{-1}$ we deduce that $A \subseteq P^{-1}$. We show now that $P^{-1} \neq A$. Let $c \in P$, $c \neq 0$, hence by (a) there exist prime ideals $P_1, \ldots, P_r$ such that $P \supseteq Ac \supseteq P_1 \ldots P_r$, with $P_i \supseteq Ac$ for every $i = 1, \ldots, r$. We may choose r to be the minimum possible, for which this property holds. Then P must contain one of these prime ideals, say $P \supseteq P_1$, hence by the hypothesis $P = P_1$, since every non-zero prime ideal is maximal. By the minimality of r, we have $Ac \nsupseteq P_2, \ldots, P_r$, so there exists $a \in P_2 \ldots P_r$, $a \notin Ac$, that is $a/c \notin A$. However $a/c \in P^{-1}$, because $(a/c)P \subseteq (1/c)P_1P_2 \ldots P_r \subseteq A$, so $A \subset P^{-1}$.

Next, we prove that $PP^{-1} = A$. At any rate $P = P \cdot A \subseteq PP^{-1} \subseteq A$, and PP^{-1} is an ideal of A, so either $PP^{-1} = A$ or $PP^{-1} = P$. If this second case takes place, then $PP^{-2} = (PP^{-1})P^{-1} = PP^{-1} = P$ and similarly $PP^{-n} = P$ for every $n \geq 1$. Thus, if $a \in P$, $b \in P^{-1}$ then $ab^n \in PP^{-1} = P$ for every $n \geq 1$. The ideal $J = \sum_{n=0}^{\infty} Aab^n$ is finitely generated, because A is a Noetherian ring, so there exists $n > 1$ and elements $c_0, \ldots, c_{n-1} \in A$ such that $ab^n = \sum_{i=0}^{n-1} c_i ab^i$, hence $b^n - \sum_{i=0}^{n-1} c_i b^i = 0$; this shows that b is integral over A, hence $b \in A$ because A is integrally closed. Thus $P^{-1} \subseteq A$, against the fact that $A \subset P^{-1}$. Therefore, $PP^{-1} = A$, proving (b).

Now, we may prove the implication $(1) \rightarrow (2)$.

If $J = A$ the assertion is trivially verified.

If J is a non-trivial ideal of A, let $P_1, \ldots, P_r$ be prime ideals of A such that $P_i \supseteq J$ ($i = 1, \ldots, r$) and $J \supseteq P_1 \ldots P_r$ (by (a)); we may assume that r is the minimum possible, $r \geq 1$.

---

* By examining the proof, we have actually shown the following fact, to be used in Chapter 10. *If A is a Noetherian ring, if J is an ideal of A, $J \neq A$, there exist prime ideals $P_1, \ldots, P_r$ of A, such that $P_i \supseteq J$ for all $i = 1, \ldots, r$, but $J \supseteq P_1P_2 \ldots P_r$.*

We shall prove by induction on r that J is expressible as a product of prime ideals.

If $r = 1$ then $J \supseteq P_1$, hence $J = P_1$.

If the statement holds for ideals containing a product of at most $r - 1$ prime ideals, let P be a prime ideal of A, such that $P \supseteq J^*$; from $P \supseteq J \supseteq P_1 \ldots P_r$, it follows that P contains some of the ideals P_i, say $P \supseteq P_r$ hence $P = P_r$; from (b), we have $A = PP^{-1} \supseteq JP_r^{-1} \supseteq P_1 \ldots P_r P_r^{-1} = P_1 \ldots P_{r-1}$. By induction, $JP_r^{-1} = P_1' \ldots P_s'$ where each P_i' is a prime ideal, and so $J = JP_r^{-1}P_r = P_1' \ldots P_s'P_r$.

Now, we prove the uniqueness of the decomposition into product of prime ideals. Let $J = P_1 \ldots P_r = P_1' \ldots P_s'$, where P_i, P_j' are prime ideals of A and $r \geqslant 1$, $s \geqslant 1$. We shall proceed by induction on the minimum of r, s.

If $r = 1$ (or $s = 1$) then $P_1 = P_1' \ldots P_s'$, so for example $P_1 \supseteq P_1'$, hence $P_1 = P_1'$ and therefore if $s \geqslant 2$ then $A = P_2' \ldots P_s'$; but this is impossible, since $P_2' \ldots P_s' \subseteq P_2' \neq A$. So $s = 1$.

Let us assume now that $r > 1$, $s > 1$. From $P_1 \supseteq P_1 \ldots P_r = P_1' \ldots P_s'$ we conclude, for example, that $P_1 = P_1'$ hence $P_2 \ldots P_r = P_2' \ldots P_s'$, and by induction, we must have $r = s$ and, after some permutation of indices, $P_2 = P_2', \ldots, P_r = P_r'$.

$(2) \rightarrow (3)$ This statement is trivial.

$(3) \rightarrow (4)$ It will be enough to show that every non-zero prime ideal of A is invertible. In fact, if J is any non-zero fractional ideal of A, let $d \in A$, $d \neq 0$ be such that $d \cdot J \subseteq A$. Then $J = (Ad)^{-1} \cdot (Ad \cdot J)$ where $(Ad)^{-1} = Ad^{-1}$ is the inverse of the principal ideal Ad. By the hypothesis, $Ad \cdot J$ and Ad are products of non-zero prime ideals; if each non-zero prime ideal of A is invertible, then the same holds for J. Thus the set of all non-zero fractional ideals of A is a multiplicative group.

Now, let P be a non-zero prime ideal, and let $a \in P$, $a \neq 0$; by hypothesis Aa is a product of prime ideals, so $P \supseteq Aa = P_1 \cdots P_r$. Since Aa is invertible then each ideal P_i is invertible. But P is prime, hence it contains some ideal P_i. The proof will be finished if we show that P_i is a maximal ideal, thus $P = P_i$ is invertible.

So, we are led to prove:

(c) If P is an invertible prime ideal of A, then P is maximal.

For this purpose, we shall require:

(d) Let $\bar{S}$ be an arbitrary domain, let $P_1, \ldots, P_r$ be invertible prime ideals, let $J = P_1 \cdots P_r$. Then if $J = P_1' \ldots P_s'$, with P_j' prime ideals, we have $s = r$ and after some permutation of indices, $P_j' = P_j$, for all $j = 1, \ldots, r$.

---

* It is known that in any ring A, every ideal $J \neq A$ is contained in a maximal ideal; in particular, this fact holds even more easily in the case where A is a Noetherian ring.

We first prove (d). Let P_1 be a minimal ideal among $P_1, P_2, \ldots, P_r$ (relative to inclusion). From $P_1 \supseteq J = P_1' \ldots P_s'$, it follows that there exists an index j, say $j = 1$, such that $P_1 \supseteq P_1'$. Similarly $P_1' \supseteq J = P_1 \ldots P_r$, hence $P_1' \supseteq P_i$ (for some i); thus $P_1 \supseteq P_i$ and by the minimality $i = 1$ and $P_1 = P_1'$.

Now, since P_1 is invertible, multiplying the relation $P_1 \ldots P_r = J = P_1' \ldots P_s'$ by P_1^{-1}, we have $P_2 \ldots P_r = P_1^{-1}J = P_2' \ldots P_s'$ where $P_1^{-1}J$ is still an integral ideal. We proceed inductively and arrive at the required conclusion.

Now, we turn to the proof of (c), and this will establish the implication.

Let $a \in A$, $a \notin P$ and consider the ideals $J = P + Aa$, $J' = P + Aa^2$. By hypothesis (3), $J = P_1 \ldots P_m$, $J' = P_1' \ldots P_n'$, where P_i, P_j' are prime ideals which must contain J, J' respectively, so contain P. In the domain $\bar{A} = A/P$, we consider the image $\bar{a}$ of a. Then $\bar{A} \cdot \bar{a} = \bar{P}_1 \ldots \bar{P}_m$, $\bar{A} \cdot \overline{a^2} = \bar{P}_1' \ldots \bar{P}_n'$, where $\bar{P}_i = P_i/P$, $\overline{P_j'} = P_j'/P$ are invertible prime ideals of $\bar{A}$ (because $\bar{a} \neq 0$ so $\bar{A}\bar{a}$, $\bar{A}\bar{a}^2$ are invertible ideals). Since $\bar{A} \cdot \overline{a^2} = (\bar{A}\bar{a})^2$, by (d), we must have $2m = n$ and, after renumbering, we have $\overline{P_{2i-1}'} = \overline{P_{2i}'} = \bar{P}_i$ for every $i = 1, \ldots, m$; therefore $P_{2i-1}' = P_{2i}' = P_i$ and so $(P + Aa)^2 = P + Aa^2$, thus $P \subseteq P + Aa^2 = (P + Aa)(P + Aa) \subseteq P^2 + Aa$. This implies that every element $x \in P$ may be written in the form $x = y + za$ with $y \in P^2$, $z \in A$. So $za = x - y \in P$ and since $a \notin P$ then $z \in P$. So, actually $P \subseteq P^2 + P \cdot Aa \subseteq P$, therefore $P = P(P + Aa)$ and since P is invertible then $A = P + Aa$. This is true for every $a \in A$, $a \notin P$, so P is a maximal ideal.

(4) $\to$ (1) We show first that A is a Noetherian ring. Let J be any (integral) ideal of A, $J \neq 0$; let J^{-1} be its inverse, so $A = JJ^{-1}$, therefore there exist elements $x_1, \ldots, x_n \in J$ and $y_1, \ldots, y_n \in J^{-1}$ such that $1 = \sum_{i=1}^n x_i y_i$. If $a \in J$ we have $a = \sum_{i=1}^n x_i(y_i a)$ with $y_i a \in J^{-1}J = A$, so the elements $x_1, \ldots, x_n$ generate the ideal J.

Next, we prove that A is integrally closed. Let $x = b/c \in K$ be a root of a monic polynomial, with coefficients in A: $X^m + a_1 X^{m-1} + \cdots + a_m$. Then $x^m = -(a_1 x^{m-1} + \cdots + a_{m-1}x + a_m)$ belongs to the fractional ideal J generated by $1, x, \ldots, x^{m-1}$. From this fact, we have $J^2 = J$, and since $J \neq 0$ is invertible, then $J = A$; in particular, $x \in A$ and so A is integrally closed.

Finally, let P be a non-zero prime ideal of A, let $a \in A$, $a \notin P$ and consider the ideal $J = P + Aa$. Since J is invertible, then $J' = J^{-1}P$ is such that $JJ' = P$, so $J' \supseteq P$.

On the other hand, if $y \in J'$, then $ay \in P$; from $a \notin P$ it follows that $y \in P$, showing the other inclusion $J' \subseteq P$. Thus $J' = P$ and $JP = P$ implies that $J = A$. This is true for every $a \in A$, $a \notin P$, so P is a maximal ideal. ∎

The unique factorization of ideals of the ring of algebraic integers was discovered by Dedekind, so this fact is usually known as Dedekind's theorem. Hurwitz gave a direct algebraic proof, and later Noether found which purely algebraic properties of a domain imply the unique factorization of ideals into prime ideals; Krull improved the form of Noether's properties, showing that (1) implies (2). Matusita proved the equivalence of conditions (2) and (3).

Since the domains with the properties of the theorem are so important, we introduce a definition:

Definition 2.   A domain A is said to be a *Dedekind domain* whenever it satisfies the equivalent properties (1), (2), (3), (4) of Theorem 1.

Many of the results which we may prove for the ring of algebraic integers are actually valid for arbitrary Dedekind domains, and when no further difficulties arise, we shall state these results in full generality. The reader may consider Dedekind domains as a natural generalization of principal ideal domains and as a counterpart of the unique factorization domains.

As a bonus of Theorem 1, we arrive at several interesting consequences.

Let us look at prime elements of a domain A; p is a *prime* in A if $p = ab$, with $a,b \in A$ implies that p is associated with either c or b; so the principal ideal Ap is indecomposable as a product of distinct principal ideals. Thus, a reasonable generalization would be the following: an integral ideal P of the domain A is said to be *indecomposable* when it is not equal to a product $P = J . J'$, where J,J' are integral ideals different from P. But, actually, these are no new kinds of ideals:

B.   *Let A be a Dedekind domain. An ideal of A is indecomposable if and only if it is prime.*
Proof: The following implication holds for an arbitrary domain. Let P be an ideal which is not indecomposable. Then we may write $P = JJ'$, where J,J' are integral ideals different from P. From $J \mid P$, $J' \mid P$, it follows that $P \subset J$, $P \subset J'$, so there exist elements $a \in J$, $a' \in J'$ such that $a \notin P$, $a' \notin P$, however, $aa' \in JJ' = P$. Thus, P is not a prime ideal.

Conversely, if A is a Dedekind domain, and if J is an indecomposable ideal, either $J = 0$ (hence it is prime) or $J \neq 0$, and then J is a product of non-zero prime ideals; but since J is indecomposable, it must be a prime ideal. ∎

Next:

C.   *If P is a prime ideal in a Dedekind domain A, if J,J' are ideals of A and $P \mid JJ'$ then $P \mid J$ or $P \mid J'$.*

Proof: By the uniqueness of the prime ideal decomposition of JJ', from $JJ' = PI$ (where I is an integral ideal), it follows that P is present among the prime ideals in the decomposition of JJ'. Hence, either $P \mid J$ or $P \mid J'$. ∎

The divisibility between ideals means the reverse inclusion:

D. *Let M, M' be non-zero fractional ideals of a Dedekind domain. Then M divides M' if and only if $M \supseteq M'$.*
Proof: We have already shown that if $M \mid M'$ then $M \supseteq M'$. Conversely, if $M \supseteq M'$, since the non-zero fractional ideals form a group, there exists the fractional ideal M^{-1} and $M'M^{-1} \subseteq MM^{-1} = A$, so $M'M^{-1}$ is an integral ideal such that $(M'M^{-1})M = M'$; that is, M divides M'. ∎

E. *If J is any integral ideal of a Dedekind domain A, then there exist only finitely many integral ideals of A which divide J.*
Proof: Let $J = \prod_{i=1}^{r} P_i^{e_i} \subseteq A$. Then the integral ideals dividing J are precisely those of the form $\prod_{i=1}^{r} P_i^{f_i}$, where $0 \leqslant f_i \leqslant e_i$ for every $i = 1, \ldots, r$. Hence, there are only finitely many integral ideals dividing J. ∎

We may now consider the greatest common divisor and least common multiple of fractional ideals of a Dedekind domain.

F. *If M, M' are non-zero fractional ideals of the Dedekind domain A, then $M + M'$ is the greatest common divisor of M and M' and $M \cap M'$ is the least common multiple of M, M'.*
Proof: We begin by observing that if M, M' are non-zero fractional ideals then $M + M'$ and $M \cap M'$ are also non-zero fractional ideals (for example, if $b, b' \in A$ are non-zero elements such that $bM \subseteq A$ and $b'M' \subseteq A$ then $bb'(M + M') \subseteq A$; also, if $a \in M$, $a' \in M'$ are non-zero elements then $0 \neq (ba)(b'a') \in AM \cap AM' \subseteq M \cap M'$).

By (**D**), $M + M'$ divides both M and M' and if $M'' \mid M$, $M'' \mid M'$ then $M'' \supseteq M$, $M'' \supseteq M'$, so $M'' \supseteq M + M'$, that is M'' divides $M + M'$; thus $M + M'$ is the greatest common divisor of M and M'.

The proof is analogous for the least common multiple. ∎

G. *A domain which is both a Dedekind and a unique factorization domain must be a principal ideal domain, and conversely.*
Proof: The converse has already been mentioned, so let us assume that the Dedekind domain A is also a unique factorization domain.

It is enough to prove that every non-zero prime ideal P is principal. Let $a \in P$, $a \neq 0$. Since a is not a unit, it is a product of prime elements, $a = \prod_{i=1}^{r} p_i^{e_i}$ with $e_i \geqslant 1$, $r > 0$.

From $\Pi_{i=1}^r p_i^{e_i} = a \in P$ it follows that $p_i \in P$ (for some index i), hence $Ap_i \subseteq P$. But Ap_i is a prime ideal, because if $a,b \in A$, $ab \in Ap_i$ then $p_i \mid ab$, thus $p_i \mid a$ or $p_i \mid b$, noting that A is a unique factorization domain.

By Theorem 1, Ap_i is a maximal ideal, so $P = Ap_i$ is a principal ideal. ∎

Thus, we have justified our previous remark, namely, the ring of integers of an algebraic number field is a unique factorization domain if and only if it is a principal ideal domain.

Another useful fact follows:

H. *If A is a Dedekind domain, for every non-zero fractional ideal M there exists a fractional ideal M', such that MM' is any given non-zero principal ideal.*

Proof: Given $Aa \neq 0$, since the non-zero fractional ideals form a group, there exists M' such that $MM' = Aa$. ∎

Since a Dedekind domain is Noetherian, all ideals are finitely generated. Much better, we may prove:

I. *If A is a Dedekind domain, then every fractional ideal of A may be generated by at most two elements: one of these elements may be arbitrarily chosen.*

Proof: It is clearly enough to prove this statement for non-zero integral ideals J. We shall prove that if J' is another non-zero integral ideal then there exists an element $b \in A$ such that $A = J' + Ab \cdot J^{-1}$.

This is enough, because if $a \in J$, $a \neq 0$, we let $J' = Aa \cdot J^{-1} \subseteq A$ and from $A = Aa \cdot J^{-1} + Ab \cdot J^{-1}$, it follows that $J = Aa + Ab$.

Let $J' = P^e, e \geqslant 1$, P being a prime ideal. Let $b \in J$, $b \notin JP$, so P does not divide the integral ideal $Ab \cdot J^{-1}$, hence the greatest common divisor of $P^e = J'$ and $Ab \cdot J^{-1}$ is $A = J' + Ab \cdot J^{-1}$.

More generally, let $J' = \Pi_{i=1}^r P_i^{e_i}$, where $r > 1$, $e_i \geqslant 1$.

For every $i = 1, \ldots, r$, let $J_i = JP_1 \ldots P_r P_i^{-1}$; by our previous argument, if $b_i \in J_i$, $b_i \notin J_i P_i$, then $A = P_i^{e_i} + Ab_i J_i^{-1}$; finally, let $b = b_1 + \cdots + b_r$, then $b \in A$ and we shall show that $b \notin JP_i$ for all $i = 1, \ldots, r$. In fact, $b_j \in J_j \subseteq JP_i$ for every $i \neq j$. On the other hand, from $b_j \notin J_j P_j = JP_1 \ldots P_r$ we deduce that $J^{-1}Ab_j \nsubseteq P_1 \ldots P_r = \cap_{i=1}^r P_i$, and by the above we must have $J^{-1}Ab_j \nsubseteq P_j$, so $b_j \notin JP_j$. Thus $b = b_1 + \cdots + b_r \notin JP_i$ for every $i = 1, \ldots, r$.

It follows that AbJ^{-1} is not contained in any of the prime ideals P_i, so the greatest common divisor of $J' = \Pi_{i=1}^r P_i^{e_i}$ and $Ab \cdot J^{-1}$ is $A = J' + Ab \cdot J^{-1}$. ∎

One of the basic theorems for Dedekind domains is the generalization of the Chinese remainder theorem (Chapter 3, Theorem 1 and (**C**)):

Theorem 2. *Let A be a Dedekind domain, let J be an integral ideal, $J = \prod_{i=1}^{r} P_i^{e_i}$ (where $r \geqslant 1$, $e_i \geqslant 1$, P_i are distinct non-zero prime ideals of A). Then*

$$A/J \cong \prod_{i=1}^{r} A/P_i^{e_i}.$$

Before proving this theorem, we shall derive a more general result, to be used later; besides, in this manner, we recognize which are the essential facts:

J. *Let A be a ring, let J be any ideal of A and assume that $J = \cap_{i=1}^{r} Q_i$, where the ideals Q_i are distinct from A and satisfy the relation $Q_j + \cap_{i \neq j} Q_i = A$ for every $j = 1, \ldots, r$. Then $A/J \cong \prod_{i=1}^{r} A/Q_i$.*

Proof: Let $\theta: A \to \prod_{i=1}^{r} A/Q_i$ be the mapping defined as follows:

$\theta(a) = (\theta_1(a), \ldots, \theta_r(a))$ where $\theta_i(a)$ denotes the image of a by the natural mapping $A \to A/Q_i$.

θ is a ring-homomorphism, whose kernel is equal to $\cap_{i=1}^{r} Q_i = J$. It remains to prove that θ maps A onto $\prod_{i=1}^{r} A/Q_i$; in other words, given elements $x_i \in A$ $(i = 1, \ldots, r)$, we shall show the existence of an element $x \in A$ such that $x - x_i \in Q_i$ for all $i = 1, \ldots, r$.

Special Case: There exists an index j such that $x_i = 0$ for all $i \neq j$, $1 \leqslant i \leqslant r$.

Since $Q_j + \cap_{i \neq j} Q_i = A$, we may write $x_j = y + z$ with $y \in Q_j$, $z \in \cap_{i \neq j} Q_i$, hence $z - x_j = -y \in Q_j$, and $z - x_i = z \in Q_i$ for $i \neq j$. We may therefore choose $x = z$.

General Case: For every $j = 1, \ldots, r$, we have shown the existence of an element $z_j \in A$ such that $z_j - x_j \in Q_j$, $z_j \in Q_i$ for all $i \neq j$. Let $x = \Sigma_{j=1}^{r} z_j$, then $x - x_i = \Sigma_{j \neq i} z_j + (z_i - x_i) \in Q_i$ for all $i = 1, \ldots, r$. ∎

Proof of the theorem: It is enough to note that $\prod_{i=1}^{r} P_i^{e_i} = \cap_{i=1}^{r} P_i^{e_i}$ and $P_j^{e_j} + \prod_{i \neq j} P_i^{e_i} = A$, when $P_1, \ldots, P_r$ are distinct prime ideals of the Dedekind domain A. This follows at once from (**F**). ∎

We may sharpen slightly the main assertion of the preceding theorem:

K. *Let A be a Dedekind domain, let $P_1, \ldots, P_r$ be distinct non-zero prime ideals of A; let $x_1, \ldots, x_r \in A$, let $e_1, \ldots, e_r$ be integers, $e_i \geqslant 0$ for every index i. Then there exists an element $x \in A$ such that*

$$x - x_i \in P_i^{e_i}, \quad x - x_i \notin P_i^{e_i+1} \quad \text{for all} \quad i = 1, \ldots, r.$$

Proof: For every $i = 1, \ldots, r$, we have $P_i^{e_i} \supset P_i^{e_i+1}$, so there exists an element $a_i \in P_i^{e_i}$, $a_i \notin P_i^{e_i+1}$. By Theorem 2, there exists $x \in A$ such that $x - (x_i + a_i) \in P_i^{e_i+1}$ for every $i = 1, \ldots, r$. Therefore, $x - x_i = [x - (x_i + a_i)] + a_i \in P_i^{e_i}$ but $x - x_i \notin P_i^{e_i+1}$. ∎

An interesting corollary concerns Dedekind domains which have only finitely many prime ideals:

L. *If a Dedekind domain has only finitely many prime ideals then it is a principal ideal domain.*

Proof: Since every non-zero ideal of the Dedekind domain A is a product of non-zero prime ideals, it is enough to show that these are principal ideals.

Thus, let $P_1, \ldots, P_r$ be the non-zero prime ideals of A. By **(K)**, for every h, $1 \leqslant h \leqslant r$, there exists an element $y_h \in A$, such that $y_h \in P_h$, $y_h \notin P_h^2$, and $y_h \notin P_j$ for all $j \neq h$. This is just the statement of **(K)**, with $x_1 = \cdots = x_r = 0$, $e_h = 1$, $e_j = 0$ for $j \neq h$.

Thus $P_h \mid Ay_h$, P_h^2 does not divide Ay_h and P_j does not divide Ay_h (for $j \neq h$). Hence the decomposition of Ay_h into prime ideals is $Ay_h = P_h$, proving that each prime ideal P_h is principal. ∎

We conclude by studying the integral closure of a Dedekind domain in a separable extension of its field of quotients:

M. *Let A be a Dedekind domain, and K its field of quotients. Let $L \mid K$ be a separable extension of degree n and let B be the integral closure of A in L. Then B is a Dedekind domain,*

Proof: B is integrally closed, by definition.

From Chapter 6, **(A)** we know that B is a submodule of a free A-module M of rank n; since A is a Noetherian ring then M is a Noetherian A-module (Chapter 6, **(G)**) hence, B is a Noetherian A-module; therefore, B is a Noetherian B-module, that is, a Noetherian ring.

Finally, if Q is a non-zero prime ideal of B, then $Q \cap A = P$ is a prime ideal of A and $Q \cap A \neq 0$ (Chapter 5, **(B)**), so P is a maximal ideal of A and by Chapter 5, **(G)**, Q is a maximal ideal of B.

By Theorem 1, B is a Dedekind domain. ∎

EXERCISES

1. Let A be the ring of algebraic integers of an algebraic number field. Prove that A has infinitely many prime ideals.

2. Let A be the ring of algebraic integers of an algebraic number field K. Prove successively the following facts:

   (a) If x is a root of $h \in A[X]$ then $h/(X - x)$ has coefficients in A.

Hint: Let d be the leading coefficient of h and m the degree of h; first note that dx is an algebraic integer; proceed by induction on m, considering the polynomial $h - dX^{m-1}(X - x)$.

(b) If $h = d(X - x_1)(X - x_2) \cdots (X - x_m) \in A[X]$ then $dx_1x_2 \cdots x_k$ is an algebraic integer (for every $k, 1 \leqslant k \leqslant m$).

(c) If $f, g \in A[X]$, if $c \in A$, $c \neq 0$, divides every coefficient of fg then, for every coefficient a of f and b of g, c divides ab.

Hint: Express f, g as products of the leading coefficient and linear factors, consider $fg/c \in A[X]$; apply (b) to this polynomial and conclude noting that the coefficients of any polynomial are expressible in terms of elementary symmetric functions of its roots.

3. Apply the previous exercise to obtain another proof, from Hurwitz, that the ring A of algebraic integers of K is a Dedekind domain. Prove successively the following facts:

(a) For every non-zero ideal I of A there exists a non-zero ideal J of A such that IJ is a principal ideal.

(b) If I, I', I'' are non-zero ideals of A and $I . I' = I . I''$ then $I' = I''$.

(c) If I, J are ideals of A such that $I \supseteq J$ then there exists an ideal I' of A such that $I . I' = J$ (that is, I divides J).

(d) An ideal P of A is prime if and only if $P = 0$ or P is maximal.

(e) If I, J are ideals of A such that $I . J$ is a multiple of the prime ideal P of A then either P divides I or P divides J.

(f) Every non-zero ideal I of A is equal to a product of prime ideals of A; this representation is unique, except for the order of the factors.

4. Prove, using transfinite induction (or Zorn's lemma) that every proper ideal I of a ring A (with unit element) is contained in a maximal ideal. Assuming that A is Noetherian, indicate a simpler proof for the same fact.

5. Prove *Hilbert basis theorem:* if A is a Noetherian ring then the ring $A[X_1, \ldots, X_n]$ of polynomials in n indeterminates is also Noetherian.

6. Prove that if a ring A' is a homomorphic image of a Noetherian ring A then A' is also a Noetherian ring.

7. Let B be a ring, A a Noetherian subring of B and let $x_1, \ldots, x_n \in B$. Prove that the subring $A[x_1, \ldots, x_n]$ of B generated by A and $x_1, \ldots, x_n$ is also a Noetherian ring.

8. Let A be a domain, and let I, J be ideals of A. The *conductor of J into I* is the set of all elements $x \in A$ such that $x . J \subseteq I$. It is denoted by $I : J$. Prove:

(a) $I : J$ is an ideal of A containing I.

(b) If J' is an ideal of A, then $J . J' \subseteq I$ if and only if $J' \subseteq I : J$.

For any ideals of A, show:

(c) $\left(\bigcap_{i=1}^{n} I_i\right) : J = \bigcap_{i=1}^{n} (I_i : J)$.

(d) $I : \left(\sum_{j=1}^{n} J_j\right) = \bigcap_{j=1}^{n} (I : J_j)$.

(e) $I : JJ' = (I : J) : J'$.

9. Let A be a domain, and let I be an ideal of A. The *root* $\sqrt{I}$ of the ideal I is the set of all elements $x \in A$ having some power $x^n (n \geqslant 1)$ in I.
Prove:

(a) $\sqrt{I}$ is an ideal containing I.

(b) $\sqrt{\sqrt{I}} = \sqrt{I}$.

(c) If I, J are ideals of A and there exists $n \geqslant 1$ such that $I^n \subseteq J$ then $\sqrt{I} \subseteq \sqrt{J}$.

(d) $\sqrt{I . J} = \sqrt{I \cap J} = \sqrt{I} \cap \sqrt{J}$.
$\sqrt{I + J} = \sqrt{\sqrt{I} + \sqrt{J}}$.

10. An ideal I of a ring A is said to be *primary* if the following condition is satisfied: If $a, b \in A$, $ab \in I$, $a \notin I$, then there exists an integer $m \geqslant 1$ such that $b^m \in I$.
Prove:

(a) Every prime ideal is a primary ideal.

(b) The root $\sqrt{I}$ of a primary ideal I is a prime ideal P; in this situation, if $a, b \in A$, $ab \in I$, $a \notin I$ then $b \in P$.

(c) If I is a primary ideal of A, if J, J' are ideals of A such that $J . J' \subseteq I$, $J \nsubseteq I$, then $J' \subseteq \sqrt{I}$.

(d) If A is a Noetherian ring and I is a primary ideal of A, show that there exists an integer $m \geqslant 1$ such that $I^m \subseteq \sqrt{I}$.

11. Let I, P be ideals of a ring A and suppose that:

(a) $I \subseteq P \subseteq \sqrt{I}$.

(b) If $ab \in I$, $a \notin I$ then $b \in P$.

Show that I is a primary ideal and $P = \sqrt{I}$.

12. Let B be a ring, and A a subring of B. Let I, I' be ideals of A, and J, J' ideals of B. For every ideal I of A let $B . I$ denote the ideal of B generated by I. Prove the following facts:

(a) $B . I$ is the set of sums $\Sigma_{i=1}^{n} b_i x_i$, with $n \geqslant 1$, $x_i \in I$, $b_i \in B$.

(b) $B(J \cap A) \subseteq J$, $B(J \cap A) \cap A = J \cap A$.

(c) $B(B . I \cap A) = B . I$.

(d) $(J + J') \cap A \supseteq J \cap A + J' \cap A$, and $B(I + I') = B . I + B . I'$.

(e) $(J \cap J') \cap A = (J \cap A) \cap (J' \cap A)$, and $B(I \cap I') \subseteq B \cdot I \cap B \cdot I'$.

(f) $J \cdot J' \cap A \supseteq (J \cap A)(J' \cap A)$ and $B \cdot (I \cdot I') = (B \cdot I)(B \cdot I')$.

(g) $(J : J') \cap A \subseteq (J \cap A : J' \cap A)$ and $B(I : I') \subseteq (B \cdot I) : (B \cdot I')$.

(h) $\sqrt{J} \cap A = \sqrt{J \cap A}$ and $B\sqrt{I} \subseteq \sqrt{B \cdot I}$.

(i) If J is a primary ideal of B, then $J \cap A$ is a primary ideal of A and $\sqrt{J \cap A} = \sqrt{J} \cap A$.

13. Prove that the powers of any maximal ideal of a ring A are primary ideals.

14. Show that the primary ideals of any Dedekind domain A are precisely the powers of the prime ideals of A.

15. Let I, J be integral ideals of the Dedekind domain A. Prove that there exists an integral ideal I' such that $I \cdot I'$ is a principal ideal and $J + I' = A$.

16. Let A be a Dedekind domain, and let I be a non-zero integral ideal of A. Show that the ring A/I satisfies the *descending chain condition for ideals* (every strictly descending chain $J_1 \supset J_2 \supset \cdots \supset J_n \supset \cdots$ of ideals is finite).

17. Let A be a domain satisfying the following properties:

(a) A is a Noetherian ring.

(b) A is integrally closed.

(c) for every non-zero ideal I of A the ring A/I satisfies the descending chain condition for ideals. Prove that A is a Dedekind domain.

18. Let A be a Dedekind domain. Show that the ideals of A satisfy the following distributive laws:

$$I \cap (J + J') = (I \cap J) + (I \cap J')$$

and

$$I + (J \cap J') = (I + J) \cap (I + J').$$

19. Prove the following *general form of the Chinese remainder theorem:* Let A be a Dedekind domain, and let $I_1, \ldots, I_n$ be ideals of A. The system of congruences $x \equiv a_i \pmod{I_i}$ (for $i = 1, \ldots, n$) has a solution in A if and only if $a_i \equiv a_j \pmod{I_i + I_j}$ for any indices i, j.

20. Let A be a domain. A subset S of A is said to be a *multiplicative subset* of A when $1 \in S$, $0 \notin S$ and if $a, b \in S$, then $ab \in S$. If I is an ideal of A let $A \setminus I$ denote the complement of I in A. Show that $A \setminus I$ is a multiplicative subset of A if and only if I is a prime ideal. If S is a multiplicative subset of A, let $\sqrt{S}$ denote the set of elements $x \in A$ such that there exist $n \geqslant 1$ for which $x^n \in S$. Show that $\sqrt{S}$ is again a multiplicative subset of A.

21. Let A be a domain, and let S be a multiplicative subset of A. In $A \times S$ we define the relation: $(a,s) \equiv (a',s')$ when $s'a = sa'$. Prove that this is an equivalence relation. Let a/s denote the equivalence class of (a,s) and let $S^{-1}A = A_S$ be the set of such classes. Define the following operations between elements of A_S:

$$\frac{a_1}{s_1} + \frac{a_2}{s_2} = \frac{s_2 a_1 + s_1 a_2}{s_1 s_2}, \qquad \frac{a_1}{s_1} \cdot \frac{a_2}{s_2} = \frac{a_1 a_2}{s_1 s_2}.$$

Prove the following:

(a) A_S is a ring.

(b) The mapping $\theta : A \to A_S$, defined by $\theta(a) = a/1$ is an isomorphism of the ring A into A_S. It is customary to identify A with its image $\theta(A)$.

(c) Every element of $\theta(S)$ is invertible in the ring A_S.

(d) $a/s = \theta(a) \cdot (1/\theta(s))$ for every $a \in A$, $s \in S$.

A_S is called the *ring of fractions of A by the multiplicative subset S*. In the particular case where $S = A \backslash P$ (for a prime ideal P of A) we use the notation A_P (see also Chapter 10, (**4**)).

22. Let A be a domain, having the following property: Every element of A is the product of finitely many prime elements (for example this holds when A is a unique factorization domain). Prove that there exists a one-to-one correspondence between the sets of prime elements of A and the ring of fractions of A (with respect to multiplicative subsets).

23. Let A be a Euclidean domain (see Chapter 1, Exercise 26), and let K be its field of quotients. Prove that if B is a subring of K containing A then there exists a multiplicative subset S of A such that $B = A_S$.

24. Let A be a domain, S a multiplicative subset of A, and A_S the corresponding ring of fractions, which contains A. Let I,I' denote ideals of A and J,J' ideals of A_S. Prove:

(a) $A_S I \cap A$ is the set of all elements $a \in A$ such that there exists $s \in S$ satisfying $as \in I$.

(b) $J = A_S(J \cap A)$.

(c) In order that $I = A_S I \cap A$ it is necessary and sufficient that if $s \in S$ then $I : As = I$.

(d) $A_S I \neq A_S$ if and only if $I \cap S = \varnothing$.

(e) If A is a Noetherian ring then A_S is also a Noetherian ring.

(f) If I is a primary ideal of A, $I \cap S = \varnothing$, then $\sqrt{I} \cap S = \varnothing$ and $A_S I$ is a primary ideal of A_S, $\sqrt{A_S I} = A_S \sqrt{I}$.

(g) The mapping $P \to A_S P$ establishes a one-to-one correspondence between the set of prime ideals of A, disjoint from S, and the set of prime ideals of A_S.

25. Let M be an additive subgroup of $\mathbb{Q}$. Prove that the following statements are equivalent:
- (a) M is a finitely generated additive group.
- (b) M is a fractional ideal of $\mathbb{Z}$.
- (c) M is a principal fractional ideal of $\mathbb{Z}$.

26. Let p be a prime number. For every $x \in \mathbb{Q}$, $x \neq 0$, let $v_p(x)$ be defined as in Chapter 1, Exercise 20. If I is any non-zero fractional ideal of $\mathbb{Z}$ then by the previous exercise $I = \mathbb{Z}x$, $x \in \mathbb{Q}$, $x \neq 0$. Define $v_p(I) = v_p(x)$.
Prove:
- (a) For every $I \neq 0$ we have $v_p(I) = 0$ except at most for a finite number of primes p.

If I, J are non-zero fractional ideals of $\mathbb{Z}$ then:
- (b) $I \subseteq J$ if and only if $v_p(I) \geqslant v_p(J)$ for every prime p.
- (c) $v_p(I + J) = \inf\{v_p(I), v_p(J)\}$.
- (d) $v_p(I \cap J) = \sup\{v_p(I), v_p(J)\}$.
- (e) $v_p(I \cdot J) = v_p(I) + v_p(J)$.

27. If M is a non non-zero additive subgroup of $\mathbb{Q}$, and p a prime number, define $v_p(M) = \inf\{v_p(x) \mid x \in M\}$. Show:
- (a) $v_p(M) \in \mathbb{Z} \cup \{-\infty\}$; for every M we have $v_p(M) \leqslant 0$ except at most for a finite number of primes p.
- (b) $M = \{x \in \mathbb{Q} \mid v_p(x) \geqslant v_p(M) \text{ for every prime } p\}$.

28. Let $\mathscr{S}$ be the set of sequences $v = (n_p)_{p \text{ prime}}$, $n_p \in \mathbb{Z} \cup \{-\infty\}$, such that $n_p \leqslant 0$ except at most for a finite number of primes p. Let $\mathscr{M}$ denote the set of all non-zero additive subgroups of $\mathbb{Q}$, and let $\theta : \mathscr{M} \to \mathscr{S}$ be the mapping defined by $\theta(M) = (v_p(M))_{p \text{ prime}}$.
If $M, M' \in \mathscr{M}$ show:
- (a) $M \subseteq M'$ if and only if $v_p(M) \geqslant v_p(M')$ for every prime p.
- (b) θ is a one-to-one mapping.
- (c) θ maps $\mathscr{M}$ onto $\mathscr{S}$.
- (d) M is a non-zero fractional ideal of $\mathbb{Z}$ if and only if $\theta(M) = (n_p)_p$ is such that each n_p is an integer and $n_p = 0$ except at most for a finite number of primes p.
- (e) M is a non-zero integral ideal of $\mathbb{Z}$ if and only if $\theta(M) = (n_p)_p$ satisfies the above condition and moreover $n_p \geqslant 0$ for every prime p.
- (f) M is a subring of $\mathbb{Q}$ if and only if $v_p(M) = 0$ or $-\infty$.
- (g) $\mathbb{Q}$ has $2^{\aleph_0}$ distinct subrings.
- (h) For every non-zero additive subgroup M of $\mathbb{Q}$ there exists a largest subring R of $\mathbb{Q}$ for which M is a R-module.

29. Let A be a Dedekind domain, P any non-zero prime ideal of A. If I is a non-zero fractional ideal of A, let $I = \Pi_P P^{v_P(I)}$ be its decomposition into prime ideals (where $v_P(I) \in \mathbb{Z}$ and $v_P(I) = 0$ except at most for finitely many prime ideals, depending on I). Define also $v_P(x) = v_P(Ax)$ for every $x \in A$, $x \neq 0$; finally, let $v_P(0) = \infty$. If I,J are non-zero fractional ideals of A, show:

(a) $v_P(xy) = v_P(x) + v_P(y)$.

(b) $v_P(x + y) \geqslant \min\{v_P(x), v_P(y)\}$.

(c) If $v_P(x) \neq v_P(y)$ then $v_P(x + y) = \min\{v_P(x), v_P(y)\}$.

(d) I divides J if and only if $v_P(I) \leqslant v_P(J)$ for every prime ideal $P \neq 0$ of A.

(e) $v_P(I) = \min\{v_P(x) \mid x \in I\}$.

(f) $v_P(I . J) = v_P(I) + v_P(J)$.

(g) $v_P(I + J) = \min\{v_P(I), v_P(J)\}$.

(h) $v_P(I \cap J) = \max\{v_P(I), v_P(J)\}$.

30. Let A be a Dedekind domain, K its field of quotients, and M any A-submodule of K; let P be any non-zero prime ideal of A. Define $v_P(M) = \inf\{v_P(I) \mid$ for every fractional ideal $I \subseteq M\}$. Show:

(a) $v_P(M) \in \mathbb{Z} \cup \{-\infty\}$; for every M we have $v_P(M) \leqslant 0$ except at most for a finite number of prime ideals P.

(b) $M = \{x \in K \mid v_P(x) \geqslant v_P(M)$ for every non-zero prime ideal $P\}$.

31. Let $\mathscr{P}$ be the set of non-zero prime ideals of the Dedekind domain A, and let $\mathscr{M}$ denote the set of all non-zero A-submodules of the quotient field K of A. Let $\mathscr{S}$ be the set of all sequences $(n_P)_{P \in \mathscr{P}}$, where $n_P \in \mathbb{Z} \cup \{-\infty\}$ and $n_P \leqslant 0$ except at most for a finite number of prime ideals P. Finally, let $\theta: \mathscr{M} \to \mathscr{S}$ be the mapping defined by $\theta(M) = (v_P(M))_{P \in \mathscr{P}}$. If $M, M' \in \mathscr{M}$ show:

(a) $M \subseteq M'$ if and only if $v_P(M) \geqslant v_P(M')$ for every $P \in \mathscr{P}$.

(b) θ is a one-to-one mapping.

(c) θ maps $\mathscr{M}$ onto $\mathscr{S}$.

(d) M is a non-zero fractional ideal of A if and only if $\theta(M) = (n_P)_P$ where each $n_P \in \mathbb{Z}$ and $n_P = 0$ except at most for a finite number of prime ideals $P \in \mathscr{P}$.

(e) M is a non-zero integral ideal of A if and only if the preceding condition is satisfied and moreover $n_P \geqslant 0$ for every $P \in \mathscr{P}$.

(f) M is a subring of K if and only if $v_P(M) = 0$ or $-\infty$ for every $P \in \mathscr{P}$.

(g) If A is a Dedekind domain with countably many prime ideals then K has $2^{\aleph_0}$ distinct subrings.

(h) For every non-zero additive subgroup M of K there exists a largest subring R of K for which M is a R-module.

CHAPTER 8

The Norm and Classes of Ideals

We know already that the ring A of integers of an algebraic number field K need not be a principal ideal domain. In this section, we associate with every field K a numerical invariant h, which measures the extent to which A deviates from being a principal ideal domain. h will be equal to 1 if and only if A is a principal ideal domain.

It is quite natural how to proceed.

Let $\mathscr{F}$ denote the multiplicative Abelian group of non-zero fractional ideals of A, let $\mathscr{P}i$ be the subgroup of non-zero principal fractional ideals. We may therefore consider the quotient group $\mathscr{F}/\mathscr{P}i$.

Explicitly, two non-zero fractional ideals M, M' are said to be *equivalent* when there exists $x \in K$, $x \neq 0$, such that $M' = Ax \cdot M$. We write $M \sim M'$.

This is clearly an equivalence relation, and if $M_1 \sim M_2$, $M_1' \sim M_2'$, then $M_1 \cdot M_1' \sim M_2 \cdot M_2'$. $\mathscr{P}i$ is precisely the subgroup of those ideals equivalent to the unit ideal A. Each element of $\mathscr{F}/\mathscr{P}i$ is called an *ideal class* of K, and $\mathscr{F}/\mathscr{P}i$ is the *ideal class group* of K. Sometimes it is also denoted by $\mathscr{C}(K)$. Roughly speaking, the larger the group $\mathscr{F}/\mathscr{P}i$ is, the more the ring A fails to be a principal ideal domain. So, a natural question is the following: is it possible that $\mathscr{F}/\mathscr{P}i$ be an infinite group?

This section is devoted to the answer of this question. It will be an outstanding theorem that $\mathscr{F}/\mathscr{P}i$ is finite; the number h of its elements will therefore be a numerical invariant of K, which will play a significant role in later work.

To begin, we must define the concept of the norm of an ideal.

In Chapter 7, (A), we have seen that if P is any non-zero prime ideal of the ring A of integers of an algebraic number field K, then the quotient domain A/P (which is a field) is necessarily finite. If $P \cap \mathbb{Z} = \mathbb{Z}p$, then $\#(A/P)$ is a power of p.

More generally:

A. *If P is a non-zero prime ideal of A, $e \geqslant 1$, then $\#(A/P^e) = \#(A/P)^e$.*
Proof: The result is true for $e = 1$. We shall proceed by induction on e, assuming it true for $e - 1$. P^{e-1}/P^e is an ideal of the ring A/P^e, hence by the isomorphism theorem for rings, we have

$$(A/P^e)/(P^{e-1}/P^e) \cong A/P^{e-1}.$$

119

It follows that $\#(A/P^e) = \#(A/P^{e-1}) \cdot \#(P^{e-1}/P^e)$. By induction, it suffices to show that $\#(P^{e-1}/P^e) = \#(A/P)$. $\#(P^{e-1}/P^e)$ is a vector space over the field A/P, with scalar multiplication so defined: if $\tilde{a} \in A/P$, $\bar{x} \in P^{e-1}/P^e$ then $\tilde{a}\bar{x} = \overline{ax}$ (the reader may easily check the details). Thus it is enough to show that P^{e-1}/P^e has dimension at most 1, since we know that $P^{e-1} \supset P^e$ therefore, the dimension of P^{e-1}/P^e is at least 1; from $P^{e-1}/P^e \cong A/P$ we deduce the required relation.

Now, let $x \in P^{e-1}$, $x \notin P^e$; we shall prove that $P^{e-1} \subseteq Ax + P^e$, and this implies that if $\bar{y} \in P^{e-1}/P^e$ then $\bar{y} = \tilde{a}\bar{x}$, for some element $a \in A$.

From $P^{e-1} \supseteq Ax + P^e$ we have $P^{e-1} \mid Ax + P^e$, hence there exists an integral ideal J such that $P^{e-1}J = Ax + P^e$; we shall prove that $J = A$. If not, let P' be a prime ideal such that $P' \supseteq J \supseteq P^{e-1}J \supseteq P^e$ hence $P' \mid P^e$ so $P' = P$ and $J = PJ'$, where J' is an integral ideal. Thus $P^e \supseteq P^eJ' = Ax + P^e \supseteq Ax$ hence $x \in P^e$, against its choice. Thus $J = A$ and $P^{e-1} = Ax + P^e$. ∎

Let us note incidentally the following useful result:

B.   *Let P be a non-zero prime ideal of A, $e \geqslant 1$. Let S be a system of representatives of A modulo P, such that $0 \in S$; let $t \in P$, $t \notin P^2$. Then*

$$R = \{s_0 + s_1 t + \cdots + s_{e-1}t^{e-1} \mid s_i \in S \text{ for } i = 0, 1, \ldots, e - 1\}$$

is a system of representatives of A modulo P^e.

Proof: The result is proved by induction on e; it is trivial for $e = 1$, and we now assume it true for $e - 1$. We show that if $r = s_0 + tr_1$, $r' = s_0' + tr_1'$ are distinct elements of R then $r - r' \notin P^e$.

Indeed,   if $r - r' = (s_0 - s_0') + (r_1 - r_1')t \in P^e$ then $s_0 - s_0' \in -(r_1 - r_1')t + P^e \subseteq P$, hence $s_0 = s_0'$ and $(r_1 - r_1')t \in P^e$. By hypothesis, $At = P \cdot J$ where P does not divide the ideal J. So we have an ideal J' of A such that $P^e \cdot J' = A(r_1 - r_1') \cdot At = A(r_1 - r_1') \cdot P \cdot J$ and therefore P^{e-1} divides $A(r_1 - r_1') \cdot J$; since P does not divide J then $r_1 - r_1' \in P^{e-1}$. By induction, we must have $r_1 = r_1'$ showing that $r = r'$.

Thus R contains already $\#(A/P)^e$ different representatives of A modulo P^e. From (**A**) it follows that R is a system of representatives of A modulo P^e. ∎

By means of Theorem 2 in Chapter 7, we deduce:

C.   *If J is any non-zero integral ideal of A, $J = \Pi_{i=1}^r P_i^{e_i}$, then*

$$\#(A/J) = \Pi_{i=1}^r \#(A/P_i)^{e_i}.$$

Proof: We just combine (**A**) with the above mentioned theorem. ∎

Thus, we have associated with every non-zero integral ideal a positive integer:

Definition 1. The *norm* of an ideal J of the ring A of integers of an algebraic number field is defined to be the positive integer $N(J) = \#(A/J)$.

By means of **(C)** we obtain:

D. *If J,J' are ideals of A, then $N(JJ') = N(J) \cdot N(J')$.*

Proof: If $J = \Pi_{i=1}^{r} P_i^{e_i}$, $J' = \Pi_{i=1}^{r} P_i^{e'_i}$, where $e_i \geqslant 0$, $e'_i \geqslant 0$, then $JJ' = \Pi_{i=1}^{r} P_i^{e_i+e'_i}$, so by means of **(C)** we deduce that $N(JJ') = N(J) \cdot N(J')$. ∎

The above property is usually called the *multiplicativity of the norm*.

Previously, we have defined the norm of an element $x \in K$ (relative to the extension $K \mid \mathbb{Q}$); in the case where $x \in A$, $x \neq 0$, we shall compare $N_{K|\mathbb{Q}}(x)$ and $N(Ax)$.

For this purpose we need an important relation between the norm of an ideal and the discriminant of any basis of this ideal.

E. *Let J be a non-zero integral ideal of A. Let $\{y_1, \ldots, y_n\}$ be any basis of the free Abelian group J, and δ the discriminant of the field K. Then:*

$$N(J)^2 = \frac{\mathrm{discr}_{K|\mathbb{Q}}(y_1, \ldots, y_n)}{\delta}$$

Proof: By Chapter 6, **(L)**, there exists an integral basis $\{x_1, \ldots, x_n\}$ and integers $f_1, \ldots, f_n$ such that $\{f_1 x_1, \ldots, f_n x_n\}$ is a basis of the free Abelian group J. Thus, $A = \mathbb{Z}x_1 \oplus \cdots \oplus \mathbb{Z}x_n$, while $J = \mathbb{Z}f_1 x_1 \oplus \cdots \oplus \mathbb{Z}f_n x_n$. Then the quotient Abelian group is $A/J \cong \Pi_{i=1}^{n} \mathbb{Z}/\mathbb{Z}f_i$, the isomorphism being obviously the one which comes from the mapping

$$A \to \prod_{i=1}^{n} \mathbb{Z}/\mathbb{Z}f_i, \qquad \sum_{i=1}^{n} a_i x_i \to (a_i \bmod f_i)_{1 \leq i \leq n},$$

by noting that the kernel is J. Therefore $N(J) = \#(A/J) = \Pi_{i=1}^{n} |f_i|$.

Since $\mathrm{discr}(f_1 x_1, \ldots, f_n x_n) = (\Pi_{i=1}^{n} f_i)^2 \cdot \mathrm{discr}(x_1, \ldots, x_n)$, noting that $\mathrm{discr}(x_1, \ldots, x_n) = \delta$ (discriminant of the field K), and

$$\mathrm{discr}_{K|\mathbb{Q}}(f_1 x_1, \ldots, f_n x_n) = \mathrm{discr}_{K|\mathbb{Q}}(y_1, \ldots, y_n)$$

for any basis $\{y_1, \ldots, y_n\}$ of the Abelian group J, we deduce that

$$N(J)^2 = \frac{\mathrm{discr}_{K|\mathbb{Q}}(y_1, \ldots, y_n)}{\delta}.$$

Now, we have:

F. *For every element $y \in A$, $y \neq 0$:*

$$N(Ay) = |N_{K|\mathbb{Q}}(y)|.$$

Proof: Let $\{x_1, \ldots, x_n\}$ be an integral basis, so $\{yx_1, \ldots, yx_n\}$ is a basis of the Abelian group Ay. Hence, by (**E**).

$$N(Ay)^2 = \frac{\text{discr}_{K|\mathbb{Q}}(yx_1, \ldots, yx_n)}{\delta}.$$

However, by Chapter 2, (**11**),

$$\text{discr}_{K|\mathbb{Q}}(yx_1, \ldots, yx_n) = \det(\sigma_i(yx_j))^2$$
$$= \det(\sigma_i(y\delta_{ij}))^2 . \det(\sigma_i(x_j))^2 = (N_{K|\mathbb{Q}}(y))^2 . \delta$$

(where $\delta_{ij} = 1$ when $i = j$, $\delta_{ij} = 0$ when $i \neq j$). Therefore, $N(Ay)^2 = (N_{K|\mathbb{Q}}(y))^2$, and since $N(Ay) > 0$ then $N(Ay) = |N_{K|\mathbb{Q}}(y)|$. ∎

One of the main facts about norms is the following:

G. *If J is a non-zero integral ideal of A, then J divides the principal ideal $A . N(J)$. For every integer $m > 0$ there are only finitely many ideals with norm equal to m.*

Proof: Let $N(J) = m = \#(A/J)$; thus in the quotient group A/J the order of every element divides m; therefore, if $x \in A$ then $mx \in J$. In particular $m = m . 1 \in J$, so J divides Am.

Since the ideal Am has only finitely many divisors (by Chapter 7, (**E**)) then there exists only finitely many ideals J with norm equal to m. ∎

Now, we are ready to prove that the group of classes of ideals is finite.

We first prove the existence of a certain uniform bound. Precisely, if J is any non-zero integral ideal of A, if $a \in J$, $a \neq 0$, then $J \mid Aa$, hence by (**D**) and (**F**), $N(J)$ divides $|N_{K|\mathbb{Q}}(a)|$, but the quotient $|N_{K|\mathbb{Q}}(a)|/N(J)$ is conceivable large.

However, we shall see that there exists an element $a \neq 0$ in J for which the above quotient becomes small, actually at most equal to some integer μ which is independent of J.

We prove it now:

H. *For every algebraic number field K there exists an integer $\mu > 0$ such that for every non-zero integral ideal J of A there exists an element $a \in J$, $a \neq 0$, such that*

$$|N_{K|\mathbb{Q}}(a)| \leqslant N(J) . \mu.$$

Proof: Let $\{x_1, \ldots, x_n\}$ be an integral basis of the field K. Let k be an integer such that $k^n \leq N(J) < (k + 1)^n$. Consider the set S of all elements $\sum_{i=1}^n d_i x_i$ where $0 \leq d_i \leq k$. Since $\#(S) = (k + 1)^n > \#(A/J)$ there must exist $b, c \in S$, $b \neq c$, such that $a = b - c = \sum_{i=1}^n a_i x_i \in J$.

We note that $|a_i| \leq k$. Now, we compute the absolute value of the norm of a; let $x_i = x_i^{(1)}, \ldots, x_i^{(n)}$ denote the conjugates of $x_i \in A$ (in the extension $K \mid \mathbb{Q}$). Thus:

$$|N_{K|\mathbb{Q}}(a)| = \prod_{j=1}^{n} \left| \sum_{i=1}^{n} a_i x_i^{(j)} \right| \leqslant \prod_{j=1}^{n} k \left(\sum_{i=1}^{n} |x_i^{(j)}| \right) = k^n \cdot \prod_{j=1}^{n} \left(\sum_{i=1}^{n} |x_i^{(j)}| \right).$$

The element $\mu = \Pi_{j=1}^{n}(\sum_{i=1}^{n} |x_i^{(j)}|)$ is an algebraic integer (because if y is an algebraic integer so is its complex conjugate $\bar{y}$, and $\sqrt{y\bar{y}} = |y|$ is also an algebraic integer); moreover μ is invariant under all conjugations (since this would only give rise to a change in the order of the product). Thus $\mu \in \mathbb{Z}$, $\mu > 0$ and from $k^n \leq N(J)$ it follows that $|N_{K|\mathbb{Q}}(a)| \leq N(J) \cdot \mu$. $\blacksquare$

Theorem 1. *The number of classes of ideals of an algebraic number field is finite.*

Proof: The norm of every non-zero integral ideal is a positive integer. Given the integer μ, defined in (**H**), by (**G**) there exists only finitely many non-zero ideals $J_1, \ldots, J_k$ such that $N(J_i) \leq \mu$.

We shall prove that if I is any non-zero ideal of A, then I is equivalent to some ideal J_i; therefore the number of classes of ideals is at most k, therefore finite.

Now, let I^{-1} denote the fractional ideal inverse of I, so there exists an element $c \in A$, $c \neq 0$, such that cI^{-1} is an integral ideal.

By (**H**), there exists an element $b \in cI^{-1}$, $b \neq 0$, such that $N(Ab) \leq N(cI^{-1}) \cdot \mu$. Multiplying by $N(I)$ and observing that $Ibc^{-1} \subseteq A$ we obtain

$$N(Ibc^{-1}) \cdot N(Ac) = N(Ibc^{-1} \cdot Ac) = N(Ib)$$

$$= N(Ab) \cdot N(I) \leq N(cI^{-1}) \cdot N(I)\mu = N(Ac)\mu,$$

thus: $N(Ibc^{-1}) \leq \mu$, so $Ibc^{-1} = J_i$ for some index i. $\blacksquare$

We introduce the following numerical invariant:

Definition 2. The number of classes of ideals of an algebraic number field K is called the *class number* of K and denoted by $h = h_K$.

An easy corollary follows:

I. *If J is any non-zero fractional ideal of the ring A then J^h is a principal fractional ideal of A.*

Proof: Just note that h is the order of the multiplicative group $\mathscr{F}/\mathscr{P}\iota$ of classes of ideals; thus the hth power of every fractional ideal is principal. $\blacksquare$

The following result will give rise to a justification of the word "ideal":

J. *Let K be an algebraic number field with class number h. There exists an extension K' of degree at most h over K with the following property: for every non-zero fractional ideal J of K there exists an element $x \in K'$ such that $J = K \cap A'x$ (where A' denotes the ring of algebraic integers of K').*

Proof: The Abelian multiplicative group $\mathscr{F}/\mathscr{P}\imath$ of classes of ideals has order h and by the structure theorem for finite Abelian groups (Chapter 3, Theorem 3), $\mathscr{F}/\mathscr{P}\imath$ is the product of cyclic groups $\mathscr{C}_1, \ldots, \mathscr{C}_m$ having orders $h_1, \ldots, h_m$ respectively; moreover $h = h_1 h_2 \ldots h_m$. Let $J_1, \ldots, J_m$ be non-zero fractional ideals whose ideal-classes are the generators of $\mathscr{C}_1, \ldots, \mathscr{C}_m$; thus if J is any non-zero fractional ideal we may write

$$J = Ac \cdot J_1^{e_1} \ldots J_m^{e_m},$$

where $c \in K^{\cdot}$, $0 \leq e_i < h_i$, $i = 1, \ldots, m$.

Since the ideal class of J_i has order h_i, then $J_i^{h_i} = Aa_i$, for some $a_i \in K^{\cdot}$. Writing $h = h_i h_i'$ $(i = 1, \ldots, m)$, we have $J^h = Ac^h \cdot J_1^{he_1} \ldots J_m^{he_m} = A(c^h a_1^{h_1'e_1} \ldots a_m^{h_m'e_m})$, thus $J^h = Aa$ with $a = c^h a_1^{h_1'e_1} \ldots a_m^{h_m'e_m}$.

For every index i let x_i be a root of the polynomial $X^{h_i} - a_i$, let $K' = K(x_1, \ldots, x_m)$, A' the ring of algebraic integers of K', thus $[K':K] \leq h_1 \ldots h_m = h$.

Then the element $x = cx_1^{e_1} \ldots x_m^{e_m} \in K'$ is such that $J = K \cap A'x$. In fact, we begin noting that $x^h = c^h x_1^{he_1} \ldots x_m^{he_m} = a$. Now, if $y \in J$ then $y^h \in J^h = Aa$ so $(y/x)^h = y^h/a \in A$, thus $y/x \in A'$, since it is integral over A and belongs to K'. Thus $y \in K \cap A'x$.

Conversely, if $y \in K \cap A'x$ then $y/x \in A'$, and $y^h/a = (y/x)^h \in A' \cap K = A$, so $(Ay \cdot J^{-1})^h$ is an integral ideal, hence $Ay \cdot J^{-1}$ is also an integral ideal of the Dedekind ring A, showing that $y \in J$. ∎

The preceding result tells that for every ideal J, not necessarily principal in K, there exists some element x in a field K' of degree at most h over K, such that J is the set of multiples in K of this element x; when J is not principal then x does not belong to K, so if we restrict our attention only to K, the element x is an "ideal element" and its multiples form the "ideal" J in K. This is the origin of this terminology, now so widespread.

It was Dirichlet who first observed that there exist rings of algebraic integers which are not unique factorization domains, that is, not principal ideal domains; he considered these questions while examining an alleged "proof" by Kummer of what is called "Fermat's last theorem."

Since not every ring of algebraic integers is a principal ideal domain, the possibility still remains that in any particular case we may enlarge the field K to another field K' whose ring A' of integers is a principal ideal domain.

In other words, is it always possible to find a finite extension K' of K such that the class-number of K' is equal to 1?

In *class field theory*, which is an advanced branch of the theory of algebraic numbers, with every algebraic number field K is associated an extension K' of degree h (the class number of K) over K. Among many important properties, for every fractional ideal J of K the ideal $A'J$ of K' is principal. However, there may exist fractional ideals in K', not generated by ideals in K, which are not principal. The next natural thing to do is to repeat the procedure just indicated, obtaining the tower of class fields: $K \subset K' \subset K'' \subset \cdots$ and the question arises whether, after a finite number of steps one reaches a field with class number 1; that is, whether the above chain is finite. This has been referred to as the "class field tower problem."

In 1966, Shafarevič and Golod showed that there exist infinite class field towers. At the present, there is reason to believe that this behavior is normal; in other words, it is perhaps even true that there exist only finitely many fields of algebraic numbers with class number equal to 1.

EXERCISES

1. Let A be the ring of algebraic integers of the field K, let I be a non-zero integral ideal of A, and $a,b \in A$. We say that a,b are congruent modulo I, and we write $a \equiv b \pmod{I}$ when $a - b \in I$. Prove:

(a) If $Aa + I = A$ there exists $x \in A$ such that $ax \equiv b \pmod{I}$; moreover, if $y \in A$ satisfies $ay \equiv b \pmod{I}$ then $x \equiv y \pmod{I}$.

(b) If $I_1, \ldots, I_n$ are integral ideals such that $I_i + I_j = A$ for $i \neq j$, if $a_1, \ldots, a_n \in A$, then there exists $x \in A$ such that $x \equiv a_i \pmod{I_i}$ for $i = 1, \ldots, n$.

2. Let I be an integral ideal of the ring A of algebraic integers of a field K. Let $\varphi(I)$ denote the number of congruence classes modulo I of elements $a \in A$ such that $Aa + I = A$. Prove:

(a) If the ideals I,J are such that $I + J = A$, then $\varphi(I \cdot J) = \varphi(I) \cdot \varphi(J)$.

(b) If P is a prime ideal, then $\varphi(P) = N(P) - 1$.

(c) If $I = P^e (e \geqslant 1)$, then

$$\varphi(I) = N(I)\left[1 - \frac{1}{N(P)}\right]$$

(d) If $I = P_1^{e_1} \ldots P_r^{e_r}$ (where $P_1, \ldots, P_r$ are distinct prime ideals), then

$$\varphi(I) = N(I) \cdot \prod_{i=1}^{r}\left[1 - \frac{1}{N(P_i)}\right]$$

(Note that in the particular case where $K = \mathbb{Q}$ then φ is identical with Euler function).

3. Let I be a non-zero ideal of the ring A of algebraic integers of K. Show that if $a \in A$ and $Aa + I = A$ then

$$a^{\varphi(I)} \equiv 1 \pmod{I}.$$

Deduce *Fermat's little theorem* for ideals: If P is a non-zero prime ideal of A, if $a \in A$, $a \notin P$, then

$$a^{N(P)-1} \equiv 1 \pmod{P}.$$

4. Let P be a non-zero prime ideal of the ring A of algebraic integers of the field K. Show that the set of residue classes $\bar{a} = a + P$ of elements $a \in A$, $a \notin P$ forms a cyclic multiplicative group (of order $N(P) - 1$).

5. Let P be a non-zero prime ideal of the ring A of algebraic integers of K, let $a \in A$. Show that there exists an integer $m \in \mathbb{Z}$ such that $a \equiv m \pmod{P}$ if and only if $a^p \equiv a \pmod{P}$ where $P \cap \mathbb{Z} = \mathbb{Z}p$.

6. Let P be a non-zero prime ideal of the ring A of algebraic integers of a field K. Let $a_1, \ldots, a_m \in A$. Show that there exists at most m pairwise non-congruent integers $x \in A$ such that

$$x^m + a_1 x^{m-1} + \cdots + a_{m-1}x + a_m \equiv 0 \pmod{P}.$$

7. Let I be an ideal of the ring A of algebraic integers of an algebraic number field K. Show that there exists $a, b \in I$ such that

$$N(I) = gcd(N_{K|\mathbb{Q}}(a), N_{K|\mathbb{Q}}(b)).$$

8. Let P be a prime ideal of A. Show that there exists $a \in A$ with the following property: For every $r \geqslant 1$ and $b \in A$ there exists $g \in \mathbb{Z}[X]$ such that $b \equiv g(a)$ $\pmod{P^r}$.

9. Show that if P is a prime ideal, $N(P) = p^f$, if $a \in A$ is the element with the property indicated in the preceding exercise, then there exists $g \in \mathbb{Z}[X]$, $\deg(g) = f$ such that $P = (p, g(a))$.

10. Let I be an ideal of the ring A of algebraic integers of the field K. Show that in every class of ideals of K there exists an integral ideal I such that $I + J = A$.

11. Let J be an integral ideal of the algebraic number field K. Show that if there exists $a \in A$ such that $N(J) = |N_{K|\mathbb{Q}}(a)|$ then $J = Aa$.

12. Let J be a non-zero ideal of the ring of integers A of the real quadratic field $K = \mathbb{Q}(\sqrt{d})$. We assume that $J^2 = Aa$, $N_{K|\mathbb{Q}}(a) < 0$ and $N_{K|\mathbb{Q}}(u) > 0$ for every unit u of A. Show that J is not a principal ideal of A.

13. Let h be the class number of the field K. Let I, J be non-zero fractional ideals of K, p a prime number not dividing h. Prove that if $I^p \sim J^p$ then $I \sim J$.

14. Let $x, y, x', y' \in K$, algebraic number field. Show that the ideals $J = (x, y)$ and $J' = (x', y')$ coincide if and only if there exist element $a, b, c, d \in K$ such that $ad - bc = 1$ and

$$\begin{cases} x' = ax + by \\ y' = cx + dy. \end{cases}$$

15. Let K be an algebraic number field, A the ring of integers of K. We say that two non-zero fractional ideals I, J are *strictly equivalent* when there exists $a \in A$ such that $I = Aa \cdot J$ and $N_{K|\mathbb{Q}}(a) > 0$. We write $I \approx J$ to express this fact. Let $\mathscr{F}$ be the multiplicative group of all non-zero fractional ideals of A, and let $\mathscr{P}\imath_0$ be the set of all non-zero principal fractional ideals Aa where $N_{K|\mathbb{Q}}(a) > 0$. Show:
  (a) If $I \approx J$ then $I \sim J$.
  (b) $\mathscr{P}\imath_0$ is a subgroup of $\mathscr{F}$.
The elements of $\mathscr{F}/\mathscr{P}\imath_0$ are called the *strict equivalence classes of ideals.*

16. Let p be an odd prime, ζ a primitive pth root of unity, $K = \mathbb{Q}(\zeta)$, and A the ring of integers of K. Show that if $x \in A$ there exists $m \in \mathbb{Z}$ such that $x^p \equiv m \pmod{A(1 - \zeta)^p}$.

17. With the same notations as in the previous exercise, let $x \in A$, $x \notin A(1 - \zeta)$. Show that there exists a positive integer e such that $\zeta^e x \equiv m \pmod{A(1 - \zeta)^2}$ for some $m \in \mathbb{Z}$.

18. Let K be an algebraic number field, let $\mathscr{C}(K)$ be the ideal class group of K. Show that there exists an integer $q \geqslant 0$ and ideal classes $\bar{I}_1, \ldots, \bar{I}_q$ such that every ideal class $\bar{I}$ of K may be written in unique way in the form $\bar{I} = \bar{I}_1^{k_1} \ldots \bar{I}_q^{k_q}$, with $0 \leqslant k_i \leqslant h_i - 1$, where $h_1 h_2 \ldots h_q = h$ (class number of K), each h_i is the power of a prime number, $\gcd(h_i, h_j) = 1$ for $i \neq j$, and $\bar{I}_i^{h_i}$ is the unit element of $\mathscr{C}(K)$. Moreover q is uniquely defined; it is called the *rank* of the class group $\mathscr{C}(K)$.

19. Let J be an integral ideal of the algebraic number field K. If there exists $x \in J$ such that $|N_{K|\mathbb{Q}}(x)| = N(J)$ show that J is the principal ideal generated by x.

CHAPTER 9

Units and Estimations for the Discriminant

As we have said, two elements of a domain are associated precisely when they generate the same ideal. Thus, by considering ideals, we ignore the units. However, it will become apparent that a number of arithmetic properties are intimately tied up with the units of the ring of integers A of the algebraic number field K.

We shall first consider the simplest type of units.

Any root of unity, that is, any root of a polynomial $X^m - 1$ (with $m \geqslant 1$) is an algebraic integer. If $\zeta^m = 1$ then $\zeta^{-m} = 1$, so ζ^{-1} is also a root of unity. Thus any root of unity in K is a unit of A.

Let $U = U_K$ denote the group of units of A and let $W = W_K$ denote the subgroup of U consisting of roots of unity. W is a non-trivial subgroup of U, since $1, -1 \in W$.

How large is W? Can it be infinite? What is the structure of the group W? Is W necessarily equal to U? If not, how can one determine the structure of U?

These are the main questions we are faced with in trying to study the units. We shall begin by describing rather accurately the group W, and afterwards we shall examine the units of quadratic fields and certain cyclotomic fields, to gain some insight into the possibilities for a reasonable structure theorem.

First, we prove a remarkable fact:

A. *Let c be any positive real number, and K an algebraic number field. Then there exist only finitely many algebraic integers x in K such that $|x^{(i)}| \leqslant c$ for all conjugates $x^{(i)}$ of x.*

Proof: We shall determine a finite set S, depending only on the constant c, and prove that if x is an algebraic integer of K such that $|x^{(i)}| \leqslant c$ for all its conjugates $x^{(i)}$, then $x \in S$.

Let $[K:\mathbb{Q}] = n$, let $s_1, s_2, \ldots, s_n$ be the elementary symmetric polynomials in n variables, namely, $s_1 = X_1 + X_2 + \cdots + X_n$, $s_2 = \Sigma_{i<j} X_i X_j, \ldots$, $s_n = X_1 . X_2 \ldots X_n$.

129

Let c' be a sufficiently large real number, for example

$$c' = \max \left\{ nc, \binom{n}{2}c, \ldots, \binom{n}{k}c^k, \ldots, c_n \right\}.$$

Let F be the set of all monic polynomials of degree at most n, whose coefficients are integers a such that $|a| \leqslant c'$; F is a finite set. Let S be the set of elements of K which are roots of some polynomial belonging to F; S is also a finite set.

If $|x^{(i)}| \leqslant c$ for all conjugates of $x \in K$, then $|s_k(x^{(1)}, x^{(2)}, \ldots, x^{(n)})| \leqslant c'$ for every $k = 1, \ldots, n$. Since x is an algebraic integer, then $s_k(x^{(1)}, \ldots, x^{(n)}) \in \mathbb{Z}$, and therefore the polynomial $\Pi_{i=1}^n (X - x^{(i)})$ belongs to F; that is $x \in S$. ∎

An immediate corollary is a characterization of the roots of unity in K:

B. *x is a root of unity in K if and only if x is an algebraic integer of K such that $|x^{(i)}| = 1$ for all conjugates $x^{(i)}$ of x.*
Proof: One implication is easily established, for if x is a root of unity, the same holds for all its conjugates. From $x^m = 1$ it follows that $|x|^m = 1$, so $|x| = 1$; the same holds for every conjugate of x.

Conversely, by **(A)** there are only finitely many algebraic integers x in K such that $|x^{(i)}| = 1$ for all conjugates of x.

Since $x, x^2, x^3, \ldots$ all share this property, then there exist integers $r, s, r < s$, such that $x^r = x^s$. Hence $x^{s-r} = 1$, showing that x is a root of unity. ∎

Combining **(A)** and **(B)**, we have the structure of the group W:

C. *The group W of roots of unity in K is a finite multiplicative cyclic group.*
Proof: By **(A)** and **(B)**, W must be finite.

Let h be the maximum of the orders of the elements in W. By Chapter 3, Lemma 2, the order of every element in W divides h, so W is contained in the group of hth roots of unity. By Chapter 2, **(8)**, this last group is cyclic, and therefore W itself is cyclic. ∎

The number of elements of W will be denoted by w. It is another numerical invariant attached to the field K. Since $1, -1$ are roots of unity belonging to any algebraic number field, then w is even. We shall prove in Chapter 10, **(7.L)** that w divides $2\delta_K$.

Now we pause and determine the units in special cases.

1. UNITS OF QUADRATIC FIELDS

Let $K = \mathbb{Q}(\sqrt{d})$, where $d < 0$ (d is a square-free integer). It is pretty easy to find all units:

D.   *If $d < 0$, $d \neq -1$, $d \neq -3$, then the units of $\mathbb{Q}(\sqrt{d})$ are $1, -1$.*

The units of $\mathbb{Q}(\sqrt{-1})$ are $1, -1, i, -i$.

The units of $\mathbb{Q}(\sqrt{-3})$ are 1, -1, $(1 + \sqrt{-3})/2$, $(1 - \sqrt{-3})/2$, $(-1 + \sqrt{-3})/2$, $(-1 - \sqrt{-3})/2$.

In all cases, every unit is a root of unity.

Proof: If $d \equiv 2 \pmod 4$ or $d \equiv 3 \pmod 4$, then the integers of $\mathbb{Q}(\sqrt{d})$ are of the form $x = a + b\sqrt{d}$ with $a, b \in \mathbb{Z}$; the conjugate of x is $x' = a - b\sqrt{d}$ and the norm is $N(x) = xx' = a^2 - b^2 d$. In order that x be a unit, it is necessary and sufficient that $N(x) = \pm 1$ (Chapter 5, (**M**)). Since $d < 0$, this means that $a^2 - b^2 d = 1$.

The only possible solutions are $a = \pm 1$, $b = 0$, except when $d = -1$, where we have another solution: $a = 0$, $b = \pm 1$.

If $d \equiv 1 \pmod 4$ then the integers of $\mathbb{Q}(\sqrt{d})$ are of the form $x = (a + b\sqrt{d})/2$, where $a, b \in \mathbb{Z}$ have the same parity. With the same argument, we are led to solve $a^2 - b^2 d = 4$. The only possible solutions are $a = \pm 2$, $b = 0$, except when $d = -3$, where we have another solution: $a = \pm 1$, $b = \pm 1$.

It is also quite clear that all units are roots of unity, those of $\mathbb{Q}(\sqrt{-3})$ being 6th roots of unity.   ∎

Let us consider now the more interesting case where $d > 0$, so $\mathbb{Q}(\sqrt{d})$ is contained in the field of real numbers. Thus, the only roots of unity in $\mathbb{Q}(\sqrt{d})$ are $1, -1$. We shall show that there exist other units in $\mathbb{Q}(\sqrt{d})$, when $d > 0$.

Lemma 1.   *If $\alpha > 0$ is an irrational real number, then for every integer $m > 0$ there exist integers a, b, not both equal to 0, such that $|a| \leqslant m$, $|b| \leqslant m$ and $|a + \alpha b| \leqslant (1 + \alpha)/m$.*

Proof: Let $f = X + \alpha Y$ and consider the set S of values $f(a,b) = a + \alpha b$, when $0 \leqslant a \leqslant m$, $0 \leqslant b \leqslant m$. Since α is irrational, if $(a,b) \neq (a',b')$ then $f(a,b) \neq f(a',b')$; therefore $\#S = (m + 1)^2$; all elements in S belong to the interval $[0, (1 + \alpha)m]$. We divide this interval into m^2 equal parts; $[0, (1 + \alpha)/m]$, $[(1 + \alpha)/m, 2(1 + \alpha)/m], \ldots$. Since there are more elements in S then sub-intervals, there exist at least two elements in S in the same subinterval:

$$\frac{r(1 + \alpha)}{m} \leqslant a_1 + \alpha b_1 < a_2 + \alpha b_2 \leqslant \frac{(r + 1)(1 + \alpha)}{m}.$$

Therefore, letting $a = a_2 - a_1$, $b = b_2 - b_1$, we have $|a + \alpha b| \leqslant (1 + \alpha)/m$, with a, b not both equal to 0, and $|a| \leqslant m$, $|b| \leqslant m$.   ∎

E. *If d is a positive square-free integer, then the group U of units of the field* $\mathbb{Q}(\sqrt{d})$ *is* $U \cong \{1,-1\} \times C$, *where C is an infinite multiplicative cyclic group.*
Proof: As we have already said, $W = \{1,-1\}$. To show that there exists another unit in $\mathbb{Q}(\sqrt{d})$, we shall make use of Lemma 1, with $\alpha = \sqrt{d}$. For every m, let S_m be the set of all couples of integers (a,b), with a,b not both 0, such that $|a| \leqslant m$, $|b| \leqslant m$, $|a + b\sqrt{d}| \leqslant (1 + \sqrt{d})/m$. By the lemma, each set S_m is non-empty. Let us write $S_m = S_m^+ \cup S_m^- \cup S_m^0$, where $S_m^+ = \{(a,b) \in S_m \,|\, a > 0\}$, $S_m^- = \{(a,b) \in S_m \,|\, a < 0\}$, $S_m^0 = \{(a,b) \in S_m \,|\, a = 0\}$. If $(a,b) \in S_m^+$ then $(-a,-b) \in S_m^-$ and vice-versa; also, if $m = 1$ then $S_1^0 = \{(0,1),(0,-1)\}$ and if $m \geqslant 2$ then $S_m^0 = \varnothing$ because $|b|\sqrt{d} \leqslant (1 + \sqrt{d})/m$ implies $|b| \leqslant (1/m)(1 + (1/\sqrt{d})) \leqslant (1/2)(1 + (1/\sqrt{d})) < 1$ which is impossible since b cannot be zero.

If $\cup_{m \geqslant 1} S_m$ is a finite set then there exists m_0 such that $1/m_0 < |a + b\sqrt{d}|$ for all $(a,b) \in \cup_{m \leqslant 1} S_m$; however, if m is sufficiently large and if $(a,b) \in S_m$ then $|a + b\sqrt{d}| < (1 + \sqrt{d})/m \leqslant 1/m_0$, which is a contradiction. Thus, $\cup_{m \geqslant 1} S_m$ is an infinite set, hence $\cup_{m \geqslant 1} S_m^+$ is also an infinite set (otherwise from $\#S_m^- = \#S_m^+$ for every m, it would follow that $\cup_{m \geqslant 1} S_m$ is finite).

From $|a| \leqslant m$, $|b| \leqslant m$ it follows that $|a - b\sqrt{d}| \leqslant |a| + |b|\sqrt{d} \leqslant m(1 + \sqrt{d})$ so

$$0 \neq |a^2 - b^2 d| = |a - b\sqrt{d}| \cdot |a + b\sqrt{d}|$$

$$\leqslant m(1 + \sqrt{d})\frac{1 + \sqrt{d}}{m} = (1 + \sqrt{d})^2$$

for every $(a,b) \in \cup_{m \geqslant 1} S_m$, hence also for every $(a,b) \in \cup_{m \geqslant 1} S_m^+$. Therefore, there exists some integer n, $0 < |n| \leqslant (1 + \sqrt{d})^2$ such that $a^2 - b^2 d = n$ for infinitely many couples of integers (a,b), where $a > 0$.

Consider $n^2 + 1$ among these couples (a,b). Let us define $(a_1,b_1) \equiv (a_2,b_2)$ when $a_1 \equiv a_2 \pmod{n}$, $b_1 \equiv b_2 \pmod{n}$; thus we have at most n^2 equivalence classes. Since the number of couples is greater than n^2, there exist at least two distinct couples (a_1,b_1), (a_2,b_2) in the same equivalence class.

Let $x_1 = a_1 + b_1\sqrt{d}$, $x_2 = a_2 + b_2\sqrt{d}$, and consider $u = x_1/x_2$. Since $N(x_1) = N(x_2) = n$ then $N(u) = 1$, and $u \neq \pm 1$, because $x_1 \neq x_2$, $x_1 \neq -x_2$ (since $a_1 > 0$, $a_2 > 0$).
But

$$u = \frac{x_1}{x_2} = 1 + \frac{x_1 - x_2}{x_2} = 1 + \frac{(x_1 - x_2)x_2'}{N(x_2)}$$

$$= 1 + \left(\frac{a_1 - a_2}{n} + \frac{b_1 - b_2}{n}\sqrt{d}\right)(a_2 - b_2\sqrt{d})$$

and noting that $(a_1 - a_2)/n$, $(b_1 - b_2)/n$ are integers, and multiplying out, we obtain $u = a + b\sqrt{d}$, with $a,b \in \mathbb{Z}$. Thus, u is a unit, different from $1, -1$.

There must exist some unit u of $\mathbb{Q}(\sqrt{d})$ such that $1 < u$; in fact, if u is a unit, then $u, -u, u^{-1}, -u^{-1}$ are units and the largest of these real numbers is greater than 1.

Now, we shall prove that among all units $u > 1$ there exists a smallest possible; it is enough to show that for every real number $c > 1$ there exists only finitely many units u such that $1 < u < c$. Now, if u is such a unit, then from $N(u) = uu' = \pm 1$, it follows that $1/c < u' < 1$ or $-1 < u' < -1/c$ and at any rate $|u'| < c$. By (A) the set of such units must be finite.

Let u_1 be the smallest unit such that $1 < u_1$. We shall prove that every positive unit u is a power of u_1. In fact, there exists an integer m such that $u_1{}^m \leqslant u < u_1^{m+1}$; then $u/u_1{}^m$ is again a unit such that $1 \leqslant u/u_1{}^m < u_1$, hence necessarily $u = u_1{}^m$. Similarly, all negative units are of the form $-u_1{}^m$, for $m \in \mathbb{Z}$.

Let C be the multiplicative cyclic group generated by u_1.

The mapping $U \to \{1,-1\} \times C$, defined by $u_1{}^m \to (1,u_1{}^m)$, $-u_1{}^m \to (-1,u_1{}^m)$ is clearly an isomorphism. ∎

The smallest unit $u_1 > 1$ is called the *fundamental unit* of $\mathbb{Q}(\sqrt{d})$.

A crude method of determining the fundamental unit is the following. First let $d \equiv 2 \pmod 4$ or $d \equiv 3 \pmod 4$.

If $u = a + b\sqrt{d}$ is a unit, $u \neq \pm 1$, so are $-u$, u^{-1}, $-u^{-1}$ and only the largest of these numbers is larger than 1; since these numbers are exactly $\pm a \pm b\sqrt{d}$, then $a + b\sqrt{d} > 1$ when $a > 0$, $b > 0$.

If $u_1 = a_1 + b_1\sqrt{d}$ is the fundamental unit, if $u_m = u_1{}^m = a_m + b_m\sqrt{d}$, then $b_{m+1} = a_1 b_m + a_m b_1$, so we have $b_1 < b_2 < b_3 < \ldots$. From $\pm 1 = N(u_1) = a_1{}^2 - b_1{}^2 d$, we have $b_1{}^2 d = a_1{}^2 \mp 1$; thus if we write the sequence d, $4d$, $9d$, $16d$, $\ldots$, then b_1 is the smallest integer such that $0 < b_1$, $b_1{}^2 d$ is a square plus or minus 1.

For example, let $d = 3$, then $b_1 = 1$, $a_1 = 2$, so $2 + \sqrt{3}$ is the fundamental unit of $\mathbb{Q}(\sqrt{3})$.

Similarly, $1 + \sqrt{2}$, $5 + 2\sqrt{6}$, $8 + 3\sqrt{7}$, are the fundamental units of the fields $\mathbb{Q}(\sqrt{2})$, $\mathbb{Q}(\sqrt{6})$, $\mathbb{Q}(\sqrt{7})$ respectively.

Now, if $d \equiv 1 \pmod 4$, by a similar argument $u_1 = a_1/2 + (b_1/2)\sqrt{d}$ with a_1, b_1 positive integers of the same parity; also $\pm 1 = N(u_1) = (a_1{}^2 - b_1{}^2 d)/4$ hence $b_1{}^2 d = a_1{}^2 \mp 4$ and we have to find the smallest integer $b_1 > 0$ such that $b_1{}^2 d$ is a square plus or minus 4. For example $1/2 + (1/2)\sqrt{5}$, $3/2 + (1/2)\sqrt{13}$ are fundamental units of $\mathbb{Q}(\sqrt{5})$, $\mathbb{Q}(\sqrt{13})$ respectively.

2. UNITS OF CYCLOTOMIC FIELDS

Let p be an odd prime, ζ a primitive pth root of unity, and let $K = \mathbb{Q}(\zeta)$. Hence K has degree $p - 1$ over $\mathbb{Q}$. As we have seen in Chapter 5, (U), the ring of integers of K is $A = \mathbb{Z}[\zeta]$ and $p = u(1 - \zeta)^{p-1}$, where u is a unit of A.

Now we prove the main result about units in $\mathbb{Q}(\zeta)$:

F.   *The multiplicative group W of roots of unity of $\mathbb{Q}(\zeta)$ is*

$$W = \{1, \zeta, \zeta^2, \ldots, \zeta^{p-1}, -1, -\zeta, -\zeta^2, \ldots, -\zeta^{p-1}\},$$

so $w = 2p$. Every unit of $\mathbb{Q}(\zeta)$ may be written as $u = \zeta^k v$ where v is a positive real unit of A.

Proof: By **(C)**, W is a cyclic group of order w. Since $-\zeta \in W$ and $-\zeta$ has order $2p$ (because p is odd) then $2p$ divides w.

Now, let $x \in W$ be an element of order w. Since $x \in K$ then $\mathbb{Q}(x) \subseteq K$ so $\varphi(w) = [\mathbb{Q}(x): \mathbb{Q}]$ divides $p - 1 = [K:\mathbb{Q}]$. But $w = p^r \cdot m$ where $r \geqslant 1$, $m \geqslant 2$, p does not divide m. So $\varphi(w) = p^{r-1}(p - 1) \cdot \varphi(m)$ and since it divides $p - 1$ then $r = 1$, $\varphi(m) = 1$ so $m = 2$ and therefore $w = 2p$. Thus, $W = \{1, \zeta, \zeta^2, \ldots, \zeta^{p-1}, -1, -\zeta, -\zeta^2, \ldots, -\zeta^{p-1}\}$.

Now, let u be any unit of $\mathbb{Q}(\zeta)$, so $u = a_0 + a_1\zeta + \cdots + a_{p-2}\zeta^{p-2}$ with $a_i \in \mathbb{Z}$. The complex conjugate of u, which is also a unit (because $uv = 1$ implies $\bar{u}\bar{v} = 1$), is given by

$$\bar{u} = a_0 + a_1\zeta^{-1} + a_2\zeta^{-2} + \cdots + a_{p-2}\zeta^{-(p-2)};$$

then $u' = u \cdot \bar{u}^{-1}$ is also a unit. Moreover, if $u^{(k)} = a_0 + a_1\zeta^k + a_2\zeta^{2k} + \cdots + a_{p-2}\zeta^{k(p-2)}$ (for $k = 1, 2, \ldots, p - 1$) are the conjugates of u, then $\overline{u^{(k)}} = \bar{u}^{(k)}$ are the conjugates of $\bar{u}$, so those of u' are $u'^{(k)} = u^{(k)} \cdot \overline{u^{(k)}}^{-1}$ and therefore, for every $k = 1, 2, \ldots, p - 1$, we have $|u'^{(k)}| = 1$. By **(B)**, the element u' is a root of unity and therefore of the form $u' = \pm\zeta^h$, with $0 \leqslant h \leqslant p - 1$, as shown above.

We must have the positive sign; if not, $u' = -\zeta^h$ then $u = -\zeta^h\bar{u}$. Let us consider the ring $R = A/A(1 - \zeta)$, and let $\theta: A \to R$ be the canonical homomorphism. Then $\theta(\zeta) = 1$, so $\theta(\zeta^k) = 1$ for every $k = 1, \ldots, p - 2$, therefore $\theta(u) = a_0 + a_1 + \cdots + a_{p-2} = \theta(\bar{u})$; from $u = -\zeta^h\bar{u}$, we have $\theta(u) = -\theta(\bar{u})$, thus $\theta(2\bar{u}) = 0$, so $2\bar{u} \in A(1 - \zeta)$, that is $1 - \zeta$ divides $2\bar{u}$, and since $\bar{u}$ is a unit, $1 - \zeta$ divides 2. Since p is associated with $(1 - \zeta)^{p-1}$ then p divides 2, so $p = 2$, against the hypothesis.

Thus, $u = \zeta^h\bar{u}$. Let k be such that $2k \equiv h \pmod{p}$, so $\zeta^h = \zeta^{2k}$, and therefore

$$\frac{u}{\zeta^k} = \zeta^k\bar{u} = \frac{\bar{u}}{\zeta^{-k}} = \overline{\left(\frac{u}{\zeta^k}\right)}$$

Letting $v = u/\zeta^k = u\zeta^{p-k}$, we see that $v \in \mathbb{R} \cap A$, v is a unit and $u = \zeta^k v$. Finally, we may take $v > 0$, by multiplying ζ^k with -1, if necessary. ∎

We may easily point out some outstanding real units of $\mathbb{Q}(\zeta)$.

For example, $(1 - \zeta^s)/(1 - \zeta)$ $(1 \leqslant s \leqslant p - 1)$ is a unit (since $1 - \zeta^s$, $1 - \zeta$ are associated elements).

By the preceding proof, there exists an even integer $2k$ such that $(1 - \zeta^s)/(1 - \zeta) = \zeta^{2k}(1 - \zeta^{-s})/(1 - \zeta^{-1})$, so

$$\frac{1 - \zeta^s}{1 - \zeta} \cdot \frac{1 - \zeta^{-s}}{1 - \zeta^{-1}} = \left(\zeta^k \frac{1 - \zeta^{-s}}{1 - \zeta^{-1}}\right)^2$$

and

$$v_s = \sqrt{\frac{1 - \zeta^s}{1 - \zeta} \cdot \frac{1 - \zeta^{-s}}{1 - \zeta^{-1}}}$$

is a real unit in $\mathbb{Q}(\zeta)$ for every $s = 1, 2, \ldots, p - 1$.

Now, if $1 \leqslant s, s' \leqslant p - 1$, $s + s' = p$ then $v_s = v_{s'}$, if $s = 1$ then $v_1 = 1$. Thus $v_2, v_3, \ldots, v_g$ $(g = (p - 1)/2)$ are $(p - 3)/2$ real units, distinct from $1, -1$, thus not roots of unity. Actually, they are distinct, because if $v_s = v_t$ then

$$\frac{1 - \zeta^s}{1 - \zeta} \cdot \frac{1 - \zeta^{-s}}{1 - \zeta^{-1}} = \frac{1 - \zeta^t}{1 - \zeta} \cdot \frac{1 - \zeta^{-t}}{1 - \zeta^{-1}},$$

so $\zeta^s + \zeta^{-s} = \zeta^t + \zeta^{-t}$.

However, $\zeta = \cos(2\pi/p) + i \sin(2\pi/p)$, so $\zeta^s = \cos(2\pi s/p) + i \sin(2\pi s/p)$, $\zeta^{-s} = \cos(2\pi s/p) - i \sin(2\pi s/p)$, so that the above relation gives

$$2 \cos(2\pi s/p) = 2 \cos(2\pi t/p).$$

Therefore, either $s = t$ or $(2\pi s/p) + (2\pi t/p) = 2k\pi$, with $k \in \mathbb{Z}$. But then $s + t = kp$, and, if $1 < s, t \leqslant (p - 1)/2$, this is impossible.

At this point we still have no way of determining even in very simple particular cases all the units of a cyclotomic field. Soon, we shall prove a theorem of Dirichlet, which indicates the structure of the group of units of an algebraic number field; this result states the existence of fundamental systems of units, which together with the roots of unity, generate all the units of the field. But even for cyclotomic fields, the determination of a fundamental system of units requires deep results and delicate computations.

Dirichlet's theorem is proved by a method devised by Minkowski, which has led to the branch of Number Theory called Geometry of Numbers.

In the n-dimensional vector space $\mathbb{R}^n$ of all n-tuples of real numbers, we shall consider certain additive subgroups.

Definition 1. A *lattice* Λ in $\mathbb{R}^n$ ($n \geqslant 1$) is the set of all linear combinations, with coefficients in $\mathbb{Z}$, of n $\mathbb{R}$-linearly independent elements $a^{(1)}, \ldots, a^{(n)} \in \mathbb{R}^n$.

In other words, Λ is a free Abelian group of rank n, contained in $\mathbb{R}^n$, having a basis which is also an $\mathbb{R}$-basis of $\mathbb{R}^n$.

The *fundamental parallelotope* of Λ is the set

$$\Pi = \left\{ \sum_{k=1}^n x_k a^{(k)} \mid x_k \in \mathbb{R}, 0 \leqslant x_k \leqslant 1 \quad \text{for } k = 1, \ldots, n \right\}.$$

The volume of Π shall be denoted by $\mu = \mu(\Pi)$.

Definition 2. A subset S of $\mathbb{R}^n$ is said to be *convex* when it satisfies the following property: If $y, y' \in S$ then the line segment joining y, y' is contained in S.

We can also phrase this property as follows: If $y, y' \in S$, if $\lambda, \lambda' \in \mathbb{R}$, $0 \leqslant \lambda, \lambda' \leqslant 1$ and $\lambda + \lambda' = 1$, then $\lambda y + \lambda' y' \in S$.

We recall now some well-known notions:

If S is any set, and $\alpha \in \mathbb{R}$, $\alpha > 0$, we define the *homothetic image* of S (with ratio α) to be the set $\alpha S = \{\alpha x = (\alpha x_1, \ldots, \alpha x_n) \mid x = (x_1, \ldots, x_n) \in S\}$. Then any homothetic image of a convex set is convex.

A subset S of $\mathbb{R}^n$ is said to be a *bounded set* when there exists a sufficiently large real number γ such that $|x| \leqslant \gamma$ for every $x \in S$.[*] In other words, S is contained in the n-dimensional sphere of centre 0, and radius $\gamma > 0$.

If S is bounded, and $\alpha > 0$ then αS is also a bounded set.

A subset S of $\mathbb{R}^n$ is said to be *closed* when it satisfies the following property: if $y^{(1)}, y^{(2)}, \ldots, y^{(n)}, \ldots$ is any sequence of elements in S then every accumulation point of this sequence still belongs to S. This means that S is a closed set in the topological space $\mathbb{R}^n$ (with its natural topology). Again, any homothetic image of a closed set is closed.

Definition 3. A non-empty convex, bounded and closed subset of $\mathbb{R}^n$ is called a *convex body*.

If S is a subset of $\mathbb{R}^n$ such that when $y \in S$ then $-y \in S$, we say that S is a *symmetric subset*.

It is very easy to picture some symmetric convex bodies. For example, the closed n-dimensional cube $S = \{x = (x_1, \ldots, x_n) \in \mathbb{R}^n \mid -1 \leqslant x_i \leqslant 1$ for $i = 1, \ldots, n\}$ and the closed n-dimensional sphere $S = \{x \in \mathbb{R}^n \mid |x| \leqslant 1\}$ are symmetric convex bodies. Similarly, if $n = 2$, if C is an ellipse of center at the origin, the closed bounded region S of $\mathbb{R}^2$ determined by C (and including the points of C) is a symmetric convex body. This is true, whatever

---

[*] If $x = (x_1, \ldots, x_n)$ then $|x|$ denotes the positive square root $\sqrt{x_1^2 + \cdots + x_n^2}$, which is the distance from x to the origin.

be the ratio between the radii of the ellipse and for any slopes of the axes; thus, we may also consider very elongated ellipses, with center at the origin and axes with irrational slopes.

The "form" of a convex body may be rather difficult to describe. However, our considerations will depend on the volume, rather than on the form of the convex body.

Here is not the place to enter into lengthy discussions on the concept of volume; let us only say that it will be used in an intuitive way.

For the convenience of the reader, we state the main pertinent facts:

(1) Let $e_1 = (1,0, \ldots, 0)$, $e_2 = (0,1, \ldots, 0), \ldots, e_n = (0,0, \ldots, 1)$ be the standard basis of $\mathbb{R}^n$; then the parallelotope determined by the origin and $e_1, \ldots, e_n$ has volume equal to 1.

(2) If φ is an invertible linear transformation of $\mathbb{R}^n$, then the volume of the parallelotope determined by 0, $\varphi(e_1), \ldots, \varphi(e_n)$ is equal to $|d|$, where $d = \det(a_{ji})$ and $\varphi(e_i) = \sum_{j=1}^{n} a_{ji} e_j$ $(i = 1, \ldots, n)$.

(3) If S_1, S_2 are subsets of $\mathbb{R}^n$ such that $S_1 \subseteq S_2$ and if their volumes are defined, then $\mathrm{vol}(S_1) \leqslant \mathrm{vol}(S_2)$.

(4) If S is a subset of $\mathbb{R}^n$, $\alpha \in \mathbb{R}$, $\alpha > 0$, let $\alpha S = \{\alpha x \mid x \in S\}$; if the volume of S is defined so is $\mathrm{vol}(\alpha S)$ and $\mathrm{vol}(\alpha S) = \alpha^n \cdot \mathrm{vol}(S)$.

(5) If S is a subset of $\mathbb{R}^n$ whose volume is defined, if $x \in \mathbb{R}^n$ and $x + S = \{x + y \mid y \in S\}$ then the volume of $x + S$ is defined and equal to the volume of S.

(6) If S is a bounded subset of $\mathbb{R}^n$ and its volume is defined, then $0 \leqslant \mathrm{vol}(S) < \infty$.

(7) If S_1, S_2 are disjoint subsets of $\mathbb{R}^n$ and their volumes are defined, so is the volume of $S_1 \cup S_2$, and $\mathrm{vol}(S_1 \cup S_2) = \mathrm{vol}(S_1) + \mathrm{vol}(S_2)$.

It will be quite clear that for all the subsets S which we shall encounter, it is possible to define the volume in an unambiguous way. More generally, a theorem due to Blaschke states that *it is possible to define the volume of every convex body*.

Before proceeding, let us note this easy fact:

G. *If y, y' are points in the interior of a convex set S, then every point of the line segment joining y, y' lies also in the interior of S.*

Proof: We recall the concept of an interior point. We say that $y \in S$ is an interior point if there exists a real number $c > 0$ such that, for every element $x \in \mathbb{R}^n$ with $|x - y| < c$ we have $x \in S$.

If y, y' are interior points of the convex set S, we may choose $c \in \mathbb{R}$, $c > 0$, so small that if $x \in \mathbb{R}^n$, $|x - y| < c$, then $x \in S$ and also if $|x - y'| < c$, then $x \in S$.

Consider now any point $\lambda y + \lambda' y'$ of the segment joining y, y', so $0 \leqslant \lambda$, $\lambda' \leqslant 1$, $\lambda + \lambda' = 1$; since S is convex then $\lambda y + \lambda' y' \in S$. Let $x \in \mathbb{R}^n$ be such that if $z = x - (\lambda y + \lambda' y')$ then $|z| < c$. We have $x = z + \lambda y + \lambda' y' = \lambda(y + z) + \lambda'(y' + z)$ with $|(y + z) - y| < c$, $|(y' + z) - y'| < c$; since y, y' are in the interior of S, then $y + z$, $y' + z$ are in S and therefore $x \in S$, showing that $\lambda y + \lambda' y'$ is an interior point of S. $\blacksquare$

Theorem 1 (*Minkowski*). *Let Λ be a lattice in $\mathbb{R}^n$ ($n \geqslant 1$), and let μ be the volume of the fundamental parallelotope of Λ.*

If S is a symmetric convex body having volume $\mathrm{vol}(S) > 2^n \mu$, then there exists a point of Λ, distinct from the origin, which belongs to the interior of S.

However if $\mathrm{vol}(S) = 2^n \mu$, it can only be said that there is a point of Λ, distinct from the origin, which is in S (but not necessarily in the interior of S).

Proof: It is enough to prove the first statement, for it will imply the second one. Indeed, let us assume that $\mathrm{vol}(S) = 2^n \mu$.

S is bounded, so there exists an integer m such that $|x| < m$ for every $x \in S$.

For every $y \in \Lambda$, let $\rho_y = \inf\{|y - x| \,|\, x \in S\}$. The distance mapping $x \to d(y,x) = |y - x|$ is a continuous function of x; since S is compact, for every $y \in \Lambda$ there exists $x(y) \in S$ such that $\rho_y = |y - x(y)|$. We shall be interested in $\rho = \inf\{\rho_y \,|\, y \in \Lambda,\, y \neq 0\}$.

Let C be the open cube with center at the origin and side $2(m + 1)$, that is C is the set of points $x = (x_1, \ldots, x_n) \in \mathbb{R}^n$ satisfying $|x_i| < m + 1$ for $i = 1, \ldots, n$. An easy geometric argument convinces us at once that for every $y \in \Lambda$ there exists $y' \in \Lambda \cap C$ such that $\rho_{y'} \leqslant \rho_y$; thus $\rho = \inf\{\rho_y \,|\, y \in \Lambda \cap C,\, y \neq 0\}$. Now, $\Lambda \cap C$ is a finite set, so there exists $y_0 \in \Lambda \cap C$, $y_0 \neq 0$, such that $\rho = \rho_0 = |y_0 - x(y_0)|$. If $\rho \neq 0$, we shall derive a contradiction, and this will prove that $y_0 = x(y_0) \in \Lambda \cap S$.

Let $\delta = \rho/2m$ and consider the homothetic convex body $(1 + \delta)S$, whose volume is $\mathrm{vol}((1 + \delta)S) = (1 + \delta)^n \mathrm{vol}(S) > 2^n \mu$. Assuming the first statement of the theorem (even in weaker form), there exists $y_1 \in \Lambda$, $y_1 \neq 0$, such that $y_1 \in (1 + \delta)S$, so $y_1 = (1 + \delta)x_1$, with $x_1 \in S$. Then

$$d(y_1, x_1) = |(1 + \delta)x_1 - x_1| = \delta |x_1| < \delta m < \rho,$$

which is a contradiction.

Let us now turn to the proof of the first assertion: if there exists no point $y \in \Lambda$, $y \neq 0$ in the interior of S then $\mathrm{vol}(S) \leqslant 2^n \mu$; that is,

$$\mathrm{vol}\left(\frac{1}{2} S\right) = \frac{1}{2^n} \mathrm{vol}(S) \leqslant \mu.$$

For every $y \in \Lambda$, let us consider the convex body $y + (1/2)S$ with center y, obtained by translation from $(1/2)S$.

If $y_1, y_2 \in \Lambda$, $y_1 \neq y_2$, then the interiors of $y_1 + (1/2)S$, $y_2 + (1/2)S$ are disjoint, or equivalently, if $y \in \Lambda$, $y \neq 0$, then the interiors of $(1/2)S$,

$y + (1/2)S$ are disjoint. In fact, if x would be in the interior of $(1/2)S$ and of $y + (1/2)S$, then $y - x$ would be in the interior of $y - (y + (1/2)S) = -(1/2)S = (1/2)S$ (since S is symmetric), hence $(1/2)y = (1/2)(y - x) + (1/2)x$ would be in the interior of $(1/2)S$ (by (**G**)), so $y \in \Lambda$, $y \neq 0$, y would be in the interior of S, against the hypothesis.

We denote by $a^{(1)}, \ldots, a^{(n)} \in \mathbb{R}^n$ the generators of the lattice Λ.

Let $m > 0$ be a sufficiently large integer, let Λ_m be the set of all lattice points $y = \sum_{i=1}^n y_i a^{(i)}$ such that $y_i \in \mathbb{Z}$, $-m \leqslant y_i \leqslant m$ for all $i = 1, \ldots, n$; thus, Λ_m consists of $(2m + 1)^n$ elements.

Let $\gamma > 0$ be a sufficiently large real number, for example, such that if $x = \sum_{i=1}^n x_i a^{(i)} \in S$, $x_i \in \mathbb{R}$, then $|x_i| \leqslant \gamma$ for all $i = 1, \ldots, n$; then for every point $y = (y_1, \ldots, y_n) \in \Lambda_m$ we have $|y_i + (1/2)x_i| \leqslant m + (1/2)\gamma$ for every point $x \in S$. Thus $\Lambda_m + (1/2)S$ is contained in the parallelotope

$$\Pi' = \left\{ \sum_{k=1}^n x_k a^{(k)} \,\middle|\, |x_k| \leqslant m + (1/2)\gamma \right\}.$$

But, $\Lambda_m + (1/2)S = \cup_{y \in \Lambda_m} [y + (1/2)S]$. Since the sets $y + (1/2)S$ have disjoint interior, by the invariance of the volume by translation, we have:

$$\text{vol}[\Lambda_m + (1/2)S] = (2m + 1)^n \cdot \text{vol}[(1/2)S] \leqslant (2m + \gamma)^n \cdot \mu$$

this last quantity being the volume of the parallelotope Π'.

The above relation is true for every sufficiently large integer m, therefore,

$$\text{vol}\left(\frac{1}{2}\right) S \leqslant \lim_{m \to \infty} \left(\frac{2m + \gamma}{2m + 1}\right)^n \cdot \mu = \mu.$$

We apply this theorem to show the existence of solutions for a system of linear inequalities; for $n = 1$ the following statement is trivial:

H. *Let $n > 1$, let $L_i = \sum_{j=1}^n a_{ij} X_j$ $(i = 1, \ldots, n)$ be n linear forms with real coefficients, such that $d = \det(a_{ij}) \neq 0$.*

If $\tau_1, \ldots, \tau_n$ are positive real numbers such that $\tau_1 \ldots \tau_n \geqslant |d|$, given any index i_0, $1 \leqslant i_0 \leqslant n$, there exist integers $x_1, \ldots, x_n \in \mathbb{Z}$, not all equal to 0, such that $|L_i(x_1, \ldots, x_n)| < \tau_i$ for $i \neq i_0$, and $|L_{i_0}(x_1, \ldots, x_n)| \leqslant \tau_{i_0}$.
Proof: For simplicity of notation we may assume that $i_0 = n$. If the assertion is not true, for all n-tuples $(x_1, \ldots, x_n)$ of integers, not all equal to 0, we have either: (1) there exists i, $1 \leqslant i \leqslant n - 1$ such that $|L_i(x_1, \ldots, x_n)| \geqslant \tau_i$; or (2) if $1 \leqslant i \leqslant n - 1$ then $|L_i(x_1, \ldots, x_n)| < \tau_i$, but $|L_n(x_1, \ldots, x_n)| > \tau_n$. Let T be the set of all n-tuples $(x_1, \ldots, x_n)$ of integers satisfying condition (2). If $T = \varnothing$ we have case (1). If $T \neq \varnothing$, let $(x_1, \ldots, x_n) \in T$ and τ_n' be such that $|L_n(x_1, \ldots, x_n)| < \tau_n'$.

Thus the subset T' of T consisting of all n-tuples $(x_1, \ldots, x_n)$ such that $|L_n(x_1, \ldots, x_n)| < \tau_n'$ is not empty; actually T' is a finite set, because the coordinates of its points are integers and the determinant of the coefficients of the linear forms is not zero. Therefore the minimum of the quantities $|L_n(x_1, \ldots, x_n)|$ when $(x_1, \ldots, x_n) \in T'$ is $\tau_n + \delta$, for some $\delta > 0$, and the same holds for all points in T. Thus, we have shown the existence of $\delta > 0$ such that if $x_1, \ldots, x_n \in \mathbb{Z}$, not all equal to 0, then either $|L_i(x_1, \ldots, x_n)| \geq \tau_i$ (for some $i = 1, \ldots, n - 1$), or $|L_n(x_1, \ldots, x_n)| \geq \tau_n + \delta$.

Let $\theta \colon \mathbb{R}^n \to \mathbb{R}^n$ be the linear transformation defined as follows: If $x = (x_1, \ldots, x_n)$ then $\theta(x) = (L_1(x), \ldots, L_n(x))$. Since $\det(\theta) = \det(a_{ij}) = d \neq 0$, then θ transforms the linearly independent vectors $e^{(1)} = (1, 0, \ldots, 0)$, $e^{(2)} = (0, 1, 0, \ldots, 0), \ldots e^{(n)} = (0, 0, \ldots, 1)$ into linearly independent vectors $\theta(e^{(1)}), \theta(e^{(2)}), \ldots, \theta(e^{(n)})$. Let Λ be the lattice defined by these vectors, and let the volume of its fundamental parallelotope be $\mu = |d|$.

Let S be the set of all elements $x = (x_1, \ldots, x_n) \in \mathbb{R}^n$ such that $|x_i| \leq \tau_i$ for $i = 1, \ldots, n - 1$ and $|x_n| \leq \tau_n + \delta$. Thus, S is a symmetric convex body, having volume $\mathrm{vol}(S) = 2\tau_1 . 2\tau_2, \ldots, 2\tau_{n-1} . 2(\tau_n + \delta) > 2^n \tau_1 \ldots \tau_n \geq 2^n |d| = 2^n . \mu$.

By Minkowski's theorem, there is a point $\theta(y) \in \Lambda$, $\theta(y) \neq 0$, belonging to the interior of S. In other words, there exist integers $y_1, \ldots, y_n$, not all zero, such that $|L_i(y)| < \tau_i$ for $i = 1, \ldots, n - 1$ and $|L_n(y)| < \tau_n + \delta$, which is a contradiction. ∎

A similar result holds for linear forms with complex coefficients:

I. *Let $n > 1$, let $L_i = \Sigma_{j=1}^n a_{ij} X_j$ $(i = 1, \ldots, n)$ be n linear forms with complex coefficients such that $d = \det(a_{ij}) \neq 0$. Let us assume that for every i there exists an index i' such that $\bar{L}_i = \Sigma_{j=1}^n \bar{a}_{ij} X_j$ (the complex conjugate form of L_i) is equal to $L_{i'}$ (for example, if L_i has real coefficients then $i' = i$).*

Let $\tau_1, \ldots, \tau_n$ be real positive numbers such that if $\bar{L}_i = L_{i'}$ then $\tau_i = \tau_{i'}$. If $\tau_1 \ldots \tau_n \geq |d|$, given any index i_0 such that $\bar{L}_{i_0} = L_{i_0}$, there exist integers $x_1, \ldots, x_n \in \mathbb{Z}$, not all equal to 0, such that $|L_i(x_1, \ldots, x_n)| < \tau_i$ for $i \neq i_0$, and $|L_{i_0}(x_1, \ldots, x_n)| \leq \tau_{i_0}$.

Proof: We begin associating real forms L_i' and constants τ_i' $(i = 1, \ldots, n)$ to the given forms L_i and constants τ_i in the following way: If L_i is a real form let $L_i' = L_i$ and $\tau_i' = \tau_i$; if L_i is not a real form and $\overline{L_i'} = \bar{L}_i$ we put $L_i' = (1/2)(L_i + L_{i'})$, $L_{i'}' = (1/2\sqrt{-1})(L_i - L_{i'})$ and $\tau_i' = \tau_{i'}' = \tau_i/\sqrt{2}$. Thus L_i', $L_{i'}'$ are real forms and $L_i = L_i' + \sqrt{-1}L_{i'}'$, $L_{i'} = L_i' - \sqrt{-1}L_{i'}'$ and $\tau_1' \ldots \tau_n' = 2^{-r_2}\tau_1 \ldots \tau_n$, where r_2 denotes the number of pairs of non-real complex conjugate forms L_i, $L_{i'}$. If d' is the determinant of the coefficients of the forms L_i' $(i = 1, \ldots, n)$ then $|d'| = 2^{-r_2}|d|$. This verification may be easily done by the reader. From $\tau_1 \tau_2 \ldots \tau_n \geq |d|$ it follows that $\tau_1' . \tau_2' \ldots \tau_n' \geq 2^{-r_2}|d| = |d'|$.

We are therefore in a position to apply (**H**), deducing the existence of integers $x_1, \ldots, x_n$, not all equal to 0, such that $|L_i'(x_1, \ldots, x_n)| < \tau_i'$ for $i \neq i_0$ and $|L_{i_0}'(x_1, \ldots, x_n)| \leqslant \tau_{i_0}'$.

It follows that if $i \neq i_0$ and $\bar{L}_i = L_i'$ with $i \neq i'$ then

$$|L_i(x)|^2 = |L_{i'}(x)|^2 = (L_i'(x))^2 + (L_{i'}'(x))^2 < \frac{\tau_i^2}{2} + \frac{\tau_i^2}{2} = \tau_i^2;$$

if $\bar{L}_i = L_i$, then obviously $|L_i(x_1, \ldots, x_n)| < \tau_i$, for $i \neq i_0$, and $|L_{i_0}(x_1, \ldots, x_n)| \leqslant \tau_{i_0}$. ∎

As a first application of these methods, we may refine the result of Chapter 8, (**H**):

J. *For every algebraic number field K, distinct from $\mathbb{Q}$, and for every non-zero integral ideal J of A there exists an element $a \in J$, $a \neq 0$, such that*

$$|N_{K|\mathbb{Q}}(a)| < N(J) \cdot \sqrt{|\delta|}$$

where δ is the discriminant of K.

Proof: Let $\{a_1, \ldots, a_n\}$ be a basis of the free Abelian group J (Chapter 6, (**K**)), and let us consider the n linear forms $L_i = \Sigma_{j=1}^n a_j^{(i)} X_j$ $(i = 1, \ldots, n)$ where $a_j^{(1)} = a_j$, $a_j^{(2)}, \ldots, a_j^{(n)}$ are the conjugates of $a_j \in K$ over $\mathbb{Q}$. The determinant of the coefficients is equal, in absolute value, to $|\det(a_j^{(i)})| = N(J) \cdot \sqrt{|\delta|} \neq 0$ (by Chapter 8, (**E**)).

Let $\tau_1 = \tau_2 = \cdots = \tau_n = |\det(a_j^{(i)})|^{1/n}$. Since $n > 1$, by (**I**) there exist integers $x_1, \ldots, x_n$, not all equal to 0, such that

$$\left| \sum_{j=1}^n a_j^{(i)} x_j \right| \leqslant |\det(a_j^{(i)})|^{1/n}, \quad \text{for all } i = 1, \ldots, n,$$

with at most one equality, hence with at least one strict inequality.

Hence, letting $a = \Sigma_{j=1}^n x_j a_j \in J$ we have $a \neq 0$ (since not all x_j are equal to 0), and

$$|N_{K|\mathbb{Q}}(a)| = \prod_{i=1}^n \left| \sum_{j=1}^n a_j^{(i)} x_j \right| < |\det(a_j^{(i)})| = N(J) \cdot \sqrt{|\delta|}. \quad ∎$$

A very important corollary follows now:

K. *If K is an algebraic number field, distinct from $\mathbb{Q}$, then $|\delta| \geqslant 2$.*

Proof: We take J equal to the unit ideal, $J = A$. Then $N(J) = 1$ and therefore $1 \leqslant |N_{K|\mathbb{Q}}(a)| < N(J) \cdot \sqrt{|\delta|} = \sqrt{|\delta|}$ for some element $a \in A$, $a \neq 0$. Since δ is an integer, then $|\delta| \geqslant 2$. ∎

Let us observe at this point, that (**J**) *is equivalent to the fact that in every class of ideals there exists a non-zero integral ideal I such that* $N(I) < \sqrt{|\delta|}$. In fact, let $J \neq 0$ be an ideal in the given class, and let $b \in A$, $b \neq 0$ be such that $bJ^{-1} \subseteq A$; applying (**J**), there exists $a \in bJ^{-1}$, $a \neq 0$, for which

$$|N_{K|\mathbb{Q}}(a)| < N(Ab \cdot J^{-1})\sqrt{|\delta|};$$

then letting $I = Aab^{-1} \cdot J \subseteq A$ we have $N(I) < \sqrt{|\delta|}$.

Conversely, let J be a non-zero integral ideal, let I be an integral ideal in the class of J^{-1} such that $N(I) < \sqrt{|\delta|}$; then there exists $a \in K$, $a \neq 0$ such that $aJ^{-1} = I \subseteq A$; hence $a \in J$ and $|N_{K|\mathbb{Q}}(a)| < N(J) \cdot \sqrt{|\delta|}$.

By Chapter 8, (**G**), there are only finitely many ideals integral I_i ($i = 1, \ldots, t$) such that $N(I_i) < \sqrt{|\delta|}$; therefore, every ideal is in the same class as some of the ideals $I_1, \ldots, I_t$; this provides a new proof of Theorem 1 of Chapter 8.

Before returning to the study of the group of units of an algebraic number field K, let us develop the method just used in order to obtain sharper estimations for the discriminant.

Let $K^{(1)}, K^{(2)}, \ldots, K^{(n)}$ be all the conjugates of the field K (over $\mathbb{Q}$). As we know, the complex conjugate of any field is a conjugate (in the algebraic sense), because if x is a root of a polynomial f, then its complex conjugate $\bar{x}$ is also a root of f. Of course, if a field is contained in the field of real numbers, then it coincides with its complex conjugate. Thus, the conjugates of K may be grouped and numbered as follows: let $r_1 \geqslant 0$ be such that $K^{(1)}, K^{(2)}, \ldots, K^{(r_1)}$ are contained in $\mathbb{R}$; let $r_2 \geqslant 0$ be such that

$$K^{(r_1+1)}, K^{(r_1+2)}, \ldots, K^{(r_1+r_2)},$$
$$K^{(r_1+r_2+1)}, K^{(r_1+r_2+2)}, \ldots, K^{(n)}$$

are non-real fields, with $\overline{K^{(r_1+i)}} = K^{(r_1+r_2+i)}$ for $i = 1, 2, \ldots, r_2$; so there are r_2 pairs of complex conjugate non-real fields and

$$n = r_1 + 2r_2.$$

It is also convenient to introduce the following notation: $l_1 = \cdots = l_{r_1} = 1$, $l_{r_1+1} = l_{r_1+2} = \cdots = l_{r_1+r_2} = 2$, so $\sum_{i=1}^{r_1+r_2} l_i = r_1 + 2r_2 = n$. If $x \in K$, we denote by $x^{(1)}, \ldots, x^{(n)}$ all its conjugates, where $x^{(i)} \in K^{(i)}$ ($i = 1, \ldots, n$).

Let $\varphi : K \to \mathbb{R}^n$ be the isomorphism of $\mathbb{Q}$-vector spaces defined by $\varphi(x) = (\xi_1, \ldots, \xi_n) \in \mathbb{R}^n$, where

$$\begin{cases} \xi_i = x^{(i)} & \text{for} \quad i = 1, \ldots, r_1 \\ \xi_j = Re(x^{(j)}) & \text{for} \quad j = r_1 + 1, \ldots, r_1 + r_2 \\ \xi_{j+r_2} = Im(x^{(j)}) & \text{for} \quad i = r_1 + 1, \ldots, r_1 + r_2 \end{cases}$$

(for every complex number $y = a + b\sqrt{-1}$ we denote $a = Re(y)$, $b = Im(y)$).

Let $D_1 = \{(\xi_1, \ldots, \xi_n) \in \mathbb{R}^n \mid \Pi_{j=1}^{r_1} |\xi_j| \cdot \Pi_{j=r_1+1}^{r_1+r_2} (\xi_j^2 + \xi_{j+r_2}^2) \leqslant 1\}$. For example, if $r_1 = 1$, $r_2 = 0$ then $n = 1$ hence D_1 is the closed interval $[-1,1]$ and $\text{vol}(D_1) = 2$. If $r_1 = 0$, $r_2 = 1$ then $n = 2$ and D_1 is the closed disk of radius 1 and center 0, hence $\text{vol}(D_1) = \pi$. If $r_1 = 2$, $r_2 = 0$ then $n = 2$ and D_1 is no more bounded; this is also true for every other case.

We may consider symmetric convex bodies D contained in D_1, and apply to them Minkowksi's theorem, as we did in (**J**).

L. *Let K be an algebraic number field of degree n, let J be a non-zero integral ideal of A. For every symmetric convex body D in $\mathbb{R}^n$, such that $D \subseteq D_1$, there exists an element $a \in J$, $a \neq 0$, such that*

$$|N_{K|}(a)| \leqslant \frac{2^{r_1+r_2}}{\text{vol}(D)} N(J) \cdot \sqrt{|\delta|}.$$

Proof: Let $\{a_1, \ldots, a_n\}$ be a basis of the free Abelian group J and consider the vectors $\varphi(a_1), \ldots, \varphi(a_n)$ in $\mathbb{R}^n$. We shall show that these vectors are linearly independent over $\mathbb{R}$, by computing the determinant of their coordinates (relative to the canonical basis of $\mathbb{R}^n$).

Let $\varphi(a_i) = (\alpha_{1i}, \ldots, \alpha_{ni})$, therefore

$$\det(\alpha_{ij}) = \det \begin{pmatrix} a_1^{(1)} & \cdots & & a_n^{(1)} \\ \cdots & & & \cdots \\ a_1^{(r_1)} & \cdots & & a_n^{(r_1)} \\ Re\, a_1^{(r_1+1)} & \cdots & & Re\, a_n^{(r_1+1)} \\ \cdots & & & \cdots \\ Re\, a_1^{(r_1+r_2)} & \cdots & & Re\, a_n^{(r_1+r_2)} \\ \cdots & & & \cdots \\ Im\, a_1^{(r_1+1)} & \cdots & & Im\, a_n^{(r_1+1)} \\ \cdots & & & \cdots \\ Im\, a_1^{(r_1+r_2)} & \cdots & & Im\, a_n^{(r_1+r_2)} \end{pmatrix}$$

To compute this determinant we make successively the following:

(1) Multiply by i each of the r_2 last rows; this introduces a factor $1/i^{r_2}$.

(2) Add the $(r_1 + j)$th row with the $(r_1 + r_2 + j)$th row.

(3) Subtract the $(r_1 + j)$th row to the double of the $(r_1 + r_2 + j)$th row and multiply the result by -1; this introduces a factor $(-1)^{r_2} \cdot (1/2^{r_2})$.

After these transformations we arrive at $(-1)^{r_2}/(2i)^{r_2} \det(a_j^{(k)})$ and therefore the absolute value of the determinant is

$$|\det(\alpha_{ij})| = \frac{1}{2^{r_2}} |\det(a_j^{(k)})| = \frac{1}{2^{r_2}} N(J) \cdot \sqrt{|\delta|} \neq 0,$$

as it was computed in Chapter 8, (E).

Hence, the vectors $\varphi(a_1), \ldots, \varphi(a_n)$ are linearly independent over $\mathbb{R}$ and define a lattice Λ, whose fundamental parallelotope has volume $\mu = |\det(\alpha_{ij})| = (1/2^{r_2})N(J) \cdot \sqrt{|\delta|}$.

Let $\rho \in \mathbb{R}$, $\rho > 0$ be such that $\rho^n = [2^{r_1+r_2}/\text{vol}(D)]N(J)\sqrt{|\delta|}$. Thus, if we consider the homothetic image $\rho D = \{\rho\xi \mid \xi \in D\}$, then

$$\text{vol}(\rho D) = \rho^n \, \text{vol}(D) = 2^{r_1+r_2}N(J) \cdot \sqrt{|\delta|} = 2^n \mu.$$

By Minkowski's theorem, there exists $\xi \neq 0$, $\xi \in D$, such that $\rho\xi \in \Lambda = \Sigma_{i=1}^n \mathbb{Z}\,\varphi(a_i) = \varphi(\Sigma_{i=1}^n \mathbb{Z}\,a_i) = \varphi(J)$; thus $\rho\xi = \varphi(a)$, $a \in J$, $a \neq 0$. We have $\varphi(a) \in \rho D \subseteq \rho D_1$, so $\varphi(a) = (\rho\beta_1, \ldots, \rho\beta_n)$ with $(\beta_1, \ldots, \beta_n) \in D_1$.

Now, let us compute $|N_{K|\mathbb{Q}}(a)| = \Pi_{j=1}^n |a^{(j)}|$. Since $a^{(r_1+r_2+j)} = \overline{a^{(r_1+j)}}$ (for $j = 1, \ldots, r_2$) then $|a^{(r_1+r_2+j)}| \cdot |a^{(r_1+j)}| = (Re\ a^{(r_1+j)})^2 + (Im\ a^{(r_1+j)})^2 = \rho^2(\beta_{r_1+j}^2 + \beta_{r_1+r_2+j}^2)$; similarly $|a^{(j)}| = \rho\,|\beta_j|$ (for $j = 1, \ldots, r_1$); thus $|N_{K|\mathbb{Q}}(a)| \leqslant \rho^n = [2^{r_1+r_2}/\text{vol}(D)]N(J) \cdot \sqrt{|\delta|}$. ∎

The statement of (L) gives an upper bound for the norms of elements of the ideal J. The larger is the volume of the convex body $D \subseteq D_1$, the smaller is the upper bound in question.

For example if $r_1 = 0$, $r_2 = 1$ and if we take $D = D_1$, we have vol $(D_1) = \pi$, and (L) states the existence of $a \in J$, $a \neq 0$, such that

$$|N_{K|\mathbb{Q}}(a)| \leqslant \frac{2}{\pi} N(J)\sqrt{|\delta|} < N(J)\sqrt{|\delta|};$$

this result was already proved in (J) for any extension $K | \mathbb{Q}$.

For larger values of r_1, r_2 we have seen that D_1 itself is not a convex body, for it is unbounded and not convex. In this situation, we reach different estimations by appropriate choices of the convex body D.

M. *Let K be an algebraic number field of degree n, and let J be a non-zero integral ideal of A. There exists an element $a \in J$, $a \neq 0$, such that*

$$|N_{K|\mathbb{Q}}(a)| \leqslant \left(\frac{4}{\pi}\right)^{r_2} \frac{n!}{n^n} \sqrt{|\delta|} \cdot N(J).$$

Proof: Let

$$D = \{(\xi_1, \ldots, \xi_n) \in \mathbb{R}^n \mid |\xi_1| + \cdots + |\xi_{r_1}|$$
$$+ 2\sqrt{\xi_{r_1+1}^2 + \xi_{r_1+r_2+1}^2} + \cdots + 2\sqrt{\xi_{r_1+r_2}^2 + \xi_n^{\;2}} \leqslant n\}.$$

We show that D is a symmetric convex body. Everything is easy to check and only the convexity requires some computation. Let $\lambda,\ \mu \in \mathbb{R},\ 0 \leqslant \lambda,\ \mu \leqslant 1,\ \lambda + \mu = 1$. If $(\xi_1, \ldots, \xi_n),\ (\eta_1, \ldots, \eta_n)$ are in D, then so is $(\lambda\xi_j + \mu\eta_j)_j$ because

$$|\lambda\xi_1 + \mu\eta_1| + \cdots + 2\sqrt{(\lambda\xi_{r_1+1} + \mu\eta_{r_1+1})^2 + (\lambda\xi_{r_1+r_2+1} + \mu\eta_{r_1+r_2+1})^2} + \cdots$$
$$\leqslant \lambda\,|\xi_1| + \mu\,|\eta_1| + \cdots + 2\lambda\sqrt{\xi_{r_1+1}^2 + \xi_{r_1+r_2+1}^2}$$
$$+ 2\mu\sqrt{\eta_{r_1+1}^2 + \eta_{r_1+r_2+1}^2} + \cdots$$
$$\leqslant \lambda\{|\xi_1| + \cdots + 2\sqrt{\xi_{r_1+1}^2 + \xi_{r_1+r_2+1}^2} + \cdots\}$$
$$+ \mu\{|\eta_1| + \cdots + 2\sqrt{\eta_{r_1+1}^2 + \eta_{r_1+r_2+1}^2} + \cdots\}$$
$$\leqslant \lambda n + \mu n = (\lambda + \mu)n = n$$

(the above inequalities are straightforward). This shows that D is convex.

Moreover $D \subseteq D_1$, because the geometric mean between positive real numbers is less than the arithmetic mean, thus

$$\left\{\prod_{j=1}^{r_1} |\xi_j| \prod_{j=r_1+1}^{r_1+r_2} (\xi_j^{\;2} + \xi_{j+r_2}^2)\right\}^{1/n} = \left\{\prod_{j=1}^{r_1} |\xi_j| \cdot \prod_{j=r_1+1}^{r_1+r_2} |\xi_j + i\xi_{j+r_2}|^2\right\}^{1/n}$$
$$\leqslant \frac{1}{n}\left\{\sum_{j=1}^{r_1} |\xi_j| + 2\sum_{j=r_1+1}^{r_1+r_2} |\xi_j + i\xi_{j+r_2}|\right\}$$
$$= \frac{1}{n}\left\{\sum_{j=1}^{r_1} |\xi_j| + 2\sum_{j=r,+1}^{r_1+r_2} \sqrt{\xi_j^{\;2} + \xi_{j+r_2}^2}\right\}.$$

We may apply **(L)**, so there exists $a \in J,\ a \neq 0$ such that

$$|N_{K|\mathbb{Q}}(a)| \leqslant \frac{2^{r_1+r_2}}{\mathrm{vol}(D)}\, N(J) \cdot \sqrt{|\delta|}.$$

It remains to compute $\mathrm{vol}(D)$, which depends on r_1, r_2. More generally, if $\rho > 0$ let $D^{(\rho)}$ be the set of vectors $(\xi_1, \ldots, \xi_n) \in \mathbb{R}^n$ such that

$$|\xi_1| + \cdots + 2\sqrt{\xi_{r_1+1}^2 + \xi_{r_1+r_2+1}^2} + \cdots \leqslant \rho.$$

We denote by $f_{r_1,r_2}(\rho)$ the volume of $D^{(\rho)}$. For example $f_{1,0}(\rho) = 2\rho$, $f_{0,1}(\rho) = \pi\rho^2/4$. For larger values of r_1, r_2 the computation is done by induction and we shall omit the details. Let us just note that

$$f_{r_1,r_2}(\rho) = 2\int_0^\rho f_{r_1-1,r_2}(\rho - t)\, dt$$

and

$$f_{r_1, r_2}(\rho) = \iint_{t^2 + u^2 \leqslant \rho^2/4} f_{r_1, r_2 - 1}(\rho - 2\sqrt{t^2 + u^2}) \, dt \, du$$

$$= 2\pi \int_0^{\rho/2} f_{r_1, r_2 - 1}(\rho - 2t) t \, dt.$$

Performing the induction, we obtain $f_{r_1, r_2}(\rho) = 2^{r_1}(\pi/2)^{r_2}\rho^n/n!$ and taking $\rho = n$ we have $\text{vol}(D) = 2^{r_1}(\pi/2)^{r_2}n^n/n!$.

We conclude that

$$|N_{K|\mathbb{Q}}(a)| \leqslant \frac{2^{r_1 + r_2}}{2^{r_1}}\left(\frac{2}{\pi}\right)^{r_2} \frac{n!}{n^n} N(J) \cdot \sqrt{|\delta|}$$

$$= \left(\frac{4}{\pi}\right)^{r_2} \frac{n!}{n^n} N(J)\sqrt{|\delta|}. \quad \blacksquare$$

The preceding result gives the following sharper estimation of the discriminant:

N. *If K is an algebraic number field of degree n then*

$$|\delta| \geqslant \left(\frac{\pi}{4}\right)^{2r_2}\left(\frac{n^n}{n!}\right)^2 \geqslant \left(\frac{\pi}{4}\right)^n \left(\frac{n^n}{n!}\right)^2.$$

Proof: We take J equal to the unit ideal; hence, applying **(M)**, we deduce:

$$1 \leqslant |N_{K|\mathbb{Q}}(a)| \leqslant \left(\frac{4}{\pi}\right)^{r_2} \frac{n!}{n^n} \sqrt{|\delta|},$$

therefore,

$$|\delta| \geqslant \left(\frac{\pi}{4}\right)^{2r_2}\left(\frac{n^n}{n!}\right)^2 \geqslant \left(\frac{\pi}{4}\right)^n \left(\frac{n^n}{n!}\right)^2,$$

since

$$\frac{\pi}{4} < 1 \quad \text{and} \quad 2r_2 \leqslant n. \quad \blacksquare$$

We may note that $(\pi/4)^n(n^n/n!)^2$ increases monotonically with n. If we express $n!$ by Stirling's formula

$$\int n! = \sqrt{2\pi n} \, n^n e^{-n + (\alpha/12n)}$$

where $0 < \alpha < 1$, then $e^{\alpha/12n} < e^{1/12} < \Sigma_{\nu=0}^{\infty} (1/12^{\nu}) = 12/11$; hence $|\delta| > (\pi e^2/4)^n(1/2\pi n)(11/12)^2$.

Since $\pi e^2/4 > 1$ then, as we consider fields of larger and larger degree, the absolute value of their discriminant must also increase; more precisely,

$$\lim_{\substack{n \to \infty \\ [K:\mathbb{Q}]=n}} \min \{|\delta_K|\} = \infty.$$

From this fact we may now deduce the following interesting consequence due to Hermite:

O. *For every integer d there exists at most finitely many fields K having discriminant* $\delta_K = d$.

Proof: Given d, there exists an integer $n_0 \geqslant 1$ such that if $n \geqslant n_0$ then $(\pi e^2/4)^n (1/2\pi n)(11/12)^2 > d$, hence if K has degree $n \geqslant n_0$, then its discriminant is greater than d. Thus, it is enough to prove that for every integer n, there exists at most finitely many fields K of degree n having discriminant equal to d; this is true for $|d| = 1$, as we have seen in **(K)**, so we may assume $|d| > 1$. Hence, $r_1 = 1$, $r_2 = 0$ is impossible. If $r_1 = 0$, $r_2 = 1$ then $n = 2$, so K is an imaginary quadratic extension, $K = \mathbb{Q}(\sqrt{a})$, with $a \in \mathbb{Z}$, a square free; by Chapter 6, **(P)**, $d = \delta_K$ is either $4a$ or a, so there exists at most one quadratic field with discriminant d.

From now on we may assume that $r_1 + r_2 > 1$. Given K, with discriminant d, we shall show the existence of a primitive element a of K which is an algebraic integer and such that:

(1) If $r_1 > 0$ then $|a^{(1)}| \leqslant \sqrt{|d|}$ and for $i \neq 1$, $|a^{(i)}| < 1$.

(2) If $r_1 = 0$ then $|a^{(1)}| = |a^{(1+r_2)}| \leqslant \sqrt[4]{|d|}$ and all other conjugates $a^{(i)}$ are such that $|a^{(i)}| < 1$.

For Case (1), we apply **(I)**; let $\{x_1, \ldots, x_n\}$ be an integral basis of K, and let $L_i = \sum_{j=1}^{n} x_j^{(i)} X_j$ (for $i = 1, \ldots, n$); these linear forms have determinant $|\det(x_j^{(i)})| = \sqrt{|d|}$, since K has discriminant d. Given $\tau_1 = \sqrt{|d|}$, $\tau_i = 1$ for $i \neq 1$, we deduce the existence of integers $m_1, \ldots, m_n \in \mathbb{Z}$ such that $|L_i(m_1, \ldots, m_n)| < \tau_i = 1$ for $i \neq 1$, $|L_1(m_1, \ldots, m_n)| \leqslant \tau_1 = \sqrt{|d|}$. This means that the element $a = \sum_{i=1}^{n} m_i x_i \in A$ is such that $|a^{(1)}| \leqslant \sqrt{|d|}$, while $|a^{(i)}| < 1$ for $i \neq 1$. From $1 \leqslant |N_{K|\mathbb{Q}}(a)| < |a^{(1)}|$, we deduce also that all the conjugates of $a^{(1)} = a$ are distinct from a and so a has n distinct conjugate and it is a primitive element of K.

For Case (2), we apply **(I)** again, with $\tau_1 = \tau_{1+r_2} = \sqrt[4]{|d|}$, $\tau_i = 1$ (for other indices i) and we arrive in the same manner at an element a with the required properties.

Now, we consider the set $\mathscr{S}$ of all fields K of degree n and discriminant d; for each such field, let a be chosen as indicated, so $K = \mathbb{Q}(a)$. We shall associate with K, or better with a, an element of a fixed finite set S, which depends only on n,d; explicitly, S is the set of n-tuples of integers $(m_1, \ldots, m_n)$, where $-\mu \leqslant m_i \leqslant \mu$ $(i = 1, \ldots, n)$ and

$$\mu = \max_{1 \leqslant i \leqslant n} \left\{ \binom{n}{i} \right\} \cdot \sqrt{|d|}.$$

This is achieved as follows. Let $f = X^n + c_1 X^{n-1} + \cdots + c_n \in \mathbb{Z}[X]$ be the minimal polynomial of the algebraic integer a, thus $c_i = (-1)^i s_i$ where s_i is the ith elementary symmetric polynomial on a and its conjugates; hence $|c_i| = |s_i| = |\Sigma \text{ (products of } i \text{ conjugates of } a)| \leqslant \Sigma \text{ |products of } i \text{ conjugates of } a| \leqslant \binom{n}{i} \sqrt{|d|} \leqslant \mu$.

With $K = \mathbb{Q}(a)$, we associate the n-tuple $(c_1, \ldots, c_n)$ which belongs to S. There are at most finitely many fields K which give rise to the same element $(c_1, \ldots, c_n)$, namely the conjugates of $K = \mathbb{Q}(a)$. Thus, $\mathscr{S}$ is a finite set, proving our statement. ∎

Now, we shall return to the study of the group of units of an algebraic number field K. We prove the following fundamental theorem, which sets forth the role of the invariant $r = r_1 + r_2 - 1$:

Theorem 2 (*Dirichlet*). *The group U of units of the ring A of algebraic integers of K has the following structure*

$$U \cong W \times C_1 \times \cdots \times C_r,$$

where W is the cyclic group of order w of roots of unity belonging to K, each C_i is an infinite multiplicative cyclic group, and $r = r_1 + r_2 - 1$.
Proof: We first put aside the trival case where $r = 0$. This means that $r_1 + r_2 = 1$, thus if $K \neq \mathbb{Q}$ then $r_1 = 0$, $r_2 = 1$, $n = 2$, so $K = \mathbb{Q}(\sqrt{-d})$. $d > 0$. By (**D**) we know that every unit of K is a root of unity, that is, $U = W$. From now on we assume that $r \geqslant 1$.

We shall require 3 lemmas in order to prove the theorem.

First we introduce the following notion. The units $u_1, \ldots, u_k$ of A are said to be *independent* whenever a relation.

$$u_1^{m_1} u_2^{m_2} \ldots u_k^{m_k} = 1, \quad \text{with} \quad m_i \in \mathbb{Z},$$

is only possible when $m_1 = \cdots = m_k = 0$. Therefore, each u_i belonging to an independent set of units cannot be a root of unity.

The above concept is of multiplicative type. We may transform it into a kind of linear independence of vectors, by considering logarithms:

Lemma 2. *The units $u_1, \ldots, u_k$ of A are independent if and only if the r-dimensional real vectors*

$$x_1 = (\log |u_1^{(1)}|, \ldots, \log |u_1^{(r)}|) \in \mathbb{R}^r$$

$$\cdot$$
$$\cdot$$
$$\cdot$$

$$x_k = (\log |u_k^{(1)}|, \ldots, \log |u_k^{(r)}|) \in \mathbb{R}^r$$

are linearly independent over $\mathbb{Q}$.

Proof: If $u_1, \ldots, u_k$ are dependent, there exist integers $m_i \in \mathbb{Z}$, not all equal to zero, such that $u_1^{m_1} \ldots u_k^{m_k} = 1$. Therefore, considering the conjugates and their absolute values, we have $|u_1^{(i)}|^{m_1} \ldots |u_k^{(i)}|^{m_k} = 1$ for $i = 1, \ldots, r$, and taking logarithms: $\Sigma_{j=1}^{k} m_j \log |u_j^{(i)}| = 0$ for $i = 1, \ldots, r$.

This means that $\Sigma_{j=1}^{k} m_j x_j = 0$, showing the linear dependence of the vectors $x_1, \ldots, x_k$ over $\mathbb{Q}$.

Conversely, if the vectors $x_1, \ldots, x_k$ are linearly dependent over $\mathbb{Q}$, by multiplying a linear dependence relation by the product of denominators of the coefficients, we obtain a relation $\Sigma_{j=1}^{k} m_j x_j = 0$, where each $m_i \in \mathbb{Z}$ and not all coefficients are zero. Thus, $\Sigma_{j=1}^{k} m_j \log |u_j^{(i)}| = 0$ for $i = 1, \ldots, r$.

Since each u_j is a unit, then

$$|N_{K|\mathbb{Q}}(u_j)| = |u_j^{(1)}| \ldots |u_j^{(r)}| \cdot |u_j^{(r+1)}| \ldots |u_j^{(n)}| = 1,$$

hence taking $l_i = 1$ for $i = 1, \ldots, r_1$ and $l_i = 2$ for $i = r_1 + 1, \ldots, r_1 + r_2$, $\Sigma_{i=1}^{r_1+r_2} l_i \log |u_j^{(i)}| = 0$ for $j = 1, \ldots, k$, after noting that $u_j^{(r_1+r_2+i)} = \overline{u_j^{(r_1+i)}}$. Thus, from the above relations we deduce:

$$l_{r_1+r_2}\left[\sum_{j=1}^{k} m_j \log |u_j^{(r_1+r_2)}|\right] = -\sum_{j=1}^{k} m_j \left(\sum_{i=1}^{r} l_i \log |u_j^{(i)}|\right)$$

$$= -\sum_{i=1}^{r} l_i \left(\sum_{j=1}^{k} m_j \log |u_j^{(i)}|\right)$$

hence,

$$\sum_{j=1}^{k} m_j \log |u_j^{(r_1+r_2)}| = 0.$$

Since $u_j^{(r_1+r_2+i)} = \overline{u_i^{(r_1+i)}}$ then we have also $\Sigma_{j=1}^{k} m_j \log |u_j^{(r_1+r_2+i)}| = 0$ for all indices $i = 1, \ldots, r_2$: thus we have shown that $\Sigma_{j=1}^{k} m_j \log |u_j^{(i)}| = 0$ for all indices $i = 1, \ldots, n$; this means that

$$\left|\left(\prod_{j=1}^{k} u_j^{m_j}\right)^{(i)}\right| = 1 \quad \text{for } i = 1, \ldots, n;$$

by **(B)**, $\Pi_{j=1}^{k} u_j^{m_j}$ is a root of unity, hence there exists $q > 0$ such that $\Pi_{j=1}^{k} u_j^{qm_j} = 1$. ∎

A much more striking fact is contained in the next lemma:

Lemma 3. *Let $u_1, \ldots, u_k$ be units of A, but not roots of unity. The r-dimensional real vectors $x_1, \ldots, x_k \in \mathbb{R}^r$ (defined in Lemma 2) are linearly independent over $\mathbb{Q}$ if and only if they are linearly independent over $\mathbb{R}$.*

Proof: Since one implication is trivial, we shall assume that $x_1, \ldots, x_k$ are linearly dependent over $\mathbb{R}$.

Each x_i is not the null vector (because u_i is not a root of unity), hence there exists q, $1 \leqslant q < k$, such that (after change in numbering, if necessary) $x_1, \ldots, x_q$ are linearly independent vectors and every vector x_s ($s > q$) depends linearly on the preceding ones. So, for every $s > q$ we have real numbers c_j such that $x_s = \Sigma_{j=1}^{q} c_j x_j$. Hence, considering the components, we have:

$$\log |u_s^{(i)}| = \sum_{j=1}^{q} c_j \log |u_j^{(i)}|, \quad \text{for } i = 1, \ldots, r. \tag{1}$$

We shall prove that $c_j \in \mathbb{Q}$ for $j = 1, \ldots, q$. For this purpose, let G be the set of all q-tuples $(a_1, \ldots, a_q) \in \mathbb{R}^q$ such that there exists a unit v of A satisfying:

$$\log |v^{(i)}| = \sum_{j=1}^{q} a_j \log |u_j^{(i)}|, \quad \text{for } i = 1, \ldots, r. \tag{2}$$

For example, $G \supseteq \mathbb{Z}^q$, because if each $a_j \in \mathbb{Z}$ then $v = \Pi_{j=1}^{q} u_j^{a_j}$ satisfies the above relation. Similarly, G is an additive subgroup of $\mathbb{R}^q$.

Besides, it has the following important property: G is a discrete subgroup of $\mathbb{R}^q$, that is, the set G_1 of all elements $(a_1, \ldots, a_q) \in G$ such that $|a_i| \leqslant 1$ for all $i = 1, \ldots, q$, is finite. Let us prove this fact. Let $\theta : G_1 \to U$ be the mapping so defined: if $(a_1, \ldots, a_q) \in G_1$ let $\theta(a_1, \ldots, a_q) = v$ be any unit satisfying the relations in (2).

Since the vectors $x_1, \ldots, x_q$ are linearly independent over $\mathbb{R}$, if $(a_1, \ldots, a_q)$, $(b_1, \ldots, b_q)$ are distinct elements in G_1 then their images v, t by θ must be distinct, otherwise $\Sigma_{j=1}^{q} (a_j - b_j) \log |u_j^{(i)}| = 0$ for $i = 1, \ldots, r$, giving a non-trivial linear combination of the vectors $x_1, \ldots, x_q$. To prove our assertion, it is enough to show that $\theta(G_1)$ is contained in a finite set. If $v \in \theta(G_1)$, from (2) we deduce by considering absolute values:

$$|\log |v^{(i)}|| \leqslant \sum_{j=1}^{q} |a_j| \cdot |\log |u_j^{(i)}|| \leqslant \sum_{j=1}^{q} |\log |u_j^{(i)}||$$

for $i = 1, \ldots, r$; thus

$$-\sum_{j=1}^{q} |\log |u_j^{(i)}|| \leqslant \log |v^{(i)}| \leqslant \sum_{j=1}^{q} |\log |u_j^{(i)}||$$

and therefore there exist non-zero positive constants β, γ such that $\beta \leqslant |v^{(i)}| \leqslant \gamma$ for $i = 1, \dots, r$.

But v is a unit, so

$$1 = |N_{K|\mathbb{Q}}(v)| = |v^{(1)}| \dots |v^{(r)}| \, |v^{(r+1)}| \dots |v^{(n)}|,$$

and from $v^{(r_1+r_2+i)} = \overline{v^{(r_1+i)}}$, we have

$$|v^{(r+1)}|^2 = \frac{1}{|v^{(1)}|^{l_1} \dots |v^{(r)}|^{l_r}}$$

and so there exists a constant $\alpha \geqslant \gamma$ such that $|v^{(i)}| \leqslant \alpha$ for every $i = 1, \dots, n$. By (A), v belongs to a finite set, showing our assertion.

Knowing that G is a discrete subgroup of $\mathbb{R}^q$ we shall conclude that $G \subseteq \mathbb{Q}^q$, or more precisely: *there exists* $b \in \mathbb{Z}, b \neq 0$ *such that if* $(a_1, \dots, a_q) \in G$ *then* $ba_i \in \mathbb{Z}$ *for* $i = 1, \dots, q$. This will establish the lemma, because the elements $c_1, \dots, c_q$ appearing in (1) must be rational numbers.

First we note that if $(a_1, \dots, a_q) \in G$, if m is an integer then $(ma_1, \dots, ma_q) \in G$; so if $m_1, \dots, m_q \in \mathbb{Z}$ then $(ma_1 - m_1, \dots, ma_q - m_q) \in G$.

Now, given an arbitrary integer m, for every $i = 1, \dots, q$, let $m_i \in \mathbb{Z}$ be such that $|ma_i - m_i| \leqslant 1/2 < 1$. This establishes a mapping from $\mathbb{Z}$ into the finite set G_1, namely $m \to (ma_1 - m_1, \dots, ma_q - m_q)$. Hence, there exist distinct integers m, m', such that $ma_i - m_i = m'a_i - m_i'$ for all $i = 1, \dots, q$; thus $a_i = (m_i - m_i')/(m - m') \in \mathbb{Q}$, proving that $G \subseteq \mathbb{Q}^q$.

To obtain the more precise statement, let $(a_1, \dots, a_q) \in G$, thus, by what we have proved, $a_i \in \mathbb{Q}$ for every $i = 1, \dots, q$. For example, $a_1 = s/t$, with $s, t \in \mathbb{Z}$, $t > 0$ and s, t relatively prime; we shall show that $t \leqslant g = \#(G_1)$.

This is true when $t = 1$. Let $t \geqslant 2$; since $gcd(s,t) = 1$, there exist integers s', t' such that $1 = s's - t't$, and therefore $1/t = s'a_1 - t'$; hence every fraction $0, 1/t, 2/t, \dots, (t-1)/t$ is of the form $h/t = s_h'a_1 - t_{1h}'$ with $s_h', t_{1h}' \in \mathbb{Z}$. Let $t_{2h}', \dots, t_{qh}' \in \mathbb{Z}$ be such that $|s_h'a_i - t_{ih}'| < 1$ for $i = 2, \dots, q$. Then the image of s_h' is $(s_h'a_1 - t_{1h}', s_h'a_2 - t_{2h}', \dots, s_h'a_q - t_{qh}') \in G_1$ (by the mapping already considered). Thus G_1 contains at least t different elements, that is $t \leqslant g = \#(G_1)$; so t divides $b = g!$ and therefore $ba_1 \in \mathbb{Z}$. Similarly, we may show that $ba_i \in \mathbb{Z}$ for every $i = 2, \dots, q$, proving the stronger statement. ∎

We shall now prove what amounts almost to the theorem:

Lemma 4. *The statement of the theoeem is true, for some integer* $k \leqslant r$: $U \cong W \times C_1 \times \cdots \times C_k$.

Proof: If every unit is a root of unity, then $U = W$ and the lemma is true. Otherwise, let $\{u_1, \dots, u_k\}$ be a maximal set of independent units (with

$k > 0$); let $x_1, \ldots, x_k$ be the corresponding vectors in $\mathbb{R}^r$; by Lemmas 2 and 3, they are linearly independent over $\mathbb{Q}$, hence also over $\mathbb{R}$, and so $k \leqslant r$.

By the maximality of the set $\{u_1, \ldots, u_k\}$, if u is any other unit, not a root of unity, then $\{u, u_1, \ldots, u_k\}$ is not independent, so there exist $a_1, \ldots, a_k \in \mathbb{Q}$, not all equal to zero, such that

$$\log |u^{(i)}| = \sum_{j=1}^{k} a_j \log |u_j^{(i)}| \quad \text{for } i = 1, \ldots, r.$$

By the final considerations in Lemma 3, there exists an integer $b \neq 0$, independent of u, such that $ba_j \in \mathbb{Z}$ for all $j = 1, \ldots, k$. Thus

$$\log |u^{(i)}|^b = \sum_{j=1}^{k} (ba_j) \log |u_j^{(i)}|,$$

hence

$$|(u^b)^{(i)}| = \left| \prod_{j=1}^{k} (u_j^{ba_j})^{(i)} \right|$$

for all $i = 1, \ldots, r$; thus $u^b / \prod_{j=1}^{k} u_j^{ba_j}$ and all its conjugates have absolute value equal to 1; by (**B**) it must be a root of unity $\zeta \in K$, that is, $u^b = \zeta \prod_{j=1}^{k} u_j^{ba_j}$. To find an expression for u, let ξ be a zero of $X^b - \zeta$, so $\xi^{bw} = 1$, where $w = \#(W)$. Let also v_j be a zero of $X^b - u_j$ (for $j = 1, \ldots, k$); ξ, v_j are not necessarily in K. Thus $u^b = (\xi \prod_{j=1}^{k} v_j^{ba_j})^b$, that is $u = \lambda \prod_{j=1}^{k} v_j^{ba_j}$, where $\lambda^{bw} = 1$.

The above computations led to an expression of u, by means of elements λ, v_j which do not necessarily belong to K. Let U' be the multiplicative group of complex numbers generated by $\lambda, v_1, \ldots, v_k$, let

$$W' = \{\lambda^m \mid m = 0, 1, \ldots bw - 1\},$$

hence $W' \cap U = W$. Since the units $v_1, \ldots, v_k$ are independent (because of the same property for $u_1, \ldots, u_k$), then the multiplicative Abelian group U'/W' is a free Abelian group of rank k, generated by the cosets of $v_1, \ldots, v_k$. From $W' \cap U = W$ it follows that the quotient group U/W may be identified to a subgroup of U'/W'; by Chapter 6, (**J**), U/W is a free Abelian group of rank at most k; but the cosets of $u_1, \ldots, u_k$ relative to W are independent, because a relation $\prod_{j=1}^{k} u_j^{f_j} = \zeta \in W$ would imply $\prod_{j=1}^{k} u_j^{f_j w} = 1$ hence, $f_j w = 0$ and $f_j = 0$ for every $j = 1, \ldots, k$. Hence U/W has rank equal to k. Explicitly, we have shown that there exist units $u_1', \ldots, u_k' \in U$, not roots of unity, such that every unit $u \in U$ may be written uniquely in the form $u = \zeta \cdot \prod_{j=1}^{k} u_j'^{f_j}$ with $\zeta \in W$, $f_j \in \mathbb{Z}$ (for $j = 1, \ldots, k$). Letting C_j be the multiplicative cyclic group generated by u_j', we have proved that $U \cong W \times C_1 \times \cdots \times C_k$. ∎

In order to complete the proof of the theorem, we still need to show the existence of sufficiently many independent units. This is basically the contents of the following lemma:

Lemma 5. *If $c_1, \ldots, c_r \in \mathbb{R}$ are not equal to zero, then there exists a unit $u \in U$ such that $\Sigma_{i=1}^r c_i \log |u^{(i)}| \neq 0$.*

Proof: We shall require Minkowski's theorem in the proof of this lemma.

Let $\{a_1, \ldots, a_n\}$ be an integral basis of K, let $d = \det(a_j^{(i)})$. Since d^2 is the discriminant of the field then $1 \leqslant d^2 \in \mathbb{Z}$ so $1 \leqslant |d|$, $0 \neq d$. Let β be a sufficiently large positive real number, for example, $\beta = (\Sigma_{i=1}^r |c_i|)\log |d| + 1$. We consider the n linear forms

$$L_i = \sum_{j=1}^n a_j^{(i)} X_j \quad (i = 1, \ldots, n).$$

Let $\tau_1, \ldots, \tau_n$ be positive real numbers satisfying the following conditions:

$$\begin{cases} \tau_{r_1+i} = \tau_{r_1+r_2+i} & \text{for } i = 1, \ldots, r_2 \\ \tau_1 \ldots \tau_n = |d|. \end{cases} \tag{1}$$

We may choose $\tau_1, \ldots, \tau_r$ arbitrarily and the above relations determine $\tau_{r_1+r_2}$ uniquely. By **(I)** there exist integers $x_1, \ldots, x_n$, not all equal to zero, such that if $y^{(i)} = L_i(x_1, \ldots, x_n) = \Sigma_{j=1}^n a_j^{(i)} x_j$ then $|y^{(i)}| \leqslant \tau_i$ (for $i = 1, \ldots, n$). In particular, $y \in A$, $y \neq 0$, and $1 \leqslant |N(y)| \leqslant |d|$ thus, $\tau_i/|d| = \tau_i/(\tau_1 \ldots \tau_n) \leqslant 1/\Pi_{j \neq i} |y^{(j)}| \leqslant |y^{(i)}| \leqslant \tau_i \leqslant \tau_i |d|$ (for every $i = 1, \ldots, n$). Letting $F(y) = \Sigma_{i=1}^r c_i \log |y^{(i)}|$ we deduce that

$$\left| F(y) - \sum_{i=1}^r c_i \log \tau_i \right| = \left| \sum_{i=1}^r c_i \log\left(\frac{|y^{(i)}|}{\tau_i}\right) \right| \leqslant \sum_{i=1}^r |c_i| \cdot \left| \log\left(\frac{|y^{(i)}|}{\tau_i}\right) \right|$$

$$\leqslant \left(\sum_{i=1}^r |c_i| \right) \log |d| < \beta$$

because $-\log |d| \leqslant \log\left(\dfrac{|y^{(i)}|}{\tau_i}\right) \leqslant 0 \leqslant \log |d|$.

Suppose now that for every $h = 1, 2, \ldots$ we choose positive real numbers $\tau_{h1}, \ldots, \tau_{hr}$ satisfying conditions in (1) and also the following condition:

$$\sum_{i=1}^r c_i \log \tau_{hi} = 2\beta h. \tag{2}$$

This is possible since there exists an index i, $1 \leqslant i \leqslant r$, for which $c_i \neq 0$. Let $y_h \in A$, $y_h \neq 0$ be obtained from $\tau_{h1}, \ldots, \tau_{hr}$ in the manner indicated above. Then $|F(y_h) - 2\beta h| = |F(y_h) - \Sigma_{i=1}^r c_i \log \tau_{hi}| < \beta$ so

$\beta(2h - 1) < F(y_n) < \beta(2h + 1)$ for all indices $h = 1, 2, \ldots$ Therefore $F(y_1) < F(y_2) < F(y_3) < \ldots$ But $N(Ay_n) = |N_{K|\mathbb{Q}}(y_n)| \leqslant |d|$, so there exist distinct indices $h \neq h'$ such that $Ay_h = Ay_{h'}$ and therefore $u = y_h{}'/y_h$ is a unit of K. Since $F(y_h) \neq F(y_{h'}) = F(uy_h) = F(u) + F(y_h)$ it follows that $F(u) \neq 0$ so

$$\sum_{i=1}^{r} c_i \log u^{(i)} \neq 0. \quad \blacksquare$$

Proof of the theorem: By Lemma 5, if $c_1 = 1$, $c_2 = \cdots = c_r = 0$ there exists a unit $u_1 \in U$ such that $\log |u_1^{(1)}| \neq 0$.

Now, given $c_1 = -\log |u_1^{(2)}|$, $c_2 = \log |u_1^{(1)}| \neq 0$, $c_3 = \cdots = c_r = 0$ there exists a unit $u_2 \in U$ such that $c_1 \log |u_2^{(1)}| + c_2 \log |u_2^{(2)}| \neq 0$; that is,

$$\det\begin{pmatrix} \log |u_1^{(1)}| & \log |u_2^{(1)}| \\ \log |u_1^{(2)}| & \log |u_2^{(2)}| \end{pmatrix} \neq 0.$$

Repeating this argument, given

$$c_1 = \det\begin{pmatrix} \log |u_1^{(2)}| & \log |u_2^{(2)}| \\ \log |u_1^{(3)}| & \log |u_2^{(3)}| \end{pmatrix}$$

$$c_2 = -\det\begin{pmatrix} \log |u_1^{(1)}| & \log |u_2^{(1)}| \\ \log |u_1^{(3)}| & \log |u_2^{(3)}| \end{pmatrix}$$

$$c_3 = \det\begin{pmatrix} \log |u^{(1)}| & \log |u_2^{(1)}| \\ \log |u_1^{(2)}| & \log |u_2^{(2)}| \end{pmatrix} \neq 0,$$

there exists a unit $u_3 \in U$ such that

$$c_1 \log |u_3^{(1)}| + c_2 \log |u_3^{(2)}| + c_3 \log |u_3^{(3)}| \neq 0;$$

that is,

$$\det\begin{pmatrix} \log |u_1^{(1)}| & \log |u_2^{(1)}| & \log |u^{(1)}| \\ \log |u_1^{(2)}| & \log |u_2^{(2)}| & \log |u_3^{(2)}| \\ \log |u_1^{(3)}| & \log |u_2^{(3)}| & \log |u_3^{(3)}| \end{pmatrix} \neq 0.$$

In this way, we determine r units $u_1, \ldots, u_r$. Since the determinant $\det(\log |u_j^{(i)}|) \neq 0$, then no column is identically zero, so each u_j is not a root of unity (by (**B**)); moreover, the column vectors are linearly independent over $\mathbb{R}$, hence the units $u_1, \ldots, u_r$ are independent. This shows that we have $k = r$ in Lemma 4, concluding the proof of the theorem. $\quad \blacksquare$

Explicitly, Dirichlet's theorem on units says that there exists a root of unity ζ and r units of infinite order $u_1, \ldots, u_r$, such that every unit u may be written uniquely in the form $u = \zeta^{e_o} u_1^{e_1} \ldots u_r^{e_r}$ with $0 \leqslant e_o < w$ and $e_1, \ldots, e_r \in \mathbb{Z}$.

Definition 4. Any set of r independent units $\{u_1, \ldots, u_r\}$ of K (with $r = r_1 + r_2 - 1$) for which the above statement holds is called a *fundamental system of units of K*.

Now, we may introduce a new numerical invariant:

Q. *Let $\{u_1, \ldots, u_r\}$, $\{v_1, \ldots, v_r\}$ be any two fundamental systems of units of K. Then*

$$|\det(\log |u_j^{(i)}|)| = |\det(\log |v_j^{(i)}|)|.$$

Proof: By the theorem, we may write $v_j = \zeta^{b_j} u_1^{a_{1j}} u_2^{a_{2j}} \ldots u_r^{a_{rj}}$ (for every $j = 1, \ldots, r$) where $b_j, a_{ij} \in \mathbb{Z}$. Similarly, we may write $u_j = \zeta^{b_j'} v_1^{a_{1j}'} v_2^{a_{2j}'} \ldots v_r^{a_{rj}'}$ (for every $j = 1, \ldots, r$). Therefore, by the uniqueness of representation of units, the matrix (a_{ij}') is the inverse of (a_{ij}), and so $\det(a_{ij}) \cdot \det(a_{ij}') = 1$, hence $|\det(a_{ij})| = |\det(a_{ij}')| = 1$. But considering the conjugates of the units, their absolute values and logarithms, we obtain

$$\log |v_j^{(i)}| = \sum_{h=1}^{r} a_{hj} \log |u_h^{(i)}|$$

and therefore

$$|\det(\log |v_j^{(i)}|)| = |\det(\log |u_j^{(i)}|)|. \quad \blacksquare$$

We introduce therefore the following concept:

Definition 5. Let $u_1, \ldots, u_r$ be any fundamental system of units of K. The positive real number

$$R = |\det(l_i \log |u_j^{(i)}|)|$$

is called the *regulator* of K.

For example, if $K = \mathbb{Q}(\sqrt{3})$, we have $r = r_1 + r_2 - 1 = 1$, and as computed before, $u_1 = 2 + \sqrt{3}$ is a fundamental unit. Then $R = |\log(2 + \sqrt{3})|$.

The computation of the regulator is difficult in practice, since it requires the knowledge of a fundamental system of units, which is usually hard to determine.

EXERCISES

1. Let K be an algebraic number field of degree 4. Determine the roots of unity of K.

2. Let K be an algebraic number field of odd degree. Show that K has only two roots of unity: $1, -1$.

3. Using properties of Euler's function, show that an algebraic number field contains only finitely many roots of unity.

Hint: If $n = [K:\mathbb{Q}]$ show that for every large d we have $\varphi(d) > n$, so K cannot contain a dth root of unity.

4. Let K be an algebraic number field which is different from $\mathbb{Q}$ and not an imaginary quadratic field. Prove that for every real number $\varepsilon > 0$, there exists an algebraic integer $x \in K$ such that $|x| < \varepsilon$.

5. Determine the fundamental units of the following quadratic fields $\mathbb{Q}(\sqrt{d})$, where:
   (a) $d = 10$
   (b) $d = 14$
   (c) $d = 19$
   (d) $d = 23$.

6. Determine the units of the field $\mathbb{Q}(\sqrt[3]{3})$.

7. Show that $u = 1 - 6\sqrt[3]{6} + 3\sqrt[3]{36}$ is a fundamental unit of $\mathbb{Q}(\sqrt[3]{6})$.

8. Construct a symmetric convex body S of volume $2^n \mu$ and having no lattice point in its interior, except the origin.

9. Develop in detail the argument in the proof of Theorem 1, where it is asserted that if $y \in \Lambda$, there exists $y' \in \Lambda \cap C$ such that $\rho_{y'} \leqslant \rho_y$.

10. Show that the homothetic image of a convex set is convex.

11. Establish the following assertions made in the proof of (N):

   (a) $f_{r_1,r_2}(\rho) = 2\int_0^\rho f_{r_1-1,r_2}(\rho - t)\, dt$.

   (b) $f_{r_1,r_2}(\rho) = \iint\limits_{t^2+u^2\leqslant\rho^2/4} f_{r_1,r_2-1}(\rho - 2\sqrt{t^2 + u^2})\, dt\, du$

   $\qquad\qquad = 2\pi \int_0^{\rho/2} f_{r_1,r_2-1}(\rho - 2t)t\, dt.$

   (c) $f_{r_1,r_2}(\rho) = 2^{r_1}\left(\dfrac{\pi}{2}\right)^{r_2} \dfrac{\rho^n}{n!}.$

12. Let α be a real number. Show that there exists a pair of relatively prime integers x,y such that $|x/y - \alpha| < 1/y^2$. Prove that if α is rational then there are only finitely many pairs (x,y) with above properties.

13. If α is a real irrational number, show that there exist an infinite number of pairs of relatively prime integers x,y such that $|x/y - \alpha| < 1/y^2$.

14. Let d be a natural number, not a square. Show that there exist infinitely many pairs of natural numbers x,y such that $|x^2 - dy^2| < 1 + 2\sqrt{d}$.

15. Let d be a natural number not a square. The equation $X^2 - dY^2 = 1$ is called *Pell's equation*. Prove:
  (a) Pell's equation has infinitely many solutions (x,y) with $x,y \in \mathbb{Z}$.
  (b) There exists a solution (x_1,y_1) with x_1,y_1 positive and such that every solution (x,y) with x,y positive satisfies $x + y\sqrt{d} = (x_1 + y_1\sqrt{d})^m$ for some integer $m \geqslant 1$.

16. Let d be a natural number not a square. Show that if $x + y\sqrt{d}$ is a unit of $\mathbb{Q}(\sqrt{d})$ such that $x > \frac{1}{2}y^2 - 1$ then u is the fundamental unit.

17. Given any natural numbers y_1 and a, let $d = a(ay_1^2 + 2)$. Show that $(1 + ay_1^2) + y_1\sqrt{d}$ is the fundamental unit of $\mathbb{Q}(\sqrt{d})$. Deduce that for every natural number y_1 there exists infinitely many real quadratic fields whose fundamental unit is of the form $x + y_1 d$.

18. Let d be a natural number, not a square. Show that if there exist integers x,y such that $x^2 - dy^2 = -1$ then every odd prime factor of d is congruent to 1 modulo 4. However, verify (for $d = 34$) that the converse is not true.

19. Let d be a natural number, not a square. Let x,y be natural numbers such that $x^2 - dy^2 = -1$. If $x',y' \in \mathbb{Z}$ are defined by $x' + y'\sqrt{d} = (x + y\sqrt{d})^2$ show that $x'^2 - dy'^2 = 1$.

20. Show that if p is a prime number, $p \equiv 1 \pmod 4$ then there exist integers x,y such that $x^2 - py^2 = -1$.

21. Let $d > 2$ be a natural number, not a square. Show that it is impossible to find a pair of integers x,y satisfying simultaneously two of the following equations:
$$X^2 - dY^2 = -1$$
$$X^2 - dY^2 = 2$$
$$X^2 - dY^2 = -2.$$

22. Let p be an odd prime. Prove that the following equations are solvable in integers:
  (a) $X^2 - pY^2 = -1$ when $p \equiv 1 \pmod 4$.
  (b) $X^2 - pY^2 = 2$ when $p \equiv 7 \pmod 8$.
  (c) $X^2 - pY^2 = -2$ when $p \equiv 3 \pmod 8$.

23. Let $m > 1$, $l > 1$ be relatively prime integers, let ζ be a primitive mth root of unity and ξ a primitive lth root of unity. Show that $1 - \zeta\xi$ is a unit in the ring of integers of the field $\mathbb{Q}(\zeta,\xi)$.

24. Show that if K is a field of degree 3 then its discriminant satisfies: $\delta < -12$ or $20 < \delta$.

25. Show that in every class of ideals of the algebraic number field K, there exists a non-zero integral ideal J such that

$$N(J) \leqslant \left(\frac{4}{\pi}\right)^{r_2} \frac{n!}{n^n} \sqrt{|\delta|}.$$

26. Let ζ be a primitive 5th root of unity. Show, with the methods of this chapter, that the class number of $\mathbb{Q}(\zeta)$ is equal to 1.

27. Let K be a quadratic field, with discriminant δ. Show that in every class of ideals of K there exists a non-zero integral ideal J such that

$$N(J) \leqslant \begin{cases} \frac{1}{2}\sqrt{\delta} & \text{when } \delta > 0 \\ \dfrac{2}{\pi}\sqrt{|\delta|} & \text{when } \delta < 0. \end{cases}$$

(Chapter 13, (7A)).

28. Let $K = \mathbb{Q}(\sqrt{d})$ be a quadratic field. Show that the number h_0 of strictly equivalent classes of ideals of K is finite (see Chapter 8, Exercise 15). Moreover, $h_0 = h$ when $d < 0$ or $d > 0$ and the fundamental unit has negative norm. Otherwise, $h_0 = 2h$.

29. Let $d > 0$, $K = \mathbb{Q}(\sqrt{d})$ and let u be a fundamental unit of K. Show that if $N(u) = -1$ then $p \equiv 1 \pmod 4$ for every odd prime divisor of δ_K.

30. Let ζ be a primitive mth root of unity, and let $K = \mathbb{Q}(\zeta)$, $K_0 = K \cap \mathbb{R}$. Show that a fundamental system of units of K_0 is a maximal independent system of units of K.

31. Let p be an odd prime, and ζ a primitive pth root of unity. Show that

$$p = (-1)^{(p-1)/2} \prod_{j=1}^{(p-1)/2} (\zeta^j - \zeta^{-j})^2.$$

Hint: Use the fact that ζ^2 is also a primitive pth root of unity and express p in terms of ζ^2 and its powers.

32. Let p, q be distinct odd primes, let ζ be a primitive pth root of unity, and A the ring of algebraic integers of $\mathbb{Q}(\zeta)$. Show that:

(a) $p^{(q-1)/2} \equiv (-1)^{\frac{p-1}{2} \cdot \frac{q-1}{2}} \prod_{j=1}^{(p-1)/2} (\zeta^{jq} - \zeta^{-jq})/(\zeta^j - \zeta^{-j}) \pmod{Aq}$.

(b) $\prod_{j=1}^{(p-1)/2} (\zeta^{jq} - \zeta^{-jq})/(\zeta^{j} - \zeta^{-j}) = (-1)^r$, where r is the number of integers kq in the set $\left\{q, 2q, \ldots, \dfrac{p-1}{2} q\right\}$ such that $kq \equiv -s$ (mod p), with $0 < s \leqslant (p-1)/2$.

(c) prove anew Gauss reciprocity law.

Hint: For (a) note that $(\zeta^{j} - \zeta^{-j})^q \equiv \zeta^{jq} - \zeta^{-jq}$ (mod Aq); for (c) make use of Euler's and Gauss' criterion (Chapter 2, (G) and (I)).

33. Show that if the algebraic number field K contains a non-real root of unity then $N_{K|\mathbb{Q}}(x) > 0$ for every $x \in K$, $x \neq 0$.

34. Let K be an algebraic number field of degree n. Let U_1 be the group of all units of K having norm $N_{K|\mathbb{Q}}(u) = 1$. Prove:

(a) If n is odd there is a fundamental system of units $\{u_1, \ldots, u_r\}$ of K, such that every unit $u \in U_1$ may be written uniquely in the form $u = u_1^{e_1} \ldots u_r^{e_r}$ (with $e_i \in \mathbb{Z}$).

(b) If n is even, if $\{u_1, \ldots, u_r\}$ is a fundamental system of units of K and $k, 0 \leqslant k \leqslant r$, is such that $N_{K|\mathbb{Q}}(u_i) = 1$ for $i = 1, \ldots, k$, $N_{K|\mathbb{Q}}(u_i) = -1$ for $i = k + 1, \ldots, r$, let $v_i = u_i$ for $i = 1, \ldots, k$, $v_i = u_i u_r$ for $i = k + 1, \ldots, r$. Then every unit $u \in U_1$ may be written uniquely in the form $u = \zeta v_1^{e_1} \ldots v_r^{e_r}$ ($e_i \in \mathbb{Z}$), where ζ is any root of unity in K.

35. Show that the fundamental unit of the real quadratic field $\mathbb{Q}(\sqrt{m^2 - 1})$ is $\varepsilon = m + \sqrt{m^2 - 1}$.

36. Let $\mathscr{S}$ be a finite set of $s \geqslant 0$ non-zero prime ideals of the ring A of algebraic integers of K. Let S be the multiplicative set, complement in A of the union $\cup_{P \notin \mathscr{S}} P$. The units of the ring A_S (Chapter 7, Exercise 21) are called the $\mathscr{S}$-*units* of A. Let U_S denote the group of $\mathscr{S}$-units of A. Prove the analogous of Dirichlet's theorem: U_S/W is a free Abelian group of rank $r + s$, where $r = r_1 + r_2 - 1$.

37. Let $n \geqslant 2$ be a power of a prime, let ζ be a primitive nth root of unity, $K = \mathbb{Q}(\zeta)$, $K_0 = K \cap \mathbb{R}$. Show that every unit of K is the product of a unit of K_0 with a root of unity.

Hint: Generalize the method of proof of (F).

CHAPTER 10

Ramification, Discriminant and Different

1. THE RAMIFICATION INDEX AND THE INERTIAL DEGREE

Let A be a Dedekind domain, K its field of quotients; let $L \mid K$ be a separable extension of degree n and B the integral closure of A in L. By Chapter 7, (**M**), B is a Dedekind domain.

If I is any ideal of A, let $B \, . \, I$ denote the ideal of B generated by I; it consists of all sums $\sum_{i=1}^{m} b_i x_i$ with $m \geqslant 1$, $b_i \in B$, $x_i \in I$ for all $i = 1, \ldots, m$.

We begin noting the following easy fact which holds in particular when A is the ring of algebraic integers of an algebraic number field:

1A. *If I is an ideal of A then $B \, . \, I \cap A = I$; in particular, if $I \neq A$ then $B \, . \, I \neq B$.*

Proof: We may assume $I \neq 0$ and we first consider the case where $L \mid K$ is a Galois extension. Let $y \in B \, . \, I \cap A$ so $y = \sum_{j=1}^{m} b_j x_j$, where $b_j \in B$, $x_j \in I$. Then for every K-isomorphism σ_i of L we have

$$y = \sigma_i(y) = \sum_{j=1}^{m} \sigma_i(b_j) x_j \qquad (i = 1, \ldots, n)$$

therefore,

$$y^n = \prod_{i=1}^{n} \left(\sum_{j=1}^{m} \sigma_i(b_j) x_j \right) = h(x_1, \ldots, x_m)$$

where h is a homogeneous polynomial in m variables of degree n, with coefficients which are invariant under the action of each σ_i $(i = 1, \ldots, n)$; hence h has coefficients in K and also in B (since $\sigma_i(B) = B$); thus

$$h \in A[X_1, \ldots, X_m]$$

and therefore $y^n \in I^n$. Since $(Ay \, . \, I^{-1})^n$ is an integral ideal of A then

$$Ay \, . \, I^{-1} \subseteq A \quad \text{so} \quad y \in I.$$

This proves the inclusion $B \, . \, I \cap A \subseteq I$. The other inclusion is trivial.

Now let L' be the smallest Galois extension of K containing L, and let B' be the integral closure of A in L'; thus, $B' \cap L = B$. We have therefore $I = B' \, . \, I \cap A = [B'(B \, . \, I) \cap B] \cap A = B \, . \, I \cap A$. ∎

Let A be the ring of integers of an algebraic number field K.

If P is a non-zero prime ideal of A then $B . P$ may be written in a unique way as a product of powers of prime ideals of B:

$$B . P = \prod_{i=1}^{g} Q_i^{e_i}.$$

We introduce the following definitions:

Definition 1. g is called the *decomposition number* of P in the extension $L \mid K$. If necessary, we shall use the notation $g_P(L \mid K)$.

We have seen in **(1A)** that $g \geqslant 1$, because $B . P \neq B$.

Definition 2.   For every $i = 1, \ldots, g$, e_i is called the *ramification index* of Q_i in $L \mid K$. If $e_i = 1$ we say that Q_i is *unramified in $L \mid K$* (or over P). If $e_1 = \cdots = e_g = 1$ we say that P *is unramified in $L \mid K$*.

We shall sometimes use the notations $e(Q_i \mid P)$ or $e_{Q_i}(L \mid K)$ for the ramification index e_i.

Let us note that $Q_i \cap A \supseteq B . P \cap A = P$ hence $Q_i \cap A = P$ and B/Q_i is a finite field containing A/P.

Definition 3.   The degree f_i of B/Q_i over A/P is called the *inertial degree* of Q_i in $L \mid K$. If $f_i = 1$ we say that Q_i is *inert in $L \mid K$* (or over P). If $f_1 = \cdots = f_g = 1$ then P *is* said to be *inert in $L \mid K$*.

Sometimes we denote the inertial degree by $f(Q_i \mid P)$ or $f_{Q_i}(L \mid K)$.

If we determine for each prime ideal P of A the decomposition number, ramification indices, and inertial degrees in $L \mid K$, then, since every non-zero ideal I of A is written in unique way as a product of prime ideals of A, then the decomposition of $B . I$ will be known.

The main theorem in this chapter will state that there are only finitely many ramified prime ideals P in $L \mid K$, precisely those dividing the relative discriminant of $L \mid K$. In particular, only finitely many prime ideals Q of B will be ramified in $L \mid K$.

We begin considering the relationship between the numbers just introduced. A first obvious fact to record is the transitivity:

1B.   *Let $K \subseteq L \subseteq L'$ be algebraic number fields, with rings of integers $A \subseteq B \subseteq B'$; let Q' be a non-zero prime ideal of B', $Q = Q' \cap B$, $P = Q \cap A$. Then $e'' = ee'$, $f'' = ff'$, where e, f are the ramification index and inertial degree of Q in $L \mid K$, and similarly e', f' correspond to Q' in $L' \mid L$, e'', f'' correspond to Q' in $L' \mid K$.*
Proof: Since Q^e divides BP, but Q^{e+1} does not divide BP, we may write $BP = Q^e . J$ where Q does not divide J. Similarly, $B'Q = Q'^{e'} . J'$ where Q' does not divide J'.

Hence $B'P = B'(BP) = (B'Q)^e \cdot (B'J) = Q'^{e'e}J'^e \cdot (B'J)$ and Q' does not divide $B'J$, otherwise $Q = Q' \cap B \supseteq B'J \cap B \supseteq J$, contrary to the assumption. Therefore $Q'^{e'e}$ is the exact power of Q' dividing $B'P$, so $e'' = ee'$.

Similarly, by definition $f = [B/Q:A/P]$, $f' = [B'/Q':B/Q]$, thus

$$f'' = [B'/Q': A/P] = ff'. \quad \blacksquare$$

In the same manner, we note that if $BP = \Pi_{i=1}^g Q_i^{e_i}$, if g_i' is the decomposition number of Q_i in $L' \mid L$, then the decomposition number of P in $L' \mid K$ is $g'' = \Sigma_{i=1}^g g_i'$.

A notable simplification arises in the important case of a Galois extension.

Let $L \mid K$ be a Galois extension of degree n, let G be its Galois group, so the elements of G leave each element of K fixed, and transform any element of L into its conjugates; in particular $\sigma(B) \subseteq B$ for every $\sigma \in G$, hence also $B = \sigma(\sigma^{-1}(B)) \subseteq \sigma(B)$, showing that $\sigma(B) = B$. If J is any ideal of B then $\sigma(J)$ is an ideal of B and $\sigma(J) \cap A = J \cap A$. Therefore σ induces a ring-isomorphism $\bar{\sigma}: B/J \rightarrow B/\sigma(J)$, namely, if $\bar{b} \in B/J$ then $\bar{\sigma}(\bar{b}) = \overline{\sigma(b)}$ (and this is indeed well-defined); $\bar{\sigma}$ leaves fixed every element of $A/(J \cap A)$.

In particular, if Q is a non-zero prime ideal of B, then $B/Q \cong B/\sigma(Q)$ and $\sigma(Q)$ is also a non-zero prime ideal of B.

More interesting is the transitivity of the action of G on the set of prime ideals Q of B, having a given intersection with A:

1C. *If Q, Q' are any prime ideals of B such that $Q \cap A = Q' \cap A \neq 0$, there exists $\sigma \in G$ such that $\sigma(Q) = Q'$.*

Proof: Let $G = \{\sigma_1, \ldots, \sigma_n\}$ be the Galois group of $L \mid K$, and let us assume that $Q' \neq \sigma_i(Q)$ for every $\sigma_i \in G$. By Chapter 7, (**K**), there exists an element $x \in B$ such that $x \notin \sigma_i(Q)$ for $i = 1, \ldots, n$, $x \in Q'$. Let

$$a = \prod_{i=1}^n \sigma_i(x),$$

then $a \in A \cap Q'$, however, $a \notin Q$ since each $\sigma_i(x) \notin Q$ for $i = 1, \ldots, n$, (otherwise $x = \sigma_i^{-1}\sigma_i(x) \in \sigma_i^{-1}(Q)$, for some index i). This is a contradiction, hence there exists $\sigma_i \in G$ such that $\sigma_i(Q) = Q'$. $\quad \blacksquare$

As a corollary of (**1C**), we have:

1D. *If $L \mid K$ is a Galois extension of degree n, if $BP = \Pi_{i=1}^g Q_i^{e_i}$ and $[B/Q_i: A/P] = f_i$, then $e_1 = \cdots = e_g$, $f_1 = \cdots = f_g$, and each B/Q_i is isomorphic with the extension of degree f_i over the finite field A/P.*

Proof: Let $BP = \Pi_{i=1}^g Q_i^{e_i}$; for every index j, $1 \leqslant j \leqslant g$, by (**1C**) there exists $\sigma \in G$ such that $\sigma(Q_1) = Q_j$. Hence from $BP = \sigma(BP) = \Pi_{i=1}^g \sigma(Q_i)^{e_i}$ and the uniqueness of the decomposition of BP into products of prime ideals, then $e_j = e_1$ for every j, $1 \leqslant j \leqslant g$. Similarly, from $B/Q_j = B/\sigma(Q_1) \cong B/Q_1$ it follows that $f_j = f_1$ for every j, $1 \leqslant j \leqslant g$.

Since A/P is finite, there exists only one extension of degree f_1, up to isomorphism, hence all the fields B/Q_i are isomorphic. ∎

We shall now derive a fundamental relation between the degree of $L \mid K$, the decomposition number and the ramification indices and inertial degrees. This is very easily done when the ground field is $\mathbb{Q}$.

Indeed, if p is a prime number and $Ap = \prod_{i=1}^{g} P_i^{e_i}$ is the decomposition of Ap into product of prime ideals of A, considering the norms of the above ideals, we have (Chapter 8, (C), (F)):

$$|N_{K|\mathbb{Q}}(p)| = \prod_{i=1}^{g} N(P_i)^{e_i},$$

and since $N_{K|\mathbb{Q}}(p) = p^n$, and $N(P_i) = p^{f_i}$ then $n = \Sigma_{i=1}^{g} e_i f_i$.

In order to generalize this relation to the case of an arbitrary ground field, we shall develop the concept of relative norm of an ideal.

In our arguments, we shall require a correspondence between polynomials in several variables and ideals; we begin establishing these facts.

E. *Let A be a ring, P a prime ideal of A. Let $P[X_1, \ldots, X_m]$ denote the set of polynomials of $A[X_1, \ldots, X_m]$ whose coefficients belong to P. Then $P[X_1, \ldots, X_m]$ is a prime ideal of $A[X_1, \ldots, X_m]$.*

Proof: With every $h \in A[X_1, \ldots, X_m]$ we associate the polynomial

$$\bar{h} \in (A/P)[X_1, \ldots, X_m]$$

obtained by reducing modulo P the coefficients of h; this defines a homomorphism from $A[X_1, \ldots, X_m]$ onto $(A/P)[X_1, \ldots, X_m]$, having kernel equal to $P[X_1, \ldots, X_m]$. Since $(A/P)[X_1, \ldots, X_n]$ is a domain then $P[X_1, \ldots, X_m]$ is a prime ideal of $A[X_1, \ldots, X_m]$. ∎

If $h \in A[X_1, \ldots, X_m]$ we define the *content of h* to be the ideal of A generated by the coefficients of h; this ideal will be denoted by $C_A(h)$ or simply by $C(h)$.

1F. *Let A be a Dedekind domain and h, $k \in A[X_1, \ldots, X_m]$. Then $C(hk) = C(h) \cdot C(k)$.*

Proof: We have to show that if P is any prime ideal of A, if P^r is the exact power of P dividing $C(h)$ and P^s the exact power of P dividing $C(k)$ (with $r \geqslant 0$, $s \geqslant 0$) then P^{r+s} is the exact power of P dividing $C(h) \cdot C(k)$.

Let $C(h) = P^r \cdot I$, where P does not divide the ideal I, and consider an element $a \in P^r$, $a \notin P^{r+1}$; then P^r divides Aa and $Aa = P^r \cdot J$, where J is an ideal of A such that P does not divide J; let $a' \in J$, $a' \notin P$, hence J divides Aa'

and we have $Aa' = J.J'$. It follows that Aa divides $Aa'.C(h)$ but P does not divide the ideal $Aa^{-1}.Aa'.C(h) = J'.I$ (because if P divides J' then it would divide Aa'). The ideal generated by the coefficients of the polynomial $a^{-1}a'h$ is $Aa^{-1}Aa.C(h) \subseteq A$, so

$$\frac{a'}{a} h \in A[X_1, \ldots, X_m], \text{ but } \frac{a'}{a} h \notin P[X_1, \ldots, X_m].$$

Similarly, let $b, b' \in A$ be elements such that $\frac{b'}{b} k \in A[X_1, \ldots, X_m]$, but $\frac{b'}{b} k \notin P[X_1, \ldots, X_m]$.

Then $g = (a'b'/ab)hk \in A[X_1, \ldots, X_m]$, $g \notin P[X_1, \ldots, X_m]$ (by the previous result) hence, $hk = (ab/a'b')g \notin P^{r+s+1}[X_1, \ldots, X_m]$ while obviously $hk \in P^{r+s}[X_1, \ldots, X_m]$ by the properties of a, a', b, b'. ∎

Let K be an algebraic number field, L an algebraic extension of degree n, let $A \subseteq K$, $B \subseteq L$ be the rings of integers, and $\sigma_1, \ldots, \sigma_n$ the K-isomorphisms of L.

1G. *If J is a fractional ideal of L then there exists a unique fractional ideal I of K such that*

$$\prod_{i=1}^{n} \sigma_i(J) = B.I.$$

If $J \subseteq B$ then $I \subseteq A$, if $J \neq 0$ then $I \neq 0$.

Proof: We consider first the case where L is a Galois extension of K and J is an integral ideal of B. Let $t_1, \ldots, t_m$ be a system of generators of the ideal J and let $h = t_1 X_1 + \cdots + t_m X_m \in B[X_1, \ldots, X_m]$; its content is $C_B(h) = J$.

For every $i = 1, \ldots, n$ let $h_i = \sigma_i(t_1)X_1 + \cdots + \sigma_i(t_m)X_m$, so $C_B(h_i) = \sigma_i(J)$. Then

$$\prod_{i=1}^{n} \sigma_i(J) = \prod_{i=1}^{n} C_B(h_i) = C_B(h_1, \ldots, h_n).$$

But $k = h_1 \ldots h_n \in A[X_1, \ldots, X_m]$ hence $C_B(k) = B.C_A(k)$, and we just take $I = C_A(k)$.

More generally, if L' is the smallest Galois extension of K containing L, and B' its ring of integers, then by what we have proved

$$B'\left(\prod_{i=1}^{n} \sigma_i(J)\right) = \prod_{i=1}^{n} \sigma_i(B'.J) = B'.I = B'(B.I);$$

it follows from (**1A**) that $\prod_{i=1}^{n} \sigma_i(J) = B.I$. The uniqueness of I follows also from (**1A**).

The case where J is a fractional ideal is easily handled; let $b \in B$, $b \neq 0$ be such that $b.J \subseteq B$. Then $\prod_{i=1}^{n} \sigma_i(Bb.J) = B.I$ for some ideal I of A.

Therefore if $a = \prod_{i=1}^{n} \sigma_i(b) \neq 0$ we have $\prod_{i=1}^{n} \sigma_i(J) = B \cdot (a^{-1}I)$. Since $a^{-1}I$ is unique with this property, then so is I. The last assertions are obvious. ∎

Definition 4. With the notations above, the fractional ideal I of K such that $B \cdot I = \prod_{i=1}^{n} \sigma_i(J)$ is called the *relative norm of J in the extension $L \mid K$.* We denote it by $N_{L \mid K}(J)$.

From the definition we have at once:

1H. *If J, J' are non-zero fractional ideals of L then*

$$N_{L \mid} (J \cdot J') = N_{L \mid K}(J) \cdot N_{L \mid K}(J').$$

If $b \in B$, $b \neq 0$, then $N_{L \mid K}(Bb) = A \cdot N_{L \mid K}(b)$. If I is any non-zero ideal of A then $N_{L \mid K}(B \cdot I) = I^n$. If $N_{L \mid K}(J)$ is a non-zero prime ideal of A then J is a non-zero prime ideal of B.

Proof: The first assertions are evident from the definition. For the last one, let $J = \prod_{i=1}^{s} Q_i^{r_i}$, where each Q_i is a non-zero prime ideal of B. Then $N_{L \mid K}(J) = \prod_{i=1}^{s} (N_{L \mid K}(Q_i))^{r_i}$. But $N_{L \mid K}(J) = P$, a non-zero prime ideal of A. By the uniqueness of the decomposition into prime ideals, we must have $N_{L \mid K}(Q_i) = P^{f_i}$ for every $i = 1, \ldots, s$ and therefore $1 = \sum_{i=1}^{s} r_i f_i$; this implies that $s = 1$, $r_1 = f_1 = 1$; that is, J is a prime ideal of B. ∎

The concept of relative norm of an ideal coincides with that of norm of an ideal, when the ground field is $K = \mathbb{Q}$. In the latter case it is also customary to refer to an *absolute norm*.

1I. *For every ideal J we have $N_{L \mid \mathbb{Q}}(J) = \mathbb{Z} \cdot N(J)$.*
Proof: Let h be the class number of the field L. From Chapter 8, **(I)** it follows that $J^h = Bb$ (for some element $b \in B$). Therefore by **(1H)** we have

$$[N_{L \mid} (J)]^h = N_{L \mid \mathbb{Q}}(Bb) = \mathbb{Z} \cdot N_{L \mid \mathbb{Q}}(b),$$

while $N(J)^h = N(Bb) = |N_{L \mid \mathbb{Q}}(b)|$. Thus $[N_{L \mid \mathbb{Q}}(J)]^h = \mathbb{Z} \cdot N(J)^h$ and therefore $N_{L \mid \mathbb{Q}}(J) = \mathbb{Z} \cdot N(J)$. ∎

1J. *Let $K \subseteq L \subseteq L'$ be algebraic number fields and J' any fractional ideal of L'. Then:*
$$N_{L \mid K}(N_{L' \mid L}(J')) = N_{L' \mid K}(J').$$

Proof: The result holds for principal ideals $B'b'$:

$$N_{L \mid K}(N_{L' \mid L}(B'b')) = N_{L \mid K}(B \cdot N_{L' \mid L}(b')) = A \cdot N_{L \mid K}(N_{L' \mid L}(b'))$$
$$= A \cdot N_{L' \mid K}(b') = N_{L' \mid K}(B'b').$$

Now if J' is any non-zero fractional ideal of L', if h is the class number of L' then $J'^h = B'b'$ and

$$N_{L|K}(N_{L'|L}(J'))^h = N_{L|K}(N_{L'|L}(B'b')) = N_{L'|K}(B'b')$$
$$= N_{L'|K}(J'^h) = N_{L'|K}(J')^h.$$

Since A is a Dedekind domain then $N_{L|K}(N_{L'|L}(J')) = N_{L'|K}(J')$. ∎

1K. *If Q is a prime ideal of B, if $Q \cap A = P$ then $N_{L|K}(Q) = P^f$, where f is the inertial degree of Q in $L \mid K$.*
Proof: We first consider the case where $L \mid K$ is a Galois extension. Let $N_{L|K}(Q) = I = \prod_{j=1}^r P_j^{t_j}$. Since

$$Q \supseteq \prod_{i=1}^n \sigma_i(Q) = B \cdot I = \prod_{j=1}^r B \cdot P_j^{t_j},$$

there exists j such that Q divides $B \cdot P_j$; hence $Q \supseteq B \cdot P_j$ and so

$$P = Q \cap A \supseteq B \cdot P_j \cap A = P_j,$$

thus $P_j = P$. Let J be an ideal of B such that $B \cdot P = Q \cdot J$. Hence

$$N_{L|K}(B \cdot P) = N_{L|K}(Q) \cdot N_{L|K}(J).$$

But $N_{L|K}(B \cdot P) = P^n$ since $\prod_{i=1}^n \sigma_i(B \cdot P) = (B \cdot P)^n = B \cdot P^n$; hence $N_{L|K}(Q)$ divides P^n and therefore we may write $N_{L|K}(Q) = P^f$ for some integer f, $1 \leqslant f \leqslant n$.

Let us consider the norms of P and Q over $\mathbb{Q}$: $N(P) = p^{f'}$, $N(Q) = p^{f''}$ where $P \cap \mathbb{Z} = \mathbb{Z}p$, $f' = [A/P : \mathbb{F}_p]$, $f'' = [B/Q : \mathbb{F}_p]$. By **(1J)** $\mathbb{Z}p^{f''} = \mathbb{Z} \cdot N(Q) = \mathbb{Z} \cdot N(N_{L|K}(Q)) = \mathbb{Z} \cdot N(P^f) = \mathbb{Z}p^{ff'}$, therefore $f'' = ff'$ and $[B/Q : \mathbb{F}_p] = f \cdot [A/P : \mathbb{F}_p]$ so $f = [B/Q : A/P]$ is the inertial degree of Q in $L \mid K$.

Now we consider the general case. Let L' be the smallest Galois extension of K containing L, let B' be the ring of integers of L', and Q' a prime ideal of B' dividing $B'.Q$; thus $Q' \cap B = Q$. By what we have just proved,

$$N_{L'|K}(Q') = P^{f'}, \text{ where } f' = [B'/Q' : A/P]. \text{ On the other hand,}$$

$$N_{L'|K}(Q') = N_{L|K}(N_{L'|L}(Q')) = N_{L|K}(Q^{f''}),$$

where $f'' = [B'/Q' : B/Q]$. Since $f' = ff''$, then $(P^f)^{f''} = (N_{L|K}(Q))^{f''}$, thus $N_{L|K}(Q) = P^f$. ∎

Having established all these properties for the relative norm of an ideal, we are able to prove in general the following fundamental relation:

1L. *If* $B \cdot P = \prod_{i=1}^{g} Q_i^{e_i}$ *and* $f_i = [B/Q_i : A/P]$ *for* $i = 1, \ldots, g$, *then*

$$\sum_{i=1}^{g} e_i f_i = n.$$

Proof: Using the preceding results we have:

$$P^n = N_{L|K}(B \cdot P) = \prod_{i=1}^{g} N_{L|K}(Q_i^{e_i}) = \prod_{i=1}^{g} P^{e_i f_i};$$

hence

$$n = \sum_{i=1}^{g} e_i f_i. \quad \blacksquare$$

In particular, if $L \mid K$ is a Galois extension then the above formula becomes: $n = efg$, where $e = e_1 = \cdots = e_g$ and $f = f_1 = \cdots = f_g$.

It follows that the decomposition number g of P in $L \mid K$ is at most equal to the degree $n = [L:K]$. If $g = n$ we say that P is *completely decomposed* in $L \mid K$.

As a corollary we deduce:

1M. *If P is a non-zero prime ideal of A then $B/B \cdot P$ is a vector space of dimension $n = [L:K]$ over A/P.*
Proof: Let $B \cdot P = \prod_{i=1}^{g} Q_i^{e_i}$ be the decomposition of $B \cdot P$ into a product of prime ideals of B; then $Q_i \cap A = P$. By Theorem 2 of Chapter 9, we have $B/B \cdot P \cong \prod_{i=1}^{g} B/Q_i^{e_i}$. Moreover,

$$\#(B/Q_i^{e_i}) = N(Q_i^{e_i}) = [N(Q_i)]^{e_i} = \#(B/Q_i)^{e_i}.$$

If $\mathbb{Z}p = P \cap \mathbb{Z}$ then the dimension over $\mathbb{F}_p$ of the corresponding vector spaces satisfy $\dim_{\mathbb{F}_p}(B/Q_i^{e_i}) = e_i \cdot \dim_{\mathbb{F}_p}(B/Q_i)$; hence,

$$\dim_{A/P}(B/Q_i^{e_i}) = e_i \cdot \dim_{A/P}(B/Q_i) \quad \text{for } i = 1, \ldots, g.$$

Therefore

$$\dim_{A/P}(B/B \cdot P) = \sum_{i=1}^{g} \dim_{A/P}(B/Q_i^{e_i}) = \sum_{i=1}^{g} e_i \cdot \dim_{A/P}(B/Q_i)$$

$$= \sum_{i=1}^{g} e_i f_i = n \quad \blacksquare$$

We shall prove soon a generalization of this result.

Before proceeding in the theory, it is instructive to examine some particular cases.

2. DECOMPOSITION OF PRIME NUMBERS IN A QUADRATIC FIELD

Let $K = \mathbb{Q}(\sqrt{d})$, where d is a square-free integer, let p be a prime number; we shall determine the decomposition of Ap into prime ideals of A.

We have $n = 2$, thus by **(1L)**, the only possibilities are the following:

(1) $g = 2, e_1 = e_2 = 1, f_1 = f_2 = 1$. In this case $Ap = P_1 P_2$, with $P_1 \neq P_2$, $N(P_1) = N(P_2) = p$. We say that p is (*totally*) *decomposed* in $\mathbb{Q}(\sqrt{d})$.

(2) $g = 1, e = 1, f = 2$. Thus $Ap = P$ is a prime ideal, $N(Ap) = p^2$, and we say that p is *inert* in $\mathbb{Q}(\sqrt{d})$.

(3) $g = 1, e = 2, f = 1$. Thus $Ap = P^2$, $N(P) = p$ and p is *ramified* in $\mathbb{Q}(\sqrt{d})$.

By Chapter 7, Theorem 2, if $Ap = P_1 P_2$ then $A/Ap \cong A/P_1 \times A/P_2$, so A/Ap is a cartesian product of two fields; in particular it has no nilpotent elements, except 0. If $Ap = P$, then A/Ap is a field. Finally, if $Ap = P^2$ then A/Ap is a ring having a non-zero ideal P/P^2, which is nilpotent. Since the above cases are mutually exclusive, we have therefore another description of the possible phenomena in terms of the ring A/Ap.

The main question now is the following: Given the prime p, for which values of d, do we have cases 1, respectively 2, 3?

A.2 *If p is a prime, $p \neq 2$: p is ramified in $\mathbb{Q}(\sqrt{d})$ if and only if p divides d; p is inert if and only if d is not a square modulo p, that is $\left(\dfrac{d}{p}\right) = -1$; p is decomposed if and only if d is a square modulo p, that is $\left(\dfrac{d}{p}\right) = 1$.*

Proof: If $d \equiv 2$ or $d \equiv 3$ (modulo 4), then the ring A of integers of $\mathbb{Q}(\sqrt{d})$ is $A = \mathbb{Z} + \mathbb{Z}\sqrt{d}$; if $d \equiv 1$ (modulo 4) then $A = \mathbb{Z} + \mathbb{Z}(1 + \sqrt{d})/2$ (by Chapter 5, **(P)**).

In both cases, $A/Ap = (\mathbb{Z} + \mathbb{Z}\sqrt{d})/Ap$. In fact, if $d \equiv 1$ (mod 4) and b is odd then

$$\left[a + b\left(\frac{1 + \sqrt{d}}{2}\right)\right] - \left[a + (b + p)\left(\frac{1 + \sqrt{d}}{2}\right)\right] \in Ap,$$

with $b + p$ even.

Now $\mathbb{Z} + \mathbb{Z}\sqrt{d} \cong \mathbb{Z}[X]/(X^2 - d)$, by the mapping induced by $\theta : \mathbb{Z}[X] \to \mathbb{Z} + \mathbb{Z}\sqrt{d}$, $\theta(h) = h(\sqrt{d})$, which has kernel $(X^2 - d)$ (principal ideal generated by $X^2 - d$ in $\mathbb{Z}[X]$).

It follows that

$$(\mathbb{Z} + \mathbb{Z}\sqrt{d})/Ap \cong \mathbb{Z}[X]/(X^2 - d, p\mathbb{Z}[X])$$
$$\cong (\mathbb{Z}[X]/p\mathbb{Z}[X])/((X^2 - d, p\mathbb{Z}[X])/p\mathbb{Z}[X])$$
$$\cong \mathbb{F}_p[X]/(X^2 - \bar{d})$$

where $(X^2 - d, p\mathbb{Z}[X])$ denotes the ideal of $\mathbb{Z}[X]$ generated by $X^2 - d$, $p\mathbb{Z}[X]$, $\bar{d}$ is the residue class of d modulo p; in deriving the above isomorphism, we have made use of the well-known isomorphism theorems for rings. [For the readers who are not familiar with these facts, we describe explicitly the isomorphism, in the following way: given $\bar{h} \in \mathbb{F}_p[X]$, let $h \in \mathbb{Z}[X]$ be such that the coefficients of $\bar{h}$ are obtained by taking the residue classes modulo p of the coefficients of h; the choice of such polynomial is of course not unique. Let $\theta(f) = h(\sqrt{d}) \in \mathbb{Z} + \mathbb{Z}\sqrt{d} \subseteq A$; denoting by $\pi: A \rightarrow A/Ap$ the canonical homomorphism, we define $\psi: \mathbb{F}_p[X] \rightarrow A/Ap$ by $\psi(\bar{h}) = \pi(\theta(h))$. If $k \in p\mathbb{Z}[X]$ then $\theta(k) \in Ap$, so if h', h are such that $\bar{h}' = \bar{h}$ then $h - h' = k \in p\mathbb{Z}[X]$ and $\pi(\theta(h)) = \pi(\theta(h'))$ hence ψ is a well-defined mapping. It is now immediate to check that ψ is a ring-homomorphism. Since $A/Ap = (\mathbb{Z} + \mathbb{Z}\sqrt{d})/Ap$, then ψ maps $\mathbb{F}_p[X]$ onto A/Ap. Clearly, $X^2 - \bar{d}$ belongs to the kernel of ψ; conversely, if $\bar{h}$ is in the kernel, then $h(\sqrt{d}) \in Ap$; but since $\mathbb{Z}[\sqrt{d}] \cap Ap = p\mathbb{Z}[\sqrt{d}]$ (as one checks easily when $d \equiv 1$ (modulo 4)) then $h(\sqrt{d}) = a + b\sqrt{d}$, with $a, b \in \mathbb{Z}p$; thus $h - (a + bX)$ vanishes on $\sqrt{d}$, so $X^2 - d$ divides $h - (a + bX)$, hence considering the residue classes modulo p of the coefficients we derive that $X^2 - \bar{d}$ divides $\bar{h}$, proving that the kernel of ψ is the ideal generated by $X^2 - \bar{d}$; hence ψ induces an isomorphism

$$\bar{\psi}: \mathbb{F}_p[X]/(X^2 - \bar{d}) \rightarrow (\mathbb{Z} + \mathbb{Z}\sqrt{d})/Ap = A/Ap$$

as we intended to show.]

If $X^2 - \bar{d}$ is irreducible in $\mathbb{F}_p[X]$, then A/Ap is isomorphic to a domain, so Ap is a prime ideal, that is, p is inert in $\mathbb{Q}(\sqrt{d})$.

If $X^2 - \bar{d} = h_1 h_2$ where h_1, h_2 are distinct irreducible polynomials, which are of degree 1, then from $(h_1 h_2) = (h_1) \cap (h_2)$, $(h_1) + (h_2) = \mathbb{F}_p[X]$, we deduce by Chapter 7, (J), that

$$A/Ap \cong \mathbb{F}_p[X]/(h_1 h_2) \cong \mathbb{F}_p[X]/(h_1) \times \mathbb{F}_p[X]/(h_2) \cong \mathbb{F}_p \times \mathbb{F}_p$$

(because $\mathbb{F}_p[X]/(h_i)$ is an algebraic extension of degree 1 over $\mathbb{F}_p$); so p is decomposed in $\mathbb{Q}(\sqrt{d})$.

If $X^2 - \bar{d}$ is the square of an irreducible polynomial, $X^2 - \bar{d} = \bar{h}^2$, then $A/Ap \cong \mathbb{F}_p[X]/(h^2)$ and the ideal $(h)/(h^2)$ is non-zero and nilpotent; therefore p is ramified in $\mathbb{Q}(\sqrt{d})$.

In this last case, letting $\bar{h} = X + \bar{a}$, we have $X^2 - \bar{d} = X^2 + 2\bar{a}X + \bar{a}^2$ and since $p \neq 2$, then $\bar{a} = \bar{0}$; that is, p divides a; thus $-\bar{d} \equiv a^2 \pmod{p}$ implies $p \mid d$. Conversely, if $p \mid d$ then $X^2 - \bar{d} = X^2$, so p is ramified.

If p does not divide d, then p is decomposed exactly when $X^2 - \bar{d} = (X + \bar{a})(X + \bar{b})$ with $\bar{a} \neq \bar{b}$, so $\bar{a} + \bar{b} = 0$, $\bar{a}\bar{b} = -\bar{d}$, hence, $\bar{a}^2 = \bar{d}$, that is, $\left(\dfrac{d}{p}\right) = 1$. Thus, by exclusion, p is inert when p does not divide d, and $\left(\dfrac{d}{p}\right) = -1$. $\blacksquare$

Let us observe at this moment that the type of decomposition of the odd prime p in $\mathbb{Q}(\sqrt{d})$ depends only on the residue class of d modulo p.

2B. *The prime 2 is ramified in $\mathbb{Q}(\sqrt{d})$ if and only if $d \equiv 2 \pmod 4$ or $d \equiv 3 \pmod 4$; 2 is inert in $\mathbb{Q}(\sqrt{d})$ if and only if $d \equiv 5 \pmod 8$; 2 is decomposed in $\mathbb{Q}(\sqrt{d})$ if and only if $d \equiv 1 \pmod 8$.*

Proof: If $d \equiv 2$ or $d \equiv 3 \pmod 4$, then the ring of algebraic integers is $A = \mathbb{Z} + \mathbb{Z}\sqrt{d}$ and $A/2A \cong \mathbb{F}_2[X]/(X^2 - \bar{d})$ (with the same proof as in (2A)). Since $\bar{d} = \bar{0}$ or $\bar{d} = \bar{1}$ then $X^2 - \bar{d}$ is a square in $\mathbb{F}_2[X]$ ($X^2 - \bar{1} = (X - \bar{1})^2$). Therefore $A/2A$ has non-zero nilpotent elements, so 2 is ramified in $\mathbb{Q}(\sqrt{d})$.

If $d \equiv 1 \pmod 4$ then

$$A = \mathbb{Z} + \mathbb{Z}\left(\frac{1 + \sqrt{d}}{2}\right) \cong \mathbb{Z}[X]\Big/\left(X^2 - X - \frac{d-1}{4}\right),$$

because $X^2 - X - (d-1)/4$ is the minimal polynomial of $(1 + \sqrt{d})/2$. Hence, $A/2A \cong \mathbb{F}_2[X]/(X^2 - X - \bar{a})$ where $a = (d-1)/4$.

If $a \equiv 1 \pmod 2$ then $d \equiv 5 \pmod 8$; in this case, since $X^2 - X - \bar{1} = X^2 + X + \bar{1} \in \mathbb{F}_2[X]$ is irreducible over $\mathbb{F}_2$, then $A/2A$ is a field, so 2 is inert in $\mathbb{Q}(\sqrt{d})$.

Finally, if $a \equiv 0 \pmod 2$ then $d \equiv 1 \pmod 8$, $X^2 - X - \bar{a} = X^2 + X = X(X + \bar{1})$, so $A/2A$ is a product of two fields and 2 is decomposed in $\mathbb{Q}(\sqrt{d})$. $\blacksquare$

We observe again that the type of decomposition of 2 in $\mathbb{Q}(\sqrt{d})$ depends only on the residue class of d modulo 8.

As an addendum, we recall that if $d \equiv 2 \pmod 4$ or $d \equiv 3 \pmod 4$ then the discriminant of $\mathbb{Q}(\sqrt{d})$ is $\delta = 4d$, and if $d \equiv 1 \pmod 4$ then $\delta = d$ (by Chapter 6, (**P**)). Thus, from (**2A**), (**2B**) we deduce: *p is ramified in $\mathbb{Q}(\sqrt{d})$ if and only if p divides the discriminant δ of $\mathbb{Q}(\sqrt{d})$.*

So for quadratic fields there exist only finitely many prime numbers p which are ramified, and these may be determined by computing the discriminant.

By means of the reciprocity law for the Jacobi symbol (see Chapter 4, (**R**), (**S**)) we deduce the following fact:

2C. *If p, p' are prime numbers and $p \equiv p'$ (mod $|\delta|$) then p, p' are of the same type in $\mathbb{Q}(\sqrt{d})$.*

Proof: If p' is ramified then p' divides $|\delta|$. From $p \equiv p'$ (mod $|\delta|$) it follows that $p = p'$.

Let p, p' be odd primes. If $d \equiv 1$ (mod 4) then $d = \delta$ and

$$\left(\frac{d}{p}\right) = \left(\frac{\delta}{p}\right) = (-1)^{\frac{\delta-1}{2}\cdot\frac{p-1}{2}}\left(\frac{p}{|\delta|}\right) = \left(\frac{p}{|\delta|}\right) = \left(\frac{p'}{|\delta|}\right) = \left(\frac{d}{p'}\right).$$

Thus by (**2A**) p is decomposed if and only if p' is decomposed in $\mathbb{Q}(\sqrt{d})$.

If $d \equiv 3$ (mod 4) then $\delta = 4d$ and

$$\left(\frac{d}{p}\right) = (-1)^{\frac{d-1}{2}\cdot\frac{p-1}{2}}\left(\frac{p}{|d|}\right) = (-1)^{\frac{p-1}{2}}\left(\frac{p}{|d|}\right).$$

But $p \equiv p'$ (mod $|\delta|$) implies $p \equiv p'$ (mod $|d|$) and $p \equiv p'$ (mod 4) so

$$(-1)^{(p-1)/2}\left(\frac{p}{|d|}\right) = (-1)^{(p'-1)/2}\left(\frac{p'}{|d|}\right) = \left(\frac{d}{p'}\right)$$

and again p, p' are primes of the same type in $\mathbb{Q}(\sqrt{d})$.

If $d \equiv 2$ (mod 4), let $d = 2d'$, d' being odd (since d has no square factor). We have $\delta = 4d = 8d'$ so $p \equiv p'$ (mod 8) and $p \equiv p'$ (mod $|d|$) hence

$$\left(\frac{d}{p}\right) = \left(\frac{2}{p}\right)\left(\frac{d'}{p}\right) = (-1)^{\frac{p^2-1}{8}+\frac{p-1}{2}\cdot\frac{d'-1}{2}}\left(\frac{p}{|d'|}\right)$$

$$= (-1)^{\frac{p'^2-1}{8}+\frac{p'-1}{2}\cdot\frac{d'-1}{2}}\left(\frac{p'}{|d'|}\right) = \left(\frac{d}{p'}\right).$$

It remains the case where $p' = 2$. From $p \equiv p'$ (mod $|\delta|$) it follows that $|\delta|$ is odd, so $d \equiv 1$ (mod 4), $d = \delta$. Now 2 is decomposed exactly when $\delta \equiv 1$ (mod 8). From

$$\left(\frac{d}{p}\right) = \left(\frac{\delta}{p}\right) = (-1)^{\frac{\delta-1}{2}\cdot\frac{p-1}{2}}\left(\frac{p}{|\delta|}\right) = \left(\frac{2}{|\delta|}\right) = (-1)^{\frac{\delta^2-1}{8}}$$

we see that $\delta \equiv 1$ (mod 8) if and only if $\left(\dfrac{d}{p}\right) = 1$; that is, p is decomposed in $\mathbb{Q}(\sqrt{d})$. ∎

The interest of this result is that it tells the type of decomposition of any prime number in $\mathbb{Q}(\sqrt{d})$ by considering its class with respect to a unique modulo, namely $|\delta|$. So, in some sense the phenomenon of decomposition of primes is already built in the residue classes modulo $|\delta|$.

3. DECOMPOSITION OF PRIME NUMBERS IN CYCLOTOMIC FIELDS

Let p be a prime number, $m = p^k > 2$ (so if $p = 2$ then $k \geqslant 2$), and let ζ be a primitive mth root of unity, $K = \mathbb{Q}(\zeta)$, A the ring of integers of K, thus $\zeta \in A$.

K is a Galois extension of degree $\varphi(m) = p^{k-1}(p - 1)$ and its Galois group $\mathscr{G}$ is isomorphic to the multiplicative group $P(m)$ of prime residue classes modulo m.

The minimal polynomial of ζ is the mth cyclotomic polynomial

$$\Phi_m = X^{p^{k-1}(p-1)} + X^{p^{k-1}(p-2)} + \cdots + X^{p^{k-1}} + 1,$$

(see Chapter 2, Exercise 5), and we have

$$\Phi_m = \prod_{\bar{a} \in (m)} (X - \zeta^a).$$

We shall indicate the decomposition into prime ideals of the ideal Aq, where q is any prime number. We first note that if a,b are non-zero integers, relatively prime to m, then $1 - \zeta^a$, $1 - \zeta^b$ are associated elements of A. In fact, we may write $b \equiv aa' \pmod{m}$ and $a \equiv bb' \pmod{m}$, thus

$$\frac{1 - \zeta^b}{1 - \zeta^a} = \frac{1 - \zeta^{aa'}}{1 - \zeta^a} = 1 + \zeta^a + \zeta^{2a} + \cdots + \zeta^{(a'-1)a} \in A$$

and similarly $[(1 - \zeta^a)/(1 - \zeta^b)] \in A$. In particular, the absolute values of the norms of the elements $1 - \zeta^a$, $1 - \zeta^b$ are equal.

Let $\xi = 1 - \zeta \in A$.

3A. $p = u\xi^{\varphi(p^k)}$, *where u is a unit of A. The principal ideal $A\xi$ is prime and* $Ap = (A\xi)^{\varphi(p^k)}$, $N(A\xi) = p$.

Proof: From $p = \Phi_m(1) = \prod_{\bar{a} \in P(m)} (1 - \zeta^a)$ and the previous observation, it follows that $p = u\xi^{\varphi(p^k)}$, where u is a unit of A. Thus $Ap = (A\xi)^{\varphi(p^k)}$. Taking norms in the extension $\mathbb{Q}(\zeta) \mid \mathbb{Q}$ we have $p^{\varphi(p^k)} = N(Ap) = (N(A\xi))^{\varphi(p^k)}$, hence $N(A\xi) = p$. By **(1H)** we conclude that $A\xi$ is a prime ideal of A. ∎

Now let q be any prime number different from p. The type of decomposition of Aq may be obtained from theorems which we shall derive in this and other chapters. However, we think it is illuminating to work out a particular case in detail. Besides, in this approach we shall meet quite interesting new methods, which will be later explained in full generality.

3B. *Let q be any prime number distinct from p. Then $Aq = Q_1 \ldots Q_g$, where the prime ideals $Q_1, \ldots, Q_g$ are distinct, $N(Q_i) = q^f$ and f is the order of $\bar{q}$ (residue class of q modulo $p^k = m$) in the group $P(m)$; moreover, $fg = \varphi(p^k)$. In particular, q is inert exactly when $q \equiv 1 \pmod{m}$.*

Proof: Let $Q_1, \ldots, Q_g$ be the distinct prime ideals of A dividing Aq. Let $\mathscr{L} = \{\sigma \in \mathscr{G} \mid \sigma(Q_1) = Q_1\}$. Then $\mathscr{L}$ is a subgroup of $\mathscr{G}$. Moreover the index of $\mathscr{L}$ in $\mathscr{G}$ is equal to g. In fact, by **(1C)** given any prime ideal Q_i there exists $\sigma \in \mathscr{G}$ such that $\sigma(Q_1) = Q_i$; if $\sigma\mathscr{L} = \tau\mathscr{L}$ then $\sigma = \tau\lambda$ with $\lambda \in \mathscr{L}$, so $\sigma(Q_1) = \tau(\lambda(Q_1)) = \tau(Q_1)$; conversely, if $\sigma(Q_1) = \tau(Q_1)$ then $\tau^{-1}\sigma(Q_1) = Q_1$, so $\tau^{-1}\sigma \in \mathscr{L}$ hence $\sigma \in \tau\mathscr{L}$, and therefore $\sigma\mathscr{L} = \tau\mathscr{L}$. Thus σ, τ are in different cosets if and only if $\sigma(Q_1) \neq \tau(Q_1)$; hence the number of cosets is equal to g.

Let Z be the field of invariants of the group $\mathscr{L}$, so $\mathscr{L} = \mathrm{Gal}(K \mid Z)$; let us consider the residue class fields $\bar{Z} = (A \cap Z)/(Q_1 \cap Z)$, which is a subfield of A/Q_1. By **(1L)** we have $\varphi(p^k) = efg$, where f is the inertial degree and e is the ramification index of Q_1 over q. By the definition of $\mathscr{L}$, it follows from **(1C)** that Q_1 is the only prime ideal of A which divides $A(Q_1 \cap Z)$; so by **(1L)** $ef = [K:Z] = e_{Q_1}(K \mid Z) \cdot f_{Q_1}(K \mid Z)$. From **(1B)** we have

$$e_{Q_1}(K \mid Z) \cdot e_{Q_1 \cap Z}(Z \mid \mathbb{Q}) = e, \qquad f_{Q_1}(K \mid Z) \cdot f_{Q_1 \cap Z}(Z \mid \mathbb{Q}) = f;$$

hence we must have $f_{Q_1}(K \mid Z) = f, f_{Q_1 \cap Z}(Z \mid \mathbb{Q}) = 1$; that is, $\bar{Z} = \mathbb{F}_q$.

With every $\sigma \in \mathscr{L}$ we associate the mapping $\bar{\sigma}: A/Q_1 \rightarrow A/Q_1$, defined by $\bar{\sigma}(\bar{x}) = \overline{\sigma(x)}$ (it is well defined, because $\sigma \in \mathscr{L}$); each $\bar{\sigma}$ belongs to the Galois group $\bar{\mathscr{G}}$ of A/Q_1 over $\mathbb{F}_q$. The kernel of the mapping $\sigma \rightarrow \bar{\sigma}$ is precisely the subgroup

$$\mathscr{T} = \{\sigma \in \mathscr{G} \mid \sigma(x) \equiv x \pmod{Q_1} \text{ for every element } x \in A\}.$$

We prove now that $\mathscr{T}$ is reduced to the identity automorphism. In fact, let $\sigma(\zeta) = \zeta^s$, where $1 < s < p^k$ and $\gcd(s, p^k) = 1$; if $\sigma \in \mathscr{T}$ then Q_1 contains the element $\sigma(\zeta) - \zeta = \zeta^s - \zeta = -\zeta(1 - \zeta^{s-1})$. If $\gcd(s - 1, p) = 1$ then we have seen that $1 - \zeta^{s-1}$ and $1 - \zeta$ are associated elements, so $\xi = 1 - \zeta \in Q_1$, that is, $Q_1 = A\xi$ and Q_1 divides Ap, which is not the case. If, however, $s - 1 = p^l t$, with $1 \leqslant l < k$, t not a multiple of p, then $\zeta^{p^{k-l}}$ is a primitive root of unity of order p^l, so

$$X^{p^l} - 1 = \prod_{a=0}^{p^l-1} (X - \zeta^{ap^{k-l}}),$$

therefore,

$$\zeta^{s-1} - 1 = \zeta^{p^l t} - 1 = \prod_{a=0}^{p^l-1} (\zeta^t - \zeta^{ap^{k-l}}) = \zeta^{p^l t} \prod_{a=0}^{p^l-1} (1 - \zeta^{ap^{k-l}-t});$$

the elements $1 - \zeta^{ap^{k-l}-t}$ are associated with $1 - \zeta$, therefore in this case we conclude also that $Q_1 = A\xi$, which is not true.

We shall show that the image of the mapping $\sigma \to \bar{\sigma}$ is equal to $\overline{\mathscr{G}}$. A/Q_1 is a finite field, hence there exists an element $x \in A$ such that $A/Q_1 = \mathbb{F}_q(\bar{x})$ (where $\bar{x}$ is the class of x modulo Q_1). Let $\rho \in \overline{\mathscr{G}}$, so $\rho(\bar{x})$ is a conjugate of $\bar{x}$. If h is the minimal polynomial of x over Z, then $h = \Pi_{\sigma \in \mathscr{L}} (X - \sigma(x))$, and each conjugate $\sigma(x)$ of x over Z is still an algebraic integer, so the coefficients of h are in $A \cap Z$. Taking the image modulo $Q_1 \cap Z$ we have

$$\bar{h} = \prod_{\sigma \in \mathscr{L}} (X - \overline{\sigma(x)}) \in (A \cap Z)/(Q_1 \cap Z) = \mathbb{F}_q.$$

For the identity automorphism we have $\bar{h}(\bar{x}) = 0$, hence $\bar{h}$ is a multiple of the minimal polynomial of $\bar{x}$ over $\mathbb{F}_q$. The conjugates of $\bar{x}$ are among the elements $\overline{\sigma(x)}$; in particular, there exists $\sigma \in \mathscr{L}$ such that $\rho(\bar{x}) = \overline{\sigma(x)} = \bar{\sigma}(\bar{x})$. This implies that $\rho = \bar{\sigma}$ showing that $\sigma \to \bar{\sigma}$ is an isomorphism between the groups $\mathscr{L}$ and $\overline{\mathscr{G}}$.

By Chapter 2, (9), $\mathscr{L} \cong \overline{\mathscr{G}}$ is a cyclic group of order f, generated by the Frobenius automorphism, which is defined by $\sigma_q(\zeta) = \zeta^q$. Thus $\zeta = \sigma_q^f(\zeta) = \zeta^{q^f}$, so $\zeta^{q^f-1} = 1$. Since ζ is a primitive p^k-th root of unity then p^k divides $q^f - 1$; on the other hand, if $f' < f$ then $\zeta^{q^{f'}} \neq \zeta$ (otherwise $\sigma_q^{f'}$ would be the identity automorphism), hence p^k does not divide $q^{f'} - 1$. This shows that the residue class of q modulo p^k has order f in the multiplicative group $P(p^k)$.

Since $(\mathscr{G}:\mathscr{L}) = g$ and $\mathscr{L}$ has order f, then $\varphi(p^k) = fg$. By the relation $[\mathbb{Q}(\zeta):\mathbb{Q}] = efg$ we deduce that $e = 1$; that is, $Aq = Q_1 \ldots Q_g$. ∎

By (3A), (3B), p is the only ramified prime in $\mathbb{Q}(\zeta)$. We have seen in Chapter 6, (R) that if $k = 1$ then $|\delta_{\mathbb{Q}(\zeta)}| = p^{p-2}$. Later in Chapter 13 we shall prove that for $m = p^k$, $k \geqslant 1$, $|\delta_{\mathbb{Q}(\zeta)}|$ is also a power of p. Thus a prime number is ramified in $\mathbb{Q}(\zeta)$ if and only if it divides the discriminant $\delta_{\mathbb{Q}(\zeta)}$.

This fact suggests again a relationship between the ramification and the discriminant. We shall take up now the algebraic preparations to introduce the relative discriminant.

The discriminant of a field was defined as a determinant of traces of elements obtained from an integral basis. This was possible because the ring of algebraic integers is a free $\mathbb{Z}$-module.

However, if the ground field K is an algebraic number field different from $\mathbb{Q}$, if $L \mid K$ is an extension of degree n, B the ring of integers of L, we may only assert that A is a Dedekind domain and B is a torsion-free A-module of rank n. In general, A need not to be a principal ideal domain, B need not to be a free A-module, thus we have no basis of B at our disposal to form the discriminant.

The way to overcome this difficulty is to associate with A certain principal ideal domains and to obtain information on the A-modules by studying modules over the associated principal ideal domains. In Chapter 7, (L) we have seen that every Dedekind domain with only one prime ideal is a principal ideal domain. We shall therefore begin constructing such domains for a given Dedekind domain. The procedure is standard in ring theory and has been developed in great generality. This will be the object of (4).

In (5) we shall consider the general concept of trace and norms of an element in a ring extension. More precisely, A is a subring of a commutative ring B and as an A-module, B is free having basis with n elements. We shall define the relative trace and norm $\mathrm{Tr}_{B|A}(x)$, $N_{B|A}(x)$, where $x \in B$. This theory belongs to Linear Algebra, but we shall also include some specific results when A,B are Dedekind domains.

In part (6) we consider the discriminant of the ring extension $B \mid A$ and study some of their purely algebraic properties.

After this preparation, we return to the situation of algebraic number fields and use this algebraic apparatus to introduce the relative discriminant and derive its relation with the ramification.

4. RINGS OF FRACTIONS

Let R be any domain, K the field of quotients of R. We recall that K consists of all equivalence classes a/b of pairs (a,b), with $a, b \in R$, $b \neq 0$, where $(a,b) \equiv (a',b')$ when $b'a = ba'$. The operations in K are defined as follows:

$$\frac{a}{b} + \frac{c}{d} = \frac{ad + bc}{bd} \qquad \frac{a}{b} \cdot \frac{c}{d} = \frac{ac}{bd}.$$

Every element a of R is identified with $a/1$, making R into a subring of K. If $a \in R$, $a \neq 0$, then a is invertible in K.

We observe that we only required that the product of two non-zero elements of R is still not equal to zero. This suggest the concept of a *multiplicative subset* S of a commutative ring R. S is a subset containing 1, not containing zero-divisors and such that if $a,b \in S$ then $ab \in S$. Hence $0 \notin S$. In spite of being mainly interested in rings of fractions of domains, it will become necessary later to consider a ring of fractions of a homomorphic image of domain, which need not be a domain anymore.

To define the ring of fractions of R by S, we consider pairs (a,s), with $a \in R$, $s \in S$, and state that $(a,s) \equiv (a',s')$ when $s'a = sa'$. Since S contains no zero-divisors, this is an equivalence relation in the above set of pairs. The equivalence class of (a,s) is denoted by a/s. The operations between equivalence classes are defined after the model of the field of quotients:

$$\frac{a}{s} + \frac{a'}{s'} = \frac{as' + a's}{ss'} \cdot \frac{a}{s} \cdot \frac{a'}{s'} = \frac{aa'}{ss'} \, .$$

It is an easy matter to check that these operations are well-defined, and that we obtain a ring, denoted by $S^{-1}R$ or R_S. It is called the *ring of fractions* of R by S.

R may be considered as a subring of $S^{-1}R$ and every element of S becomes invertible in $S^{-1}R$.

If S_0 is the multiplicative set of all elements of R which are not zero-divisors, then $S_0^{-1}R$ is called the *total ring of fractions* of R. In particular, if R is a domain, this ring $S_0^{-1}R$ is the field of quotients of R. Moreover, if R is a domain and S is any multiplicative subset then $S \subseteq S_0$ and $S^{-1}R$ is contained in the field of quotients $S_0^{-1}R$, hence it is a domain.

In the case where R is a domain and P is a non-zero prime ideal of R, the set-complement S of P in R is a multiplicative set. The ring of fractions $S^{-1}R$ is also denoted by R_P and plays an important role in the sequel.

The following proposition indicates the relationship between the ideals of R and of $S^{-1}R$:

4A. *Let $R' = S^{-1}R$*

(1) *If J' is any ideal of R' then $R'(J' \cap R) = J'$, hence the mapping $J' \to J' \cap R$, is one-to-one and preserves inclusions; if $J' \neq R'$ then $J' \cap R$ is disjoint from S.*

(2) *The mapping $P' \to P' \cap R$ applies the set of prime ideals of R' onto the set of prime ideals P of R, disjoint from S and $R'P \cap R = P$.*

(3) *In particular, if S is the set-complement of the prime ideal P in R, then by the above mapping we obtain all the prime ideals of R contained in P, and $R' = R_P$ has only one maximal ideal, namely $R'P$.*

Proof:

(1) Obviously $J' \supseteq R'(J' \cap R)$. Conversely, if $x \in J'$ then $x = a/s$, with $a \in R$, $s \in S$; hence $a = sx \in R'J' \subseteq J'$, so $a \in J' \cap R$ and

$$x = (1/s) \cdot a \in R'(J' \cap R).$$

This shows that the mapping $J' \to J' \cap R$ is one-to-one and of course it preserves inclusions. If $(J' \cap R) \cap S$ contains an element s, then

$$1 = (1/s) \cdot s \in R'(J' \cap R) = J'.$$

(2) If P' is a prime ideal of R', then clearly $P' \cap R$ is a prime ideal of R, and by (1), $P' \cap R$ is disjoint from S.

Conversely, let P be a prime ideal of R, $P \cap S = \varnothing$, and let us show that $R'P$ is a prime ideal of R' such that $R'P \cap R = P$. Every element of $R'P$ is of the form $\Sigma_{i=1}^{h} (a_i/s_i)x_i$ where $a_i \in R$, $s_i \in S$, $x_i \in P$, $h \geq 1$; this may be rewritten with a common denominator $s = s_1 \ldots s_h \in S$, as follows $\Sigma_{i=1}^{h} (a_i/s_i)x_i = \Sigma_{i=1}^{h} (b_i/s)x_i = (1/s)(\Sigma_{i=1}^{h} b_i x_i) \in R'P$ (where each $b_i \in R$); in other words, every element of $R'P$ is of the form x/s, with $x \in P$, $s \in S$.

Now, $R'P$ is a prime ideal, because if a/s, $b/t \in R'$ and $(a/s) \cdot (b/t) \in R'P$ then $(a/s) \cdot (b/t) = x/u$ with $x \in P$, $a, b \in R$, $s, t, u \in S$; thus $abu = xst \in P$; since $P \cap S = \varnothing$, then $ab \in P$, so either $a \in P$ or $b \in P$, that is $a/s \in R'P$ or $b/t \in R'P$, showing that $R'P$ is a prime ideal.

Finally, $P \subseteq R'P \cap R$; conversely, if $a \in R'P \cap R$, then $a = x/s$, with $x \in P$, $s \in S$, so $sa = x \in P$, but $s \notin P$ so $a \in P$.

(3) This assertion follows immediately from the preceding ones. ▊

As a corollary, we have:

4B. *If R is a Noetherian ring, S a multiplicative subset of R, then $R' = S^{-1}R$ is also a Noetherian ring.*
Proof: By (**4A**), Part (1), there is a one-to-one correspondence, preserving inclusions, from the set of ideals of $R' = S^{-1}R$ into the set of ideals of R. Hence, every strictly increasing chain of ideals of R' must be finite. ▊

It follows from (**4A**) that if P is a non-zero prime ideal of the domain R then $R' = R_P$ has only one maximal ideal $R'P$ and $P = R'P \cap R$; the prime ideals of R' correspond to those of R which are contained in P. Thus, to pass from R to the ring R_P amounts essentially to disregard all prime ideals P' of R which are not contained in P. This process is usually called the *localization of R at P*. It is especially important for us in the case where every non-zero prime ideal of R is maximal (for example, when R is a Dedekind domain). Then $R' = R_P$ has only one non-zero prime ideal $R'P$.

A reverse procedure is the *globalization*. We consider the family of maximal ideals P_i of the domain R; for every ideal I of R we have $I \subseteq R_{P_i}I$; but in fact, it holds:

4C. $I = \cap R_{P_i}I$ *(intersection over the set of maximal ideals P_i of the domain R).*
Proof: The result is trivial when R is a field, so we assume that this is not the case. Let $x \in \cap R_{P_i}I$. For every maximal ideal P_i of R we may write $x = a_i/b_i$ with $a_i \in I$, $b_i \in R$, $b_i \notin P_i$. Let J be the ideal of R generated by the elements b_i. Since $b_i \notin P_i$ then $J \nsubseteq P_i$ for every maximal ideal P_i. Therefore, $J = R$, because as it is known, every ideal of R, distinct from R,

is contained in a maximal ideal (this follows from Zorn's lemma; in the case where R is a Noetherian ring it is immediate by the maximal condition on ideals). In particular, 1 may be expressed in terms of the generators of $J = R$, that is, there exist elements $c_{i_1}, \ldots, c_{i_m} \in R$ such that $1 = \sum_{k=1}^{m} c_{i_k} b_{i_k}$ and so

$$x = \sum_{k=1}^{m} c_{i_k}(b_{i_k} x) = \sum_{k=1}^{m} c_{i_k} a_{i_k} \in I. \quad \blacksquare$$

In order to apply this method to rings of algebraic integers, we want to describe the behavior of the integral closure by going to rings of fractions:

4D. *Let A be a domain, let L be a field containing A; let B be the integral closure of A in L. If S is a multiplicative subset of A, then $S^{-1}B$ is the integral closure of $S^{-1}A$ in L.*

Proof: If $x \in L$ is an integral element over $S^{-1}A$, then there exist elements $a_i/s_i \in S^{-1}A$ (with $a_i \in A$, $s_i \in S$, $i = 1, \ldots, n$), such that

$$x^n + \frac{a_1}{s_1} x^{n-1} + \frac{a_2}{s_2} x^{n-2} + \cdots + \frac{a_n}{s_n} = 0.$$

Letting $s = s_1 \ldots s_n \in S$, we may rewrite

$$x^n + \frac{b_1}{s} x^{n-1} + \frac{b_2}{s} x^{n-2} + \cdots + \frac{b_n}{s} = 0,$$

with $b_i \in A$, and therefore

$$(sx)^n + b_1(sx)^{n-1} + b_2 s(sx)^{n-2} + \cdots + b_n s^{n-1} = 0.$$

This shows that $sx \in L$ is integral over A, hence $sx \in B$ and $x \in S^{-1}B$.

On the other hand, every element b/s of $S^{-1}B$ is integral over $S^{-1}A$; indeed, since $b \in B$, there exist elements $a_1, \ldots, a_n \in A$ such that

$$b^n + a_1 b^{n-1} + \cdots + a_n = 0.$$

so

$$\left(\frac{b}{s}\right)^n + \frac{a_1}{s}\left(\frac{b}{s}\right)^{n-1} + \frac{a_2}{s^2}\left(\frac{b}{s}\right)^{n-2} + \cdots + \frac{a_n}{s^n} = 0,$$

thus b/s is integral over $S^{-1}A$. $\quad \blacksquare$

4E. *If A is an integrally closed domain, if S is a multiplicative subset of A, then $S^{-1}A$ is an integrally closed domain.*

Proof: This is a particular case of (**4D**). $\quad \blacksquare$

Combining the previous results, we prove now:

4F. *If A is a Dedekind domain and S is a multiplicative subset of A, then $A' = S^{-1}A$ is a Dedekind domain. If J is an ideal of A, $J = \prod_{i=1}^{r} P_i^{e_i}$, then the decomposition of $A'J$ into prime ideals of A' is given by*

$$A'J = \prod_{P_i \cap S = \varnothing} (A'P_i)^{e_i}.$$

Proof: By (**4B**), (**4E**) A' is also a Noetherian integrally closed domain. Let us show that every non-zero prime ideal of A' is maximal. By Chapter 7, Theorem 1, this implies that A' is a Dedekind domain.

Let P' be a non-zero prime ideal of A', so $P' \cap A = P$ is a prime ideal of A such that $P \cap S = \varnothing$. Also $P \neq 0$ since $P' = A'P$. Thus, P is a maximal ideal of A, hence by (**4A**), P' is also a maximal ideal of A'.

From $J = \prod_{i=1}^{r} P_i^{e_i}$ it follows that

$$A'J = A\left(\prod_{i=1}^{r} P_i^{e_i}\right) = \prod_{i=1}^{r} (A'P_i)^{e_i} = \prod_{P_i \cap S = \varnothing} (A'P_i)^{e_i},$$

noting that if $P_i \cap S \neq \varnothing$ then $A'P_i = A'$ and if $P_i \cap S = \varnothing$ then $A'P_i$ is a prime ideal of A'. ∎

As a corollary:

4G. *If A is a Dedekind domain, if P is a non-zero prime ideal of A, then:*
(1) A_P *is a principal ideal domain, with only one non-zero prime ideal which is $A_P P$.*
(2) *Every non-zero fractional ideal of A_P is a power of $A_P P$ and $A_P P^s \cap A = P^s$ for every $s \geqslant 1$.*
(3) *An element is invertible in A_P if and only if it does not belong to $A_P P$.*
(4) *If R is a subring of the field of quotients K of A and A_P is properly contained in R then $R = K$.*
Proof:
(1) By (**4F**) A_P is a Dedekind domain; by (**4A**) it has only one non-zero prime ideal, namely $A_P P$, thus by Chapter 7, (**L**), A_P is a principal ideal domain.

(2) By (**4A**), (**4F**) every non-zero integral ideal of A_P is a power of $A_P P$; the same holds therefore for the non-zero fractional ideals of A_P.

By (**4A**) we know that $A_P P \cap A = P$. If $A_P P^s \cap A = P^s$ for $s \geqslant 1$ then from (**4A**) we have $P^s = A_P P^s \cap A \supset A_P P^{s+1} \cap A \supseteq P^{s+1}$. Since A is a Dedekind domain, there exists no ideal J of A such that $P^s \supset J \supset P^{s+1}$; thus $A_P P^{s+1} \cap A = P^{s+1}$, proving the statement.

(3) The elements of A_P which are not in the only maximal ideal $A_P P$ are precisely those which generate the unit ideal; so they are the invertible elements of A_P.

(4) Let $x \in R$, $x \notin A_P$ and let $n > 0$ be such that $A_P x = A_P t^{-n}$ where t is a generator of the principal ideal $A_P P$. If $y \in K$, $y \notin A_P$, let $m > 0$ be such that $A_P y = A_P t^{-m}$. If r is a positive integer such that $rn \geqslant m$, then $A_P x^{-r} = A_P t^{rn} \subseteq A_P t^m = A_P y^{-1}$, thus $y \in A_P x^r \subseteq A_P R \subseteq R$. This proves that $R = K$. ∎

Another useful property relates the rings of fractions to quotient rings:

4H. *Let A be a commutative ring, let S be a multiplicative subset of A, and $A' = S^{-1}A$. Let J be an ideal of A, distinct from A, such that: if $as \in J$, $a \in A$, $s \in S$ then $a \in J$. Then the image $\bar{S}$ of S by the canonical mapping $A \to A/J$ is a multiplicative subset of A/J (containing no zero divisors) and there exists a canonical isomorphism $\varphi: (A/J)_{\bar{S}} \to A'/A'J$. In particular, if all the elements of $\bar{S}$ are invertible in A/J then $A/J = A'/A'J$ (after an identification). This happens when $J = P$ is a maximal ideal and S is the complement of P in A.*
Proof: First we note that $J \cap S = \varnothing$ because if $a = 1 . a \in J \cap S$ then $1 \in J$, contrary to the hypothesis.

Similarly, $\bar{S}$ is a multiplicative subset of A/J, because $\bar{1} \in \bar{S}$, if a, $b \in S$, then $\bar{a}$, $\bar{b} \in \bar{S}$, $ab \in S$, so $\overline{ab} \in \bar{S}$; also if $\bar{a}$, $\bar{b} \in \bar{S}$ and $\overline{ab} = \bar{0}$ then $ab \in J$, with a, $b \in S$, hence by the condition on J, we must have $a \in S \cap J$, which is a contradiction.

Given any element of $(A/J)_{\bar{S}}$, which is written as $\bar{a}/\bar{s}$, with $\bar{a}$, $\bar{s} \in A/J$, $\bar{s} \in \bar{S}$, we define $\varphi(\bar{a}/\bar{s}) = \overline{(a/s)}$ (image of $a/s \in A'$ by the canonical homomorphism onto $A'/A'J$). First, we note that φ is well-defined, in other words, if $\bar{a}/\bar{s} = \overline{a'}/\overline{s'}$ then $\overline{(a/s)} = \overline{(a'/s')}$. In fact, $\overline{s'a} = \bar{s} . \bar{a'}$ so $s'a - sa' \in J$, hence

$$\frac{a}{s} - \frac{a'}{s'} = \frac{s'a - a's}{ss'} \in A'J.$$

It follows easily that φ is a ring-homomorphism. Clearly, φ maps $(A/J)_{\bar{S}}$ onto $A'/A'J$, since every element of this ring is of the type $\overline{(a/s)}$, with $a \in A$, $s \in S$. Finally, the kernel of φ is zero, because from $\overline{(a/s)} = \bar{0}$ we deduce $a/s \in A'J$ so we may write $a/s = a'/s'$ with $a' \in J$, $s' \in S$; hence, $s'a = sa' \in J$; by the hypothesis, we have $a \in J$, so $\bar{a}/\bar{s} = 0$.

For the second assertion, if P is a maximal ideal then A/P is a field. Since $A/P \subseteq (A/P)_{\bar{S}}$ and every element of $\bar{S}$ is invertible in A/P it follows that $A/P = (A/P)_{\bar{S}}$. Thus φ is an isomorphism between A/P and $A'/A'P$, and we may write $A/P = A'/A'P$, after an identification. ∎

Now we show the generalization of (**1M**):

4I. *Let A be a Dedekind domain, K its field of quotients, and let $L \mid K$ be a separable extension of degree n and B the integral closure of A in L. If P is*

*any non-zero prime ideal of A then B/BP is a vector space of dimension n
over A/P.*

Proof: Let S be the set-complement of P in A and $A' = S^{-1}A$ be the
corresponding ring of fractions. By (**4G**), A' is a principal ideal domain.
Since B is the integral closure of A, by (**4D**) $B' = S^{-1}B$ is the integral closure
of A' in L. By Chapter 6, (**B**), B' is contained in a free A'-module of rank n.
By Chapter 6, (**J**), B' is itself a free A'-module of rank at most n; from
$L = KB'$ it follows that B' has rank n over A'.

From (**1B**) we have $BP \cap A = P$ and since $B'P = B'(A'P)$ then (**1A**)
implies again that $B'P \cap A' = A'P$. Therefore, $B'/B'P$ contains the field
$A'/A'P$. We show now that it is a vector space of dimension n. In fact, if
$\{x_1, \ldots, x_n\}$ is a basis of the A'-module B', if $\bar{x}_i$ denotes the image of x_i in
$B'/B'P$ (by the natural mapping), then $\{\bar{x}_1, \ldots, \bar{x}_n\}$ generates $B'/B'P$ over
$A'/A'P$. On the other hand, if $\sum_{i=1}^{n} \bar{a}_i \bar{x}_i = \bar{0}$ (with $\bar{a}_i \in A'/A'P$) then

$$\sum_{i=1}^{n} a_i x_i \in B'P = B'(A'P)$$

and so we may write $\sum_{i=1}^{n} a_i x_i = \sum_{j=1}^{m} a_j' y_j$ with $a_j' \in A'P, y_j \in B'$; expressing
the elements y_j in terms of the generators $x_1, \ldots, x_n$ of the A'-module B'
we may write $\sum_{i=1}^{n} a_i x_i = \sum_{i=1}^{n} a_i'' x_i$ and necessarily $a_i = a_i''$ for every
$i = 1, \ldots, n$ (since $x_1, \ldots, x_n$ are linearly independent over A').

The image $\bar{S}$ of S by the natural homomorphism $A \to A/P$ is the set of
non-zero elements of the field A/P; a fortiori, the elements of $\bar{S}$ are invertible
in $B'/B'P$. By (**4H**), $A'/A'P = A/P$, $B'/B'P = B/BP$ and so B/BP is a vector
space of dimension n over A/P. ∎

5. TRACES AND NORMS IN RING EXTENSIONS

We shall consider the following general situation:

*A is a subring of the commutative ring B and B is a free A-module having a
basis with n elements.*

It is well-known that any other basis of the A-module B has also n elements.

Let $\theta: B \to B$ be any linear mapping, $\{z_1, \ldots, z_n\}$ any basis of the A-
module B. Then

$$\theta(z_j) = \sum_{i=1}^{n} a_{ij} z_i \qquad (j = 1, \ldots, n)$$

with $a_{ij} \in A$. The matrix $M(\theta) = (a_{ij})_{i,j}$ is called the matrix of θ with respect
to the basis $\{z_1, \ldots, z_n\}$.

If $\{z_1', \ldots, z_n'\}$ is any other basis of the A-module B, and $M'(\theta)$ the
corresponding matrix, $M'(\theta) = (a_{ij}')_{i,j}$, if $z_j = \sum_{i=1}^{n} c_{ij} z_i'$ $(j = 1, \ldots, n)$,

with each $c_{ij} \in A$, $C = (c_{ij})_{i,j}$, then we have:

$$\theta(z_j) = \sum_{i=1}^{n} a_{ij} z_i = \sum_{i=1}^{n} a_{ij} \left(\sum_{k=1}^{n} c_{ki} z_k' \right) = \sum_{k=1}^{n} \left(\sum_{i=1}^{n} c_{ki} a_{ij} \right) z_k',$$

and on the other hand,

$$\theta(z_j) = \sum_{i=1}^{n} c_{ij} \theta(z_i') = \sum_{i=1}^{n} c_{ij} \left(\sum_{k=1}^{n} a_{ki}' z_k' \right) = \sum_{k=1}^{n} \left(\sum_{i=1}^{n} a_{ki}' c_{ij} \right) z_k'.$$

This shows that $M'(\theta) \,.\, C = C \,.\, M(\theta)$.

Since C is the matrix of a change of basis, it is invertible, hence $M'(\theta) = C \,.\, M(\theta) \,.\, C^{-1}$. In particular, $\det(C)$ is a unit of the ring A, because

$$\det(C) \,.\, \det(C^{-1}) = 1.$$

Following the well-known method of linear algebra, we now consider the matrix $XI - M(\theta)$, where X is an indeterminate, I is the unit $n \times n$ matrix; thus, the entries of $XI - M(\theta)$ are elements of A or linear monic polynomials in X:

$$XI - M(\theta) = \begin{pmatrix} X - a_{11} & -a_{12} & \cdots & -a_{1n} \\ -a_{21} & X - a_{22} & \cdots & -a_{2n} \\ \cdot & \cdot & & \cdot \\ \cdot & \cdot & & \cdot \\ \cdot & \cdot & & \cdot \\ -a_{n1} & -a_{n2} & \cdots & X - a_{nn} \end{pmatrix}.$$

The determinant of this matrix remains unchanged, when we change basis of the A-module B. In fact,

$$\begin{aligned} XI - M'(\theta) &= XI - C \,.\, M(\theta) \,.\, C^{-1} \\ &= XC \,.\, I \,.\, C^{-1} - C \,.\, M(\theta) \,.\, C^{-1} \\ &= C(XI - M(\theta))C^{-1}. \end{aligned}$$

Hence

$$\begin{aligned} \det(XI - M'(\theta)) &= \det(C) \,.\, \det(XI - M(\theta)) \,.\, \det(C^{-1}) \\ &= \det(XI - M(\theta)). \end{aligned}$$

The element $\det(XI - M(\theta))$ depends therefore only on θ. It is called *the characteristic polynomial of* θ and denoted by

$$F(\theta) = X^n + a_1 X^{n-1} + a_2 X^{n-2} + \cdots + a_n,$$

with each a_i in A. Sometimes we use also the notation $F_{B|A}(\theta)$.

We define now the *trace of* θ as $\text{Tr}_{B|A}(\theta) = -a_1 \in A$ and the *determinant of* θ as $\det_{B|A}(\theta) = (-1)^n a_n \in A$. It is clear that the trace of θ is the sum of the elements in the diagonal of $M(\theta)$, while the determinant of θ is the determinant of the matrix $M(\theta)$.

Now we apply these general notions to the following special situation. If $x \in B$, let $\theta = \theta_x : B \to B$ be the mapping of multiplication by x; that is, $\theta_x(z) = xz$ for every $z \in B$. Then the characteristic polynomial of θ_x is called *the characteristic polynomial of x in $B \mid A$* and denoted by $F_{B \mid A}(x)$. Similarly, the trace of θ_x is called *the trace of x in $B \mid A$* and the determinant of θ_x is called *the norm of x in $B \mid A$*. They are respectively denoted by $\mathrm{Tr}_{B \mid A}(x)$, $N_{B \mid A}(x)$.

It is important to compare the notions of trace and norm with the ones known for field extensions. As a temporary notation, if $L \mid K$ is a separable field extension of degree n and $x \in L$ let $\mathrm{Tr}'_{L \mid K}(x)$ denote the sum of all conjugates of x in $L \mid K$ and similarly, $N'_{L \mid K}(x)$ shall denote the product of all conjugates of x in $L \mid K$.

5A.   *Let $x \in L$ and let $f \in K[X]$ be its minimal polynomial over K. Then the characteristic polynomial is $F_{L \mid K}(x) = f^s$ where $s = [L : K(x)]$ and $\mathrm{Tr}'_{L \mid K}(x) = \mathrm{Tr}_{L \mid K}(x)$, $N'_{L \mid K}(x) = N_{L \mid K}(x)$.*

Proof: Let $\{x_1, \ldots, x_r\}$ be a basis of the K-vector space $K(x)$, and let $\{y_1, \ldots, y_s\}$ be a basis of the $K(x)$-vector space L. So

$$\{x_1 y_1, x_2 y_1, \ldots, x_r y_1, x_1 y_2, x_2 y_2, \ldots, x_r y_2, \ldots, x_1 y_s, x_2 y_s, \ldots, x_r y_s\}$$

is a basis of the K-vector space L (with $rs = n$). Since $x \in K(x)$ we have

$$xx_j = \sum_{i=1}^{r} a_{ij} x_i \qquad \text{(for } j = 1, \ldots, r\text{)}.$$

Therefore, $T = (a_{ij})_{i,j}$ is the matrix of $\theta_x : K(x) \to K(x)$ with respect to the basis $\{x_1, \ldots, x_r\}$. It follows that

$$xx_j y_k = \sum_{i=1}^{r} a_{ij} x_i y_k \qquad \text{(for } j = 1, \ldots, r; \ k = 1, \ldots, s\text{)}.$$

Hence the matrix of $\theta_x : L \to L$, with respect to the basis of the K-vector space L, considered above, is a block diagonal $n \times n$ matrix

$$M(\theta_x) = \begin{pmatrix} T & 0 & \cdots & 0 \\ 0 & T & \cdots & 0 \\ \cdot & \cdot & & \cdot \\ \cdot & \cdot & & \cdot \\ \cdot & \cdot & & \cdot \\ 0 & 0 & \cdots & T \end{pmatrix}.$$

Therefore

$$F_{L|K}(x) = (F_{K(x)|K}(x))^s,$$

hence

$$\mathrm{Tr}_{L|K}(x) = s\,\mathrm{Tr}_{K(x)|K}(x),$$
$$N_{L|K}(x) = (N_{L|K}(x))^s.$$

As we know, we have also

$$\mathrm{Tr}'_{L|K}(x) = s\,\mathrm{Tr}'_{K(x)|K}(x) \quad \text{and} \quad N'_{L|K}(x) = (N_{K(x)|K}(x))^s.$$

Thus, it is enough to show that $F_{K(x)|K}(x) = f$, the minimal polynomial of x over K, for this implies that

$$\mathrm{Tr}'_{K(x)|K}(x) = \mathrm{Tr}_{K(x)|K}(x) \quad \text{and} \quad N'_{K(x)|K}(x) = N_{K(x)|K}(x).$$

Now, in $K(x) \mid K$ the matrix of θ_x with respect to the basis

$$\{1, x, x^2, \ldots, x^{r-1}\}$$

is the *companion matrix* of the minimal polynomial

$$f = X^n + a_1 X^{n-1} + \cdots + a_n;$$

that is,

$$M(\theta_x) = \begin{pmatrix} 0 & 0 & \cdots & 0 & -a_n \\ 1 & 0 & \cdots & 0 & -a_{n-1} \\ 0 & 1 & \cdots & 0 & -a_{n-2} \\ \cdot & \cdot & & \cdot & \cdot \\ \cdot & \cdot & & \cdot & \cdot \\ \cdot & \cdot & & \cdot & \cdot \\ 0 & 0 & \cdots & 1 & -a_1 \end{pmatrix}.$$

Hence $F_{K(x)|K}(x) = \det(XI - M(\theta_x)) = X_n + a_1 X^{n-1} + \cdots + a_n = f.$ ∎

Let us note the following algebraic properties of the trace and the norm:

5B. *If $x_1, x_2 \in B$, $a \in A$ then*

$$\mathrm{Tr}_{B|A}(x_1 + x_2) = \mathrm{Tr}_{B|A}(x_1) + \mathrm{Tr}_{B|A}(x_2), \qquad \mathrm{Tr}_{B|A}(ax_1) = a\,\mathrm{Tr}_{B|A}(x_1),$$
$$N_{B|A}(x_1 x_2) = N_{B|A}(x_1) \cdot N_{B|A}(x_2).$$

Proof: The proof is straightforward. For example, $\theta_{x_1+x_2} = \theta_{x_1} + \theta_{x_2}$, hence $M(\theta_{x_1+x_2}) = M(\theta_{x_1}) + M(\theta_{x_2})$, and considering the elements in the diagonal of these matrices, we obtain $\mathrm{Tr}_{B|A}(x_1 + x_2) = \mathrm{Tr}_{B|A}(x_1) + \mathrm{Tr}_{B|A}(x_2)$.

Similarly $\theta_{ax_1} = a\theta_{x_1}$, hence $M(\theta_{ax_1}) = aI \cdot M(\theta_{x_1})$ so

$$\mathrm{Tr}_{B|A}(ax_1) = a\,\mathrm{Tr}_{B|A}(x_1).$$

Finally, $\theta_{x_1x_2} = \theta_{x_1} \cdot \theta_{x_2}$, then $M(\theta_{x_1x_2}) = M(\theta_{x_1}) \cdot M(\theta_{x_2})$ and therefore, considering the determinants, we have $N_{B|A}(x_1x_2) = N_{B|A}(x_1) \cdot N_{B|A}(x_2)$. ∎

Now we study the behavior of the trace and norms when we consider rings of fractions. Let S be a multiplicative subset of $A \subseteq B$ and $A' = A_S$, $B' = B_S$. If $\{z_1, \ldots, z_n\}$ is an A-basis of B then it is still an A'-basis of B'.

5C. *With these notations, for every $x \in B$ we have:*

$$F_{B'|A'}(x) = F_{B|A}(x), \qquad \mathrm{Tr}_{B'|A'}(x) = \mathrm{Tr}_{B|A}(x), \qquad N_{B'|A'}(x) = N_{B|A}(x).$$

Proof: Let $\theta_x: B \to B$, $\theta_x': B' \to B'$ be the homomorphisms of multiplication by $x \in B$ and let $M(\theta_x)$, $M(\theta_x')$ be the corresponding matrices with respect to the basis $\{z_1, \ldots, z_n\}$. Obviously these matrices coincide, therefore the same happens with the characteristic polynomials, with the traces and with the norms of the element x in $B \mid A$ and in $B' \mid A'$. ∎

In particular, if B, A are domains having fields of quotients L, K respectively, then

$$F_{B|A}(x) = F_{L|K}(x), \qquad \mathrm{Tr}_{B|A}(x) = \mathrm{Tr}_{L|K}(x) \quad \text{and} \quad N_{B|A}(x) = N_{L|K}(x).$$

We shall study the characteristic polynomial, the trace and the norm when we consider Cartesian products of rings.

Let $B_1, \ldots, B_r$ be commutative rings containing the subring A and such that each ring B_i is a free A-module of finite rank. Let $B = \prod_{i=1}^r B_i$ be their Cartesian product and $\pi_i: B \to B_i$ the ith projection, which is a ring homomorphism from B onto B_i. B contains the subring $\{(a, \ldots, a) \in B \mid a \in A\}$, which is naturally isomorphic to A; hence, we may consider A as a subring of B; then B is also a free A-module of finite rank.

We may easily prove:

5D. *With these notations, if $x \in B$ then*

$$F_{B|A}(x) = \prod_{i=1}^r F_{B_i|A}(\pi_i(x))$$

$$\mathrm{Tr}_{B|A}(x) = \sum_{i=1}^r \mathrm{Tr}_{B_i|A}(\pi_i(x))$$

$$N_{B|A}(x) = \prod_{i=1}^r N_{B_i|A}(\pi_i(x)).$$

Proof: It is enough to prove the statement when $r = 2$. Let $\{t_1, \ldots, t_n\}$ be a basis of the A-module B_1, and let $\{u_1, \ldots, u_m\}$ be a basis of the A-module B_2. Then $\{(t_1,0), \ldots, (t_n,0), (0,u_1), \ldots, (0,u_m)\}$ is a basis of the A-module

$B = B_1 \times B_2$. Let $z_i = (t_i,0)$ for $i = 1, \ldots, n$ and let $z_{n+i} = (0,u_i)$ for $i = 1, \ldots, m$. If $x = (x_1,x_2) \in B_1 \times B_2$ then

$$(x_1,x_2) \cdot (t_j,0) = (x_1 t_j,0) = \left(\sum_{i=1}^{n} a_{ij} t_i,0 \right) = \sum_{i=1}^{n} a_{ij}(t_i,0)$$

where $(a_{ij})_{i,j} = M(\theta_{x_1})$, the matrix of the A-linear transformation θ_{x_1} from B_1 to B_1 (with respect to the basis $\{t_1, \ldots, t_n\}$). Similarly,

$$(x_1,x_2) \cdot (0,u_j) = \sum_{i=1}^{m} a_{ij}{}'(0,u_i)$$

where $(a_{ij}{}')_{i,j} = M(\theta_{x_2})$, the matrix of the A-linear transformation θ_{x_2} from B_2 to B_2 (with respect to the basis $\{u_1, \ldots, u_m\}$).

Thus the matrix of $\theta_x: B_1 \times B_2 \to B_1 \times B_2$ with respect to the basis $\{z_1, \ldots, z_{n+m}\}$ is

$$M(\theta_x) = \left(\begin{array}{c|c} M(\theta_{x_1}) & 0 \\ \hline 0 & M(\theta_{x_2}) \end{array} \right).$$

Hence $\det(XI - M(\theta_x)) = \det(XI - M(\theta_{x_1})) \cdot \det(XI - M(\theta_{x_2}))$; that is, $F_{B|A}(x) = F_{B_1|A}(x_1) \cdot F_{B_2|A}(x_2)$. The assertions about the trace and the norm are now immediate. ∎

We consider now the effect on the characteristic polynomial of certain ring-homomorphisms.

5E. *Let $\psi: B \to \bar{B}$ be a homomorphism from B onto the ring $\bar{B}$, and let $\psi(A) = \bar{A}$. We assume that there exists a basis $\{z_1, \ldots, z_n\}$ of the A-module B such that $\{\bar{z}_1, \ldots, \bar{z}_n\}$ is a basis of the $\bar{A}$-module $\bar{B}$ (where $\bar{z} = \psi(z)$ for every $z \in B$). If $x \in B$ then*

$$\psi(F_{B|A}(x)) = F_{\bar{B}|\bar{A}}(\bar{x}),$$

$$\psi(\mathrm{Tr}_{B|A}(x)) = \mathrm{Tr}_{\bar{B}|\bar{A}}(\bar{x}),$$

$$\psi(N_{B|A}(x)) = N_{\bar{B}|\bar{A}}(\bar{x}).$$

Proof: If $xz_j = \Sigma_{i=1}^{n} a_{ij} z_i \; (j = 1, \ldots, n)$ with $a_{ij} \in A$ then

$$\bar{x}\bar{z}_j = \sum_{i=1}^{n} \bar{a}_{ij} \bar{z}_i \qquad (j = 1, \ldots, n).$$

Thus $M(\theta_x) = (a_{ij})$, $M(\theta_{\bar{x}}) = (\bar{a}_{ij})$ (with respect to the above bases). Applying ψ to the coefficients of the characteristic polynomial $F_{B|A}(x) = \det(XI - M(\theta_x))$ we obtain $\det(XI - M(\theta_{\bar{x}})) = F_{\bar{B}|\bar{A}}(\bar{x})$. The assertions about the trace and the norm follow at once. ∎

Let R be a ring, and K a subfield of R such that R is a vector space of finite dimension over K. Let $\theta: R \to R$ be *a* K-linear transformation, and consider a strictly decreasing chain of subspaces of R,

$$R = R_0 \supset R_1 \supset R_2 \ldots \supset R_{k-1} \supset R_k = 0$$

such that $\theta(R_i) \subseteq R_i$ for every $i = 1, \ldots, k$.

The elements of the K-vector space R_{i-1}/R_i are the cosets $z + R_i$, where $z \in R_{i-1}$. θ induces a linear transformation

$$\theta_i: R_{i-1}/R_i \to R_{i-1}/R_i$$

defined as follows: $\theta_i(z + R_i) = \theta(z) + R_i$ for every $z \in R_{i-1}$. In virtue of the hypothesis on the subspaces R_j it follows that θ_i is well-defined.

For each index $i = 1, \ldots, k$ let $B_i = \{z_{i1}, \ldots, z_{im_i}\}$ be a set of elements of R_{i-1} such that the set of cosets $\{z_{i1} + R_i, \ldots, z_{im_i} + R_i\}$ forms a basis of the vector space R_{i-1}/R_i. Then for every $i = 1, \ldots, k$,

$$B_i \cup B_{i+1} \cup \cdots \cup B_k$$

constitutes a basis of the K-vector space R_{i-1}. In particular,

$$B = B_1 \cup \cdots \cup B_k$$

is a basis of R. The verification is standard and therefore omitted.

We shall consider the matrix $M(\theta)$ of θ with respect to the basis B; it may be expressed in terms of the matrices $M(\theta_i)$ of θ_i with respect to B_i:

$$M(\theta) = \begin{pmatrix} M(\theta_1) & 0 & \cdots & 0 \\ M_{21} & M(\theta_2) & \cdots & 0 \\ \cdot & \cdot & & \cdot \\ \cdot & \cdot & & \cdot \\ \cdot & \cdot & & \cdot \\ M_{k1} & M_{k2} & \cdots & M(\theta_k) \end{pmatrix}$$

where M_{ij} $(i > j)$ are matrices with entries in K of the appropriate size.

Indeed, $\theta(z_{ij}) \in R_{i-1}$ hence it may be expressed in terms of the basis $B_i \cup B_{i+1} \cup \cdots \cup B_k$ as follows:

$$\theta(z_{ij}) = \sum_{h=1}^{m_i} a_{ihj} z_{ih} + \sum_{h=1}^{m_{i+1}} a_{i+1,hj} z_{i+1,h} + \cdots + \sum_{h=1}^{m_k} a_{khj} z_{kh}$$

(with coefficients $a_{thj} \in K$). Then

$$\theta_i(z_{ij} + R_i) = \sum_{h=1}^{m_i} a_{ihj}(z_{ih} + R_i) + R_i = \sum_{h=1}^{m_i} a_{ihj} z_{ih} + R_i$$

and so $M(\theta)$ has the form indicated.

If we consider the characteristic polynomials of the linear transformations $\theta, \theta_1, \ldots, \theta_k$ then

$$F_{R|K}(\theta) = \prod_{i=1}^{k} F_{(R_{i-1}/R_i)|K}(\theta_i). \quad \blacksquare$$

5F. *With above hypotheses and notations, we assume further that:*

(1) *Each R_i is an ideal of R.*

(2) *For every $i = 1, 2, \ldots, k$ there exists no ideal R' of R such that $R_{i-1} \supset R' \supset R_i$*

(3) *If $y \in R_1$, $z \in R_{i-1}$ then $yz \in R_i$*

(4) *If $y, z \in R$, $yz \in R_i$ and $y \notin R_1$ then $z \in R_i$.*

If $x \in R$ and $\theta = \theta_x$ then for every $i = 1, \ldots, k$

$$F_{(R_{i-1}/R_i)|K}(\theta_i) = F_{R|K}(\theta)$$

and so

$$F_{R|K}(\theta) = [F_{(R/R_1)|K}(\theta_1)]^k.$$

Proof: For every $i = 1, \ldots, k$, there exists an isomorphism of R-modules $\lambda_i : R_{i-1}/R_i \to R/R_1$ such that $\theta_1 \circ \lambda_i = \lambda_i \circ \theta_i$. Indeed, let $u \in R_{i-1}$, $u \notin R_i$, hence $R_i \subset R_i + Ru \subseteq R_{i-1}$. By (1) and (2) we have $R_{i-1} = R_i + Ru$. Given any element $y + R_i \in R_{i-1}/R_i$ let $y = y'u + y''$ with $y' \in R$, $y'' \in R_i$; we put $\lambda_i(y + R_i) = y' + R_1$. This defines well the mapping

$$\lambda_i : R_{i-1}/R_i \to R/R_1.$$

In fact, if $y + R_i = z + R_i$ with $z \in R_{i-1}$ and $z = z'u + z''$ with $z' \in R$, $z'' \in R_i$, then $y - z = (y' - z')u + (y'' - z'')$ so $(y' - z')u \in R_i$. Since $u \notin R_i$, it follows from (4) that $y' - z' \in R_1$ so $y' + R_1 = z' + R_1$. It is also obvious that λ_i is a homomorphism of R-modules. Moreover, if $y = y'u + y''$ where $y' \in R_1$ then $y \in R_i$ (by (3)) hence if $\lambda_i(y + R_i) = \bar{0} \in R/R_1$ then $y + R_i = \bar{0} \in R_{i-1}/R_i$. Of course, for every $y' \in R$ if $y = y'u \in R_{i-1}$, then $\lambda_i(y + R_i) = y' + R_1$. Thus λ_i is an isomorphism.

It remains to show that $\theta_1 \circ \lambda_i = \lambda_i \circ \theta_i$. Given $y \in R_{i-1}$, if $y = y'u + y''$ with $y' \in R$, $y'' \in R_i$, then $xy = (xy')u + xy''$ with $xy' \in R$, $xy'' \in R_i$, and so we have

$$\theta_1(\lambda_i(y + R_i)) = \theta_1(y' + R_1) = \theta(y') + R_1 = xy' + R_1$$
$$= \lambda_i(xy + R_i) = \lambda_i(\theta(y) + R_i) = \lambda_i(\theta_i(y + R_i)).$$

If $\bar{B}_i$ is a basis of the K-vector space R_{i-1}/R_i and $\lambda_i(\bar{B}_i)$ is the corresponding basis of the isomorphic vector space R/R_1 then the matrices of θ_i with respect to $\bar{B}_i$ and of θ_1 with respect to $\lambda_i(\bar{B}_i)$ are the same. Hence,

$$F_{(R_{i-1}/R_i)|(K}(\theta_i) = F_{(R/R_1)|K}(\theta_1).$$

Therefore,

$$F_{R|K}(\theta) = [F_{(R/R_1)|K}(\theta_1)]^k. \quad \blacksquare$$

We apply these considerations of linear algebra to the following specific situation.

Let A be a Dedekind domain, K its field of quotients, let $L \mid K$ be a separable extension of degree n, and B the integral closure of A in L, so B is also a Dedekind domain (Chapter 7, (**M**)). If P is a non-zero prime ideal of A, let $BP = \prod_{i=1}^{g} Q_i^{e_i}$, where each Q_i is a prime ideal of B. We recall (see (**4I**)) that under these hypotheses B/BP is a vector space of dimension n over A/P. Let $\psi \colon A \to A/P = \bar{K}$, $\psi_0 \colon B \to B/BP$, $\psi_i \colon B \to B/Q_i = \bar{L}_i$ be the canonical ring homomorphisms; for every $i = 1, \ldots, g$ let

$$\pi_i \colon B/BP \to B/Q_i^{e_i}$$

be the ith projection induced by the natural isomorphism

$$B/BP \xrightarrow{\sim} \prod_{i=1}^{g} B/Q_i^{e_i};$$

explicitly, if $y \in B$ then $\psi_0(y) = y + BP$, $\pi_i(\psi_0(y)) = y + Q_i^{e_i}$. These mappings are naturally extended to the polynomials, by acting on their coefficients.

With these notations and hypotheses, we have the following relations between characteristic polynomials, traces and norms:

5G.   If $x \in B$ then $F_{L \mid K}(x) \in A[X]$ and

$$\psi(F_{L \mid K}(x)) = \prod_{j=1}^{g} [F_{L_j \mid \bar{K}}(\psi_j(x))]^{e_j},$$

$$\psi(\mathrm{Tr}_{L \mid K}(x)) = \sum_{j=1}^{g} e_j \, \mathrm{Tr}_{L_j \mid \bar{K}}(\psi_j(x)),$$

$$\psi(N_{L \mid K}(x)) = \prod_{j=1}^{g} [N_{L_j \mid \bar{K}}(\psi_j(x))]^{e_j}.$$

Proof: Since $x \in B$ its minimal polynomial over K has coefficients in A; therefore its characteristic polynomial, which is a power of the minimal polynomial (see (**5A**)) belongs also to $A[X]$.

Let S be the multiplicative set, complement of P in A, let $A' = S^{-1}A$, $B' = S^{-1}B$, $P' = A'P$, so $B'P' = B'(BP) = B'(A'P)$. By (**4H**) we have $B'/B'P = B/BP$, $A'/A'P = A/P$.

A' is a principal ideal domain, and B' is its integral closure in L (by (**4D**)); moreover, B' is a free A'-module of rank n. By the corollary of (**5C**),

$$F_{B' \mid A'}(x) = F_{L \mid K}(x).$$

Since $B'/B'P' = B/BP$ is a vector space of dimension n over $A'/A'P = A/P = \bar{K}$ (by (4I)), it follows from (5E) and (5D) that

$$\psi(F_{L|K}(x)) = \psi(F_{B'|A'}(x)) = F_{(B'/B'P')|(A'/A'P)}(\psi_0(x))$$

$$= F_{(B/BP)|\bar{K}}(\psi_0(x)) = \prod_{j=1}^{g} F_{(B/Q_j^{e_j})|\bar{K}}(\pi_j\psi_0(x)).$$

It remains now to determine these last characteristic polynomials and for this purpose we apply (5F), taking $k = e_j$,

$$R = B/Q_j^{k}, \; R_1 = Q_j/Q_j^{k}, \ldots, R_i = Q_j^{i}/Q_j^{k}, \ldots, R_{k-1} = Q_j^{k-1}/Q_j^{k}, \; R_k = 0$$

R is a ring, $\bar{K} = A/P$ is a subfield of R and R is a $\bar{K}$-space of finite dimension (equal to the inertial degree of Q_j in $L \mid K$). We have the strictly decreasing chain of $\bar{K}$-subspaces $R \supset R_1 \supset R_2 \supset \cdots \supset R_{k-1} \supset R_k = 0$; actually, we may define a scalar multiplication as follows:

$$(b + Q_j^{k}) \cdot (y + Q_j^{k}) = by + Q_j^{k}, \quad \text{where } b \in B, \quad y \in Q_j^{i};$$

then each R_i becomes an ideal of R. Since B is a Dedekind domain there exists no ideal J such that $Q_j^{i-1} \supset J \supset Q_j^{i}$ hence condition (2) of (5F) is satisfied. Condition (3) is obvious and (4) follows from the fact that B is a Dedekind domain: if $y, z \in B$, $\bar{y} = y + Q_j^{k} \in R$, $\bar{z} = z + Q_j^{k} \in R$ and $\bar{y} \cdot \bar{z} = yz + Q_j^{k} \in R_i$, but $\bar{y} \notin R_1$ then $y \notin Q_j$, $yz \in Q_j^{i}$ so $z \in Q_j^{i-1}$ and $\bar{z} \in Q_j^{i-1}$.

Thus, if $x \in B$ and $\bar{x} = x + Q_j^{k} \in R$ then

$$F_{R|\bar{K}}(\bar{x}) = [F_{(R/R_1)|\bar{K}}(\theta_1)]^{k},$$

where $\theta_1: R/R_1 \to R/R_1$ is defined by $\theta_1(\bar{y} + R_1) = \bar{x}\bar{y} + R_1$, so θ_1 is the mapping of multiplication by $\bar{x} = \pi_j\psi_0(x)$.

Now $R/R_1 = (B/Q_j^{k})/(Q_j/Q_j^{k}) \cong B/Q_j = \bar{L}_j$; the isomorphism

$$\eta: R/R_1 \to B/Q_j$$

is given explicitly as follows: If $\bar{y} + R_1 \in R/R_1$, with $y \in B$, then $\eta(\bar{y} + R_1) = \psi_j(y)$. Then we have

$$\begin{array}{ccc} R/R_1 & \xrightarrow{\;\eta\;} & \bar{L}_j \\ \theta_1 \downarrow & & \downarrow \theta_{\psi_j(x)} \\ R/R_1 & \xrightarrow{\;\eta\;} & \bar{L}_j \end{array}$$

$$\theta_{\psi_j(x)} \circ \eta(\bar{y} + R_1) = \psi_j(x)\psi_j(y) = \psi_j(xy) = (\eta\bar{x}\bar{y} + R_1) = \eta \circ \theta_1(\bar{y} + R_1).$$

Therefore, $F_{(R/R_1)|\bar{R}}(\theta_1) = F_{L_j|\bar{R}}(\psi_j(x))$. Concluding, we have shown that

$$\psi(F_{L|K}(x)) = \prod_{j=1}^{g} [F_{L_j|\bar{R}}(\psi_j(x))]^{e_j}$$

and the relations for the trace and for the norm follow at once. ∎

Now we shall prove the transitivity of the trace and norm. We have the following situation: C is a commutative ring, B, A are subrings of C, such that $C \supseteq B \supseteq A$, and we assume that B is a free A-module of rank n, while C is a free B-module of rank m. From this it follows that if $\{x_1, \ldots, x_n\}$ is an A-basis of B and $\{y_1, \ldots, y_m\}$ a B-basis of C, then

$$\{x_1y_1, x_2y_1, \ldots, x_ny_1, x_1y_2, \ldots, x_ny_m\}$$

is a A-basis of C, and so C is a free A-module of rank mn.

Thus, we may consider for every element $y \in C$, the following elements: $\text{Tr}_{C|A}(y)$ and $\text{Tr}_{B|A}(\text{Tr}_{C|B}(y))$ as well as the corresponding elements for the norm.

5H.
$$\text{Tr}_{C|A}(y) = \text{Tr}_{B|A}(\text{Tr}_{C|B}(y))$$
$$N_{C|A}(y) = N_{B|A}(N_{C|B}(y))$$

for every element $y \in C$.

Proof: Let θ be any endomorphism of the B-module C. Thus, θ satisfies also $\theta(ay) = a\theta(y)$ for every $a \in A$, that is, θ is also an endomorphism of the A-module C, and as such, it will be denoted by θ_A.

To find the matrix of θ with respect to the basis $\{y_1, \ldots, y_m\}$ we write

$$\theta(y_j) = \sum_{i=1}^{m} b_{ij}y_i, \quad \text{with } b_{ij} \in B.$$

So,

$$b_{ij} = \sum_{k=1}^{n} a_{kij}x_k$$

for all indices $i, j = 1, \ldots, m$.

To find the matrix of θ_A with respect to the basis

$$\{x_1y_1, x_2y_1, \ldots, x_ny_1, x_1y_2, \ldots x_ny_2, \ldots, x_ny_m\}$$

we note that $\theta(x_l y_j) = \sum_{i=1}^{m} x_l b_{ij} y_i$, and we write $x_l b_{ij} = \sum_{k=1}^{n} a_{kij} x_l x_k$, $x_l x_k = \sum_{h=1}^{n} a'_{lhk} x_h$ for all $l, k = 1, \ldots, n$, hence

$$x_l b_{ij} = \sum_{k=1}^{n} a_{kij} \sum_{h=1}^{n} a'_{lhk} x_h = \sum_{h=1}^{n} \left(\sum_{k=1}^{n} a_{kij} a'_{lhk} \right) x_h.$$

Thus

$$\theta(x_l y_j) = \sum_{i=1}^{m} \sum_{h=1}^{n} \left(\sum_{k=1}^{n} a_{kij} a'_{lhj} \right) x_h y_i,$$

therefore, the matrix M of θ_A, with respect to the basis considered above, has entry

$$\sum_{k=1}^{n} a_{kij} a'_{lhk}$$

at the row (h,i) and column (l,j).

On the other hand, By_j is a free A-module with basis $\{x_1 y_j, \ldots, x_n y_j\}$ and similarly for By_i. Let $\theta_{ji} \colon By_j \to By_i$ be the A-linear transformation defined by $\theta_{ji}(xy_j) = xb_{ij}y_i$ for every $x \in B$. With respect to the above bases, the matrix M_{ji} of θ_{ji} is obtained as follows:

$$\theta_{ji}(x_l y_j) = \sum_{h=1}^{n} \left(\sum_{k=1}^{n} a_{kij} a'_{lhk} \right) x_h y_i,$$

thus M_{ji} is a $n \times n$ matrix, with coefficients in A, and its entry in row h, column l is $\sum_{k=1}^{n} a_{kij} a'_{lhk}$. Therefore, M may be written as a matrix of m^2 blocks M_{ji}, each being a $n \times n$ matrix with coefficients in A:

$$M = \begin{pmatrix} M_{11} & M_{12} & \cdots & M_{1m} \\ M_{21} & M_{22} & \cdots & M_{2m} \\ \cdot & \cdot & & \cdot \\ \cdot & \cdot & & \cdot \\ \cdot & \cdot & & \cdot \\ M_{m1} & M_{m2} & \cdots & M_{mm} \end{pmatrix}$$

We now prove that the matrices M_{ji} are permutable (by multiplication). Let λ_{ji} be the endomorphism of the A-module B defined by $\lambda_{ji}(x) = xb_{ij}$; thus $\lambda_{ji}(x_l) = x_l b_{ij} = \sum_{h=1}^{n} (\sum_{k=1}^{n} a_{kij} a'_{lhk}) x_h$. With respect to the basis $\{x_1, \ldots, x_n\}$ of B, the matrix of λ_{ji} is equal to M_{ji}. Since $\lambda_{ji} \circ \lambda_{kh}(x) = xb_{kh}b_{ji} = xb_{ji}b_{kh} = \lambda_{kh} \circ \lambda_{ji}(x)$ for every $x \in B$, then the corresponding matrices satisfy $M_{ji} M_{kh} = M_{kh} M_{ji}$, as we have claimed.

Now let $\theta = \theta_y$, the B-endomorphism of C defined by multiplication with y; thus θ_A is the induced A-endomorphism of C. By definition, $\text{Tr}_{C|A}(y) = \text{Tr}(\theta_A)$, so it is equal to the sum of the elements in the diagonal of the matrix M which corresponds to θ_A; as we proved, this sum is equal to

$$\sum_{i=1}^{m} (\text{sum of diagonal elements of } M_{ii}) = \sum_{i=1}^{m} \text{Tr}(M_{ii});$$

but M_{ii} is the matrix of λ_{ii} (which is the A-endomorphism of multiplication by b_{ii}), hence $\text{Tr}(M_{ii}) = \text{Tr}_{B|A}(b_{ii})$, so $\text{Tr}_{C|A}(y) = \sum_{i=1}^{m} \text{Tr}_{B|A}(b_{ii}) =$

$\mathrm{Tr}_{B|A} (\Sigma_{i=1}^m b_{ii}) = \mathrm{Tr}_{B|A}(\mathrm{Tr}_{B|C}(y))$, because the matrix of θ with respect to the B-basis $\{y_1, \ldots, y_m\}$ of C, has diagonal elements b_{ii} $(i = 1, \ldots, m)$.

In order to prove the corresponding statement for the norm, we recall that $N_{C|A}(y) = \det(\theta_A) = \det(M)$.

We shall soon establish in a lemma that the computation of the determinant of M may be done as follows: Regard the blocks M_{ji} as if they were elements, compute the determinant obtaining a matrix with coefficients in A, and then compute the determinant of this matrix.

Now, we note the general fact: With respect to a given A-basis of B, if M', M'' are respectively the matrices of $\theta_{b'}$, $\theta_{b''}$ then $M' + M''$, $M'M''$ are the matrices of $\theta_{b'+b''}$, $\theta_{b'b''}$. From this, we deduce that if μ is the A-endomorphism of B of multiplication by $N_{C|B}(y) = \det(b_{ij})_{i,j}$, then the matrix of μ with respect to $\{x_1, \ldots, x_n\}$ is equal to $\det(M_{ji})_{j,i}$. Thus

$$\det(\mu) = \det(\det[(M_{ji})_{j,i}]) = \det(M) = N_{C|A}(y);$$

on the other hand,

$$\det(\mu) = \det(\theta_{N_{C|B}(y)}) = N_{B|A}(N_{C|B}(y)),$$

showing the formula for the norm. ∎

Now, we have to prove the lemma used above:

Lemma 1. *Let X_{ij} be m^2 indeterminates, and consider the $m \times m$ matrix $X = (X_{ij})_{i,j}$; let D be the determinant of X, $D \in \mathbb{Z}[X_{11}, \ldots, X_{mm}]$. If A is a commutative ring, if M_{ij} are $n \times n$ matrices with coefficients in A, for $i, j = 1, \ldots, m$, such that $M_{ij}M_{kh} = M_{kh} \cdot M_{ij}$ for any indices i, j, k, h, if*

$$M = \begin{pmatrix} M_{11} & M_{12} & \cdots & M_{1m} \\ M_{21} & M_{22} & \cdots & M_{2m} \\ \cdot & \cdot & & \cdot \\ \cdot & \cdot & & \cdot \\ \cdot & \cdot & & \cdot \\ M_{m1} & M_{m2} & \cdots & M_{mm} \end{pmatrix}$$

is considered as a $mn \times mn$ matrix with elements in A, then

$$\det(M) = \det(D(M_{11}, \ldots, M_{mm})).$$

Proof: The result is true when $m = 1$, and it will be proved by induction on m. In order to include the case where the ring A may have zero-divisors, we make use of the following device. Let T be a new indeterminate, for all indices i, j, let as usual δ_{ij} be 0 when $i \neq j$, and $\delta_{ii} = 1$; we denote by N_{ij} the matrix $N_{ij} = M_{ij} + \delta_{ij}TI_n$, where I_n is the unit $n \times n$ matrix.

Computing the determinant of X by considering cofactors of the elements in any column, we have the well-known relations:

$$\sum_{i=1}^{m} X_{ji} D^{ik} = \delta_{jk} D$$

where D^{ik} is the cofactor of X_{ki} in the matrix X.

Let $D^{ik}(N_{11}, \ldots, N_{mm}) = N^{ik}$, so N^{ik} is a $n \times n$ matrix with entries in $A[T]$.

If

$$P = \begin{pmatrix} N^{11} & N^{12} & \cdots & N^{1m} \\ 0 & I_n & \cdots & 0 \\ \cdot & \cdot & & \cdot \\ \cdot & \cdot & & \cdot \\ \cdot & \cdot & & \cdot \\ 0 & 0 & \cdots & I_n \end{pmatrix}$$

and

$$N = \begin{pmatrix} N_{11} & N_{12} & \cdots & N_{1m} \\ N_{21} & N_{22} & \cdots & N_{2m} \\ \cdot & \cdot & & \cdot \\ \cdot & \cdot & & \cdot \\ \cdot & \cdot & & \cdot \\ N_{m1} & N_{m2} & \cdots & N_{mm} \end{pmatrix}$$

by multiplication we have

$$PN = \begin{pmatrix} D(N_{11}, N_{12}, \ldots, N_{mm}) & 0 & \cdots & 0 \\ N_{21} & N_{22} & \cdots & N_{2m} \\ \cdot & \cdot & & \cdot \\ \cdot & \cdot & & \cdot \\ \cdot & \cdot & & \cdot \\ N_{m1} & N_{m2} & \cdots & N_{mm} \end{pmatrix}$$

Let

$$Q = \begin{pmatrix} N_{22} & \cdots & N_{2m} \\ \cdot & & \cdot \\ \cdot & & \cdot \\ \cdot & & \cdot \\ N_{m2} & \cdots & N_{mm} \end{pmatrix},$$

so Q is a $(m-1)n \times (m-1)n$ matrix with entries in $A[T]$. Since the first row of PN has only one block which is not zero,

$$\det(PN) = \det(D(N_{11}, \ldots, N_{mm})) \cdot \det(Q);$$

but on the other hand $\det(PN) = \det(P) \cdot \det(N)$, and $\det(P) = \det(N^{11})$.
Applying the induction on Q, we have:

$$\det(Q) = \det(D^{11}(N_{11}, \ldots, N_{mm})) = \det(N^{11}).$$

But $\det(N^{11})$ is a monic polynomial in T, having degree $n(m-1)$, so it is not a zero-divisor in the ring $A[T]$. Therefore, we conclude that

$$\det(D(N_{11}, \ldots, N_{mm})) = \det(N).$$

Now, letting $h: A[T] \to A$ be the homomorphism such that $h(T) = 0$ and h leaves fixed every element of A, we deduce that h induces a homomorphism $\bar{h}$ from the associated matrix rings and

$$
\begin{aligned}
\det(D(M_{11}, \ldots, M_{mm})) &= \det(D(\bar{h}(N_{11}), \ldots, \bar{h}(N_{mm}))) \\
&= \det(\bar{h}(D(N_{11}, \ldots, N_{mm}))) \\
&= h[\det(D(N_{11}, \ldots, N_{mm}))] \\
&= h(\det(N)) = \det(\bar{h}(N)) = \det(M). \quad \blacksquare
\end{aligned}
$$

6. DISCRIMINANT OF RING EXTENSIONS

Let B be a commutative ring, A a subring of B such that B is a free A-module of rank n. If $x_1, \ldots, x_n \in B$ we define the *discriminant of* $(x_1, \ldots, x_n)$ (in the ring extension $B \mid A$) as

$$\mathrm{discr}_{B\mid A}(x_1, \ldots, x_n) = \det(\mathrm{Tr}_{B\mid A}(x_i x_j));$$

that is, the determinant of the matrix whose (i,j)-entry is $\mathrm{Tr}_{B\mid A}(x_i x_j)$. Thus $\mathrm{discr}(x_1, \ldots, x_n) \in A$.

Let us note at once, if $B = L$, $A = K$ where $L \mid K$ is a separable field extension of degree n, then by (**5A**) the new concept of discriminant coincides with the one in Chapter 2, (**11**).

6A. *If $(x_1', \ldots, x_n')$ is another n-tuple of elements in B, and $x_j' = \sum_{i=1}^{n} a_{ij} x_i$ (for all $j = 1, \ldots, n$), with $a_{ij} \in A$, then*

$$\mathrm{discr}_{B\mid A}(x_1', \ldots, x_n') = [\det(a_{ij})]^2 \cdot \mathrm{discr}_{B\mid A}(x_1, \ldots, x_n)$$

Proof: The proof is standard. We first note that

$$\mathrm{Tr}_{B\mid A}(x_i' x_j') = \mathrm{Tr}_{B\mid A}\left[\left(\sum_{k=1}^{n} a_{ki} x_k\right)\left(\sum_{h=1}^{n} a_{hj} x_h\right)\right] = \sum_{k=1}^{n} \sum_{h=1}^{n} a_{ki} a_{hj} \mathrm{Tr}(x_k x_h),$$

hence letting $M = (a_{ij})$ and M' denote the transpose matrix of M, then

$$\begin{aligned}
\text{discr}_{B|A}(x_1', \ldots, x_n') &= \det(\text{Tr}_{B|A}(x_i' x_j')) \\
&= \det(M' . (\text{Tr}(x_k x_h)) . M) \\
&= \det(M') . \det(\text{Tr}(x_k x_h)) . \det(M) \\
&= [\det(a_{ij})]^2 . \text{discr}_{B|A}(x_1, \ldots, x_n). \quad \blacksquare
\end{aligned}$$

From the next result, we deduce that it is only interesting to consider the discriminant of linearly independent n-tuples:

6B. *If* $\{x_1, \ldots, x_n\}$ *is linearly dependent over the domain A then*

$$\text{discr}_{B|A}(x_1, \ldots, x_n) = 0. \quad \blacksquare$$

Proof: We assume that there exist elements $a_1, \ldots, a_n \in A$, not all equal to zero, such that $\sum_{j=1}^n a_j x_j = 0$. For example, let $a_1 \neq 0$.

Now, we consider the n-tuple $(x_1', \ldots, x_n')$, where $x_1' = 0$, $x_i' = x_i$ for $i = 2, \ldots, n$. Thus, $x_i' = \sum_{j=1}^n a_{ji} x_j$ $(i = 1, \ldots, n)$ by letting $a_{j1} = a_j$, and if $i > 1$, then $a_{ji} = 1$ for $j = i$, $a_{ji} = 0$ for $j \neq i$. By **(6A)** we have

$$0 = \text{discr}_{B|A}(0, x_2, \ldots, x_n) = [\det(a_{ij})]^2 . \text{discr}_{B|A}(x_1, \ldots, x_n).$$

Since $\det(a_{ij}) = a_1 \neq 0$ and A is a domain, then

$$\text{discr}_{B|A}(x_1, \ldots, x_n) = 0. \quad \blacksquare$$

6C. *Let A be a domain, let* $\{x_1, \ldots, x_n\}$, $\{x_1', \ldots, x_n'\}$ *be any two bases of the A-module B. Then: Either*

$$\text{discr}_{B|A}(x_1, \ldots, x_n) = \text{discr}_{B|A}(x_1', \ldots, x_n') = 0,$$

or

$$\text{discr}_{B|A}(x_1, \ldots, x_n), \qquad \text{discr}_{B|A}(x_1', \ldots, x_n')$$

are associated elements of A (see Chapter 1, (1)).

Proof: By hypothesis there exist elements $a_{ij}, a \in A$ such that

$$x_j' = \sum_{i=1}^n a_{ij} x_i,$$

for every $j = 1, \ldots, n$. By **(6A)** we have

$$\text{discr}_{B|A}(x_1', \ldots, x_n') = [\det(a_{ij})]^2 . \text{discr}_{B|A}(x_1, \ldots, x_n).$$

Since (a_{ij}) is an invertible matrix, then $\det(a_{ij})$ is a unit in the ring A; hence either both discriminants are zero or both are associated elements of A. $\quad \blacksquare$

The preceding result justifies the following definition:

Let A be a domain, let B be a commutative ring, having A as a subring, and such that B is a free A-module of rank n. If $\{x_1, \ldots, x_n\}$ is any basis of the A-module B, the principal ideal $A \cdot \mathrm{discr}_{B|A}(x_1, \ldots, x_n)$ is called the *discriminant of B relative to A*, and denoted by $\mathrm{discr}(B \mid A)$.

In the case where A is a field K, $\mathrm{discr}(B \mid K)$ is either 0 or the unit ideal of K (since K has only trivial ideals). Moreover, we shall see in (6E), that if L is an algebraic number field then $\mathrm{discr}(L \mid \mathbb{Q}) = \mathbb{Q}$ the unit ideal of $\mathbb{Q}$; so this concept does not constitute an appropriate generalization of δ_L, the discriminant of the field L, introduced in Chapter 6, Definition 4. Later, in this section, we shall explain what is the relative discriminant $\delta_{L|K}$ of an algebraic number field L over a subfield K.

One of the tools used in determining the discriminant is the following easy result:

6D. *Let $B_1, \ldots, B_r$ be commutative rings, containing the domain A and such that each ring B_i is a free A-module of finite rank. Then:*

$$\mathrm{discr}(B_1 \times \cdots \times B_r \mid A) = \prod_{i=1}^{r} \mathrm{discr}(B_i \mid A).$$

Proof: It is enough to prove the statement when $r = 2$.

Let $\{x_1, \ldots, x_n\}$ be a basis of the A-module B_1, let $\{y_1, \ldots, y_m\}$ be a basis of the A-module B_2. Then $\{(x_1, 0), \ldots, (x_n, 0), (0, y_1), \ldots, (0, y_m)\}$ is a basis of the A-module $B_1 \times B_2$. Letting $z_i = (x_i, 0)$ for $i = 1, \ldots, n$, $z_{n+i} = (0, y_i)$ for $i = 1, \ldots, m$, then $\mathrm{discr}(B_1 \times B_2 \mid A)$ is the principal ideal of A generated by $\det(\mathrm{Tr}_{B_1 \times B_2|A}(z_i z_j))$.

Now, if $t \in B_1$ then $\mathrm{Tr}_{B_1 \times B_2|A}(t, 0) = \mathrm{Tr}_{B_1|A}(t)$, as we deduce by considering the matrices of the endomorphisms $\theta_{(t,0)}$ of $B_1 \times B_2$ and θ_t of B_1, relative to the basis $\{z_1, \ldots, z_{n+m}\}$ and $\{x_1, \ldots, x_n\}$, respectively. In the same way, if $t \in B_2$ then $\mathrm{Tr}_{B_1 \times B_2|A}(0, t) = \mathrm{Tr}_{B_2|A}(t)$. Thus,

$$\det(\mathrm{Tr}_{B_1 \times B_2|A}(z_i z_j)) = \det\left(\begin{array}{c|c} (\mathrm{Tr}_{B_1|A}(x_i x_j)) & 0 \\ \hline 0 & (\mathrm{Tr}_{B_2|A}(y_i y_j)) \end{array}\right)$$

$$= \det(\mathrm{Tr}_{B_1|A}(x_i x_j)) \cdot \det(\mathrm{Tr}_{B_2|A}(y_i y_j))$$

and so this element generates the ideal $\mathrm{discr}(B_1 \mid A) \cdot \mathrm{discr}(B_2 \mid A)$. ∎

6E. *If K is a field, if B is a commutative algebra of dimension n over K†, then $\mathrm{discr}(B \mid K) = 0$ if and only if the trace in $B \mid K$ is degenerate, that is, there exists an element $x \in B$, $x \neq 0$, such that $\mathrm{Tr}_{B|K}(xy) = 0$ for every $y \in B$.*

---

† We may therefore identify K with a subring of B.

Proof: Let us assume that the trace is degenerate, with $x \in B$, $x \neq 0$, such that $\text{Tr}_{B|K}(xy) = 0$ for every $y \in B$. Let us consider a basis $\{x_1, \ldots, x_n\}$ of the vector-space B over K, such that $x_1 = x$. Then $\text{discr}(B \mid K)$ is the ideal of K generated by $\text{discr}_{B|K}(x_1, \ldots, x_n) = \det(\text{Tr}_{B|K}(x_i x_j)) = 0$.

Conversely, if $\text{discr}(B \mid K) = 0$, let $\{x_1, \ldots, x_n\}$ be a K-basis of B, hence $\text{discr}_{B|K}(x_1, \ldots, x_n) = \det(\text{Tr}_{B|K}(x_i x_j)) = 0$; thus, there exist elements $a_i \in K$, not all equal to zero, such that $\sum_{i=1}^{n} a_i \cdot \text{Tr}_{B|K}(x_i x_j) = 0$ for every $j = 1, \ldots, n$. Thus, letting $x = \sum_{i=1}^{n} a_i x_i$, we have $x \neq 0$ and for every element $y = \sum_{j=1}^{n} b_j x_j \in B$, (with $b_j \in K$) we have:

$$\text{Tr}_{B|K}(xy) = \sum_{i,j=1}^{n} a_i b_j \, \text{Tr}(x_i x_j) = 0;$$

this shows that the trace is degenerate. ∎

Let us assume now that K is a *perfect field*, that is, every algebraic extension L of K is separable; we may improve the preceding result, taking into account the fact that if $L \mid K$ is separable, there exists an element $x \in L$ such that $\text{Tr}_{L|K}(x) \neq 0$ (see Chapter 2, **(10)**). We note that every field of characteristic zero is perfect; also, every finite field is perfect.

6F. *Let K be a perfect field, let B be a commutative K-algebra of finite dimension. Then $\text{discr}(B \mid K) \neq 0$ if and only if 0 is the only nilpotent element of B.*

Proof: Let us assume that B contains the nilpotent element $x \neq 0$. Let $\{x_1, \ldots, x_n\}$ be a K-basis of B, such that $x_1 = x$. Since B is commutative, then xx_j is also nilpotent. The minimal polynomial of the endomorphism θ_{xx_j} of multiplication by xx_j is equal to X^r, for some $r > 0$; as it is known from the theory of linear transformations of vector-spaces, the characteristic polynomial of θ_{xx_j} is a multiple of the minimal polynomial, having the same irreducible factors and of degree n; thus, the characteristic polynomial is X^n, and $\text{Tr}_{B|K}(xx_j) = 0$ for every $j = 1, \ldots, n$.

Hence $\text{discr}(x_1, \ldots, x_n) = \det(\text{Tr}_{B|K}(x_i x_j)) = 0$ because the matrix of traces has the first row of zeroes. This shows that $\text{discr}(B \mid K) = 0$.

Conversely, let us assume that 0 is the only nilpotent element of B. We note that since every ideal of B is in particular a subspace of the K-space B, from the fact that B has dimension n over K, every chain of subspaces, hence also of ideals of B must be finite. Thus, B is a Noetherian ring.

We shall require the following lemma:

Lemma 2. *If B is a noetherian ring, such that 0 is the only nilpotent element, then the zero-ideal is the intersection of finitely many prime ideals.*

Assuming the lemma true, we may write $0 = P_1 \cap \cdots \cap P_r$, where each P_i is a prime ideal of B. Since $P_i \cap K$ is an ideal of K, distinct from K, then

$P_i \cap K = 0$ (for $i = 1, \ldots, r$). Thus $K \subseteq B/P_i$ (up to a natural identification), and B/P_i is a finite dimensional K-space which is also a domain. Since every element of B/P_i is integral over K (by Chapter 5, (A)) then B/P_i is a field (by Chapter 5, (F)), so P_i is a maximal ideal of B.

Now, we know that the distinct ideals $P_1, \ldots, P_r$ are maximal; hence $P_i + \cap_{j \neq i} P_j = B$, otherwise $P_i \supseteq \cap_{j \neq i} P_j$, hence $P_i \supseteq P_j$ for some $j \neq i$ and necessarily $P_i = P_j$, against the fact that these ideals are distinct. By Chapter 8, (K), we have $B = B/0 \simeq \Pi_{i=1}^r B/P_i = \Pi_{i=1}^r L_i$. Hence

$$\mathrm{discr}(B \mid K) = \prod_{i=1}^r \mathrm{discr}(L_i \mid K).$$

But the field L_i is a finite extension, thus an algebraic extension of K. Since K is a perfect field, L_i is deparable over K. As we quoted, there exists an element $x_i \in L$ such that $\mathrm{Tr}_{L_i \mid K}(x_i) \neq 0$; so the trace is not degenerate (because if there exists $x' \in L_i$, $x' \neq 0$, such that $\mathrm{Tr}_{L_i \mid K}(x'y) = 0$ for every $y \in L_i$, then from $x_i = x'(x'^{-1}x_i)$ we would have $\mathrm{Tr}_{L_i \mid K}(x_i) = 0$). By (6E), $\mathrm{discr}(L_i \mid K) \neq 0$, thus, $\mathrm{discr}(L_i \mid K) = K$; hence $\mathrm{discr}(B \mid K) = K$. ∎

Proof of the lemma: By the footnote to Theorem 1, Chapter 7, we have $0 = \Pi_{i=1}^r P_i^{e_i}$, where the prime ideals P_i are all distinct, $e_i \geqslant 1$ (for $i = 1, \ldots, r$). We show that $P_1 \cap \cdots \cap P_r = 0$. If $x \in P_1 \cap \cdots \cap P_r$ then $x^{e_1+e_2+\cdots+e_r} \in P_1^{e_1} P_2^{e_2} \ldots P_r^{e_r} = 0$, hence x is nilpotent and therefore $x = 0$. ∎

The following result is the crucial part of the main theorem to be proved soon:

6G. *Let A be a principal ideal domain, K its field of quotients: let $L \mid K$ be a separable extension of degree n and B the integral closure of A in L. Let P be a non-zero prime ideal of A such that the field A/P is perfect. Then the ring B/BP has non-zero nilpotent elements if and only if $P \supseteq \mathrm{discr}(B \mid A)$.*
Proof: B is a free A-module of rank n (Chapter 6, (B) and Theorem 1) and a Dedekind domain (by Chapter 7, (M)). By (4I) B/BP is a vector space of dimension n over the field A/P; actually, if $\{x_1, \ldots, x_n\}$ is a basis of the A-module B, then their images in B/BP form a basis $\{\bar{x}_1, \ldots, \bar{x}_n\}$ of the A/P-vector space B/BP. Thus, $\mathrm{discr}(B \mid A) = \mathrm{discr}_{B\mid A}(x_1, \ldots, x_n)$, and $\mathrm{discr}((B/BP) \mid (A/P)) = \mathrm{discr}_{(B/BP)\mid(A/P)}(\bar{x}_1, \ldots, \bar{x}_n)$.

Since A/P is a perfect field, by (6F), B/BP has non-zero nilpotent elements if and only if $\mathrm{discr}((B/BP) \mid (A/P)) = 0$; by (6E) this means that

$$0 = \mathrm{discr}_{(B/BP)\mid(A/P)}(\bar{x}_1, \ldots, \bar{x}_n) = \det(\mathrm{Tr}_{(B/BP)\mid(A/P)}(\bar{x}_i\bar{x}_j))$$

$$= \overline{\det(\mathrm{Tr}_{B\mid A}(x_ix_j))} = \overline{\mathrm{discr}_{B\mid A}(x_1, \ldots, x_n)},$$

that is $\mathrm{discr}(B \mid A) \subseteq P$. ∎

7. RELATIVE DISCRIMINANT AND DIFFERENT OF ALGEBRAIC NUMBER FIELDS

We return to the situation where K is an algebraic number field, $L \mid K$ an extension of degree n and A, B are the rings of integers of K, L respectively.

Definition 5. The *relative discriminant* of $L \mid K$ is the ideal $\delta_{L|K}$ of A generated by the elements $\mathrm{discr}_{L|K}(x_1, \ldots, x_n)$, for all possible bases $\{x_1, \ldots, x_n\}$ of $L \mid K$ such that each $x_i \in B$.

7A. *Let $\{x_1, \ldots, x_n\}$ be a basis of $L \mid K$ such that each $x_i \in B$. Then $\delta_{L|K} = A \cdot \mathrm{discr}_{L|K}(x_1, \ldots, x_n)$ if and only if B is a free A-module and $\{x_1, \ldots, x_n\}$ is an A-basis of B.*

Proof: If $\{x_1, \ldots, x_n\}$ is an A-basis of B then by definition $\delta_{L|K} \supseteq A \cdot \mathrm{discr}_{L|K}(x_1, \ldots, x_n)$. Now, if $\{x_1', \ldots, x_n'\}$ is any K-basis of L, with $x_j' \in B$ for every $j = 1, \ldots, n$, we have

$$x_j' = \sum_{i=j}^{n} a_{ij} x_i, \quad \text{with } a_{ij} \in A,$$

so

$$\mathrm{discr}_{L|K}(x_1', \ldots, x_n') = [\det(a_{ij})]^2 \cdot \mathrm{discr}_{L|K}(x_1, \ldots, x_n)$$

hence every generator of the ideal $\delta_{L|K}$ is contained in $A \cdot \mathrm{discr}_{L|K}(x_1, \ldots, x_n)$.

Conversely, let us assume that $\delta_{L|K} = A \cdot \mathrm{discr}_{L|K}(x_1, \ldots, x_n)$, where $\{x_1, \ldots, x_n\}$ is a K-basis of L contained in B. Let us show that $\{x_1, \ldots, x_n\}$ generates the A-module B.

Let P be any non-zero prime ideal of A, let S be the multiplicative set complement of P in A, $A' = S^{-1}A$, $B' = S^{-1}B$, $P' = A'P$. A' is a principal ideal domain, B' is a free A'-module of rank n; let $\{x_1', \ldots, x_n'\}$ be a basis of this module. Writing $x_i' = y_i/s_i$ with $y_i \in B$, $s_i \in S$, we deduce that $\mathrm{discr}_{L|K}(x_1', \ldots, x_n') \in A' \cdot \mathrm{discr}_{L|K}(y_1, \ldots, y_n) \subseteq A' \cdot \delta_{L|K}$. On the other hand, we have

$$x_j = \sum_{i=1}^{n} a_{ij}' x_i', \quad \text{with } a_{ij}' \in A'.$$

Hence,

$$\mathrm{discr}_{L|K}(x_1, \ldots, x_n) = [\det(a_{ij}')]^2 \cdot \mathrm{discr}_{L|K}(x_1', \ldots, x_n'),$$

thus $A' \cdot \delta_{L|K} = A' \cdot \mathrm{discr}_{L|K}(x_1', \ldots, x_n')$. From **(5C)** we have

$$\mathrm{discr}_{L|K}(x_1, \ldots, x_n) = \mathrm{discr}_{B'|A'}(x_1, \ldots, x_n)$$

and

$$\mathrm{discr}_{L|K}(x_1', \ldots, x_n') = \mathrm{discr}_{B'|A'}(x_1', \ldots, x_n').$$

Hence by (6C) these are associated elements of A', thus $[\det(a_{ij}')]^2 \in A'$ is a unit of A', and therefore $\det(a_{ij}')$ is also a unit. This means that the inverse of the matrix $(a_{ij}')_{i,j}$ has coefficients in A' and so each element x_i' belongs to the A'-module generated by $x_1, \ldots, x_n$. Since these element are linearly independent, they constitute a basis of the A'-module B'.

The above considerations held for every non-zero prime ideal P of A. It follows that $\{x_1, \ldots, x_n\}$ is a basis of the A-module B. Indeed, if $y \in B$, we may write $y = \Sigma_{i=1}^n c_i x_i$ with $c_i \in K$; this expression is unique. But for every prime ideal $P \neq 0$ we have $c_i \in A_P$ as we have just shown. Thus

$$c_i \in \cap A_P = A \text{ (intersection for all non-zero prime ideals of } A)$$

as follows from (4C) and $\{x_1, \ldots, x_n\}$ generates the A-module B, as we had to prove. ∎

This result may be applied when A is a principal ideal domain or when there exists a primitive element t of $L \mid K$ such that $B = A[t]$. It follows that $\delta_{L \mid \mathbb{Q}}$ is the principal ideal generated by the discriminant δ_L, as introduced in Chapter 6, Definition 4.

Now we come to the main theorem, connecting the ramification and the discriminant:

Theorem 1. (*Dedekind*). *The non-zero prime ideal P of A is ramified in $L \mid K$ if and only if $P \supseteq \delta_{L \mid K}$. In particular, there exist only finitely many prime ideals which are ramified in $L \mid K$.*
Proof: The second assertion follows at once from the first, by Chapter 7 (D) and (E).

We write $BP = \Pi_{i=1}^g Q_i^{e_i}$, where Q_i are distinct prime ideals of B and $e_i \geqslant 1$. From Chapter 7, Theorem 2, we have $B/BP = \Pi_{i=1}^g B/Q_i^{e_i}$.

P is ramified when some e_i is greater than 1; that is, $B/Q_i^{e_i}$ has a non-zero nilpotent element; or equivalently, B/BP has a non-zero nilpotent element. By (6F) this means that $\operatorname{discr}((B/BP) \mid (A/P)) = 0$.

If S is the set complement of P in A, if $A' = S^{-1}A$, $B' = S^{-1}B$, $P' = A'P$ it follows from (4H) that $A'/A'P = A/P$, $B'/B'P = B/P$ so the above condition is that $\operatorname{discr}((B'/B'P) \mid (A'/P')) = 0$. We know further that B' is a free module of rank n over the principal ideal domain A'; moreover, if $\{x_1', \ldots, x_n'\}$ is any basis of the A'-module B' then the images $\bar{x}_i'$ of these elements by the homomorphism $B' \to B'/B'P$ constitute a basis over $A'/A'P$ (as it was seen in the proof of (4I)).

Now, $\operatorname{discr}((B'/B'P) \mid (A'/P'))$ is the ideal generated by the elements $\operatorname{discr}_{(B'/B'P) \mid (A'/P')}(\bar{x}_1', \ldots, \bar{x}_n')$ for all possible bases $\{x_1', \ldots, x_n'\}$ of $B'/B'P$ over A'/P'. So $\operatorname{discr}((B'/B'P) \mid (A'/P') = 0$ exactly when

$$\operatorname{discr}_{(B'/B'P) \mid (A'/P')}(\bar{x}_1', \ldots, \bar{x}_n') = \overline{\operatorname{discr}_{B'/A'}(x_1', \ldots, x_n')} = \bar{0};$$

that is, $\operatorname{discr}_{B'|A}(x_1', \ldots, x_n') \in P'$ for every basis $\{x_1', \ldots, x_n'\}$ of the A'-module B'.

This last condition is actually equivalent to $\delta_{L|K} \subseteq P$. Indeed, let $\{x_1, \ldots, x_n\}$ be a K-basis of L, where each x_i belongs to B. Then it is also a basis of the A'-module B' and by **(5C)** we have

$$\operatorname{discr}_{L|K}(x_1, \ldots, x_n) = \operatorname{discr}_{B'|A'}(x_1, \ldots, x_n) \in P' \cap A = P$$

(because each $x_i \in B$). We deduce that $\delta_{L|K} \subseteq P$. Conversely, if $\delta_{L|K} \subseteq P$, if $\{x_1', \ldots, x_n'\}$ is a basis of the A'-module B', let $x_i' = x_i/s_i$ with $x_i \in B$, $s_i \in S$; thus $\{x_1, \ldots, x_n\}$ is a K-basis of L contained in B, with

$$\operatorname{discr}_{B'|A'}(x_1', \ldots, x_n') = \operatorname{discr}_{L|K}(x_1', \ldots, x_n')$$

$$= \left(\frac{1}{s_1 \ldots s_n}\right)^2 \cdot \operatorname{discr}_{L|K}(x_1, \ldots, x_n) \in A'P = P'. \quad \blacksquare$$

It will be a feature of the theory that unramified prime ideals may be handled without difficulty; thus the preceding theorem asserts that it is necessary to concentrate only on those finitely many prime ideals which ramify. In the next chapter we shall study in more detail the steps of ramification in the case of Galois extensions.

Dedekind's theorem tells which prime ideals P of A are ramified in $L \mid K$. A more precise problem is to determine the prime ideals Q of B which are ramified in $L \mid K$. Clearly, if Q is ramified and $P = Q \cap A$ then P is ramified; the converse is also true when $L \mid K$ is a Galois extension as follows from **(1D)**. However, if $L \mid K$ is not a Galois extension, it may well exist different prime ideals Q, Q' of B such that $Q \cap A = Q' \cap A = P$, and Q is ramified while Q' is not ramified in $L \mid K$.

To find out which prime ideals Q of B are ramified in $L \mid K$ we shall introduce the relative different of $L \mid K$. Besides, we shall also consider the different above a given prime ideal of A. We may treat these two cases simultaneously.

Let R be a Dedekind domain, and K its field of quotients; let $L \mid K$ be a separable field extension of degree n and T the integral closure of R in L; so T is also a Dedekind domain, L is its field of quotients, $T \cap K = R$.

The relative trace $\operatorname{Tr}_{L|K}$ induces a mapping from $L \times L$ into K, which associates with every pair $(x,y) \in L \times L$ the element $\operatorname{Tr}_{L|K}(xy) \in K$. This is a symmetric K-bilinear form.

If $x \in L$ let $\varphi_x: L \to K$ be the linear form defined by $\varphi_x(y) = \operatorname{Tr}_{L|K}(xy)$ for every $y \in L$. Thus φ_x belongs to L', the dual of the K-vector space L and since $\varphi_{ax} = a\varphi_x$, $\varphi_{x_1+x_2} = \varphi_{x_1} + \varphi_{x_2}$ (for $a \in K$, x_1, x_2, $x \in L$) we have a K-linear mapping $\varphi: L \to L'$. In order that φ_x be the zero mapping we must have $\operatorname{Tr}_{L|K}(xy) = 0$ for every $y \in L$; this means that $x = 0$ since the trace in the separable extension $L \mid K$ is non-degenerate (see Chapter 2, (**10**)). Therefore, φ is an isomorphism between the K-spaces L, L'.

If $\{x_1, \ldots, x_n\}$ is a K-basis of L, let $x_1{}^*, \ldots, x_n{}^* \in L$ be elements such that $\{\varphi_{x_1{}^*}, \ldots, \varphi_{x_n{}^*}\}$ is the dual basis; that is,

$$\varphi_{x_i{}^*}(x_j) = \mathrm{Tr}_{L \mid K}(x_i{}^* x_j) = \delta_{ij} \qquad (\delta_{ii} = 1,\ \delta_{ij} = 0 \text{ when } i \neq j).$$

Thus $\{x_1{}^*, \ldots, x_n{}^*\}$ is also a basis of L, which we call the *complementary basis* of $\{x_1, \ldots, x_n\}$. Let us note here that

$$\mathrm{discr}_{L \mid K}(x_1, \ldots, x_n) \cdot \mathrm{discr}_{L \mid K}(x_1{}^*, \ldots, x_n{}^*) = 1.$$

Indeed, if $\sigma_1, \ldots, \sigma_n$ are the K-isomorphisms of L, if $X = (\sigma_i(x_j))_{i,j}$, $X^* = (\sigma_i(x_j{}^*))_{i,j}$, if X' denotes the transpose of the matrix X then $X^{*\prime} \cdot X = (\mathrm{Tr}_{L \mid K}(x_i{}^* x_j))_{i,j}$. Therefore,

$$\det(X) \cdot \det(X^*) = 1;$$

but

$$\mathrm{discr}_{L \mid K}(x_1, \ldots, x_n) = \det(X)^2$$

and

$$\mathrm{discr}_{L \mid K}(x_1{}^*, \ldots, x_n{}^*) = \det(X^*)^2$$

so

$$\mathrm{discr}_{L \mid K}(x_1, \ldots, x_n) \cdot \mathrm{discr}_{L \mid K}(x_1{}^*, \ldots \ x_n{}^*) = 1.$$

Now we define complementary sets in L. Let M be a subset of L, then $M^* = \{x \in L \mid \mathrm{Tr}_{L \mid K}(xy) \in R \text{ for every } y \in M\}$ is called the *complementary set* of M (with respect to R).

Let us note at once the following properties:

7B. *If M is a subset of L and M^* is the complementary set then:*
(1) *M^* is a module over R; if $T \cdot M \subseteq M$ then M^* is a module over T.*
(2) *If $M_1 \subseteq M_2 \subseteq L$ then $M_2{}^* \subseteq M_1{}^* \subseteq L$.*
(3) *$T \subseteq T^*$.*
(4) *If M is a free R-module with basis $\{x_1, \ldots, x_n\}$ then M^* is a free R-module with basis $\{x_1{}^*, \ldots, x_n{}^*\}$ and $M^{**} = M$.*
Proof:
(1) Let $x_1, x_2 \in M^*$; for every $y \in M$ we have

$$\mathrm{Tr}_{L \mid K}((x_1 + x_2)y) = \mathrm{Tr}_{L \mid K}(x_1 y) + \mathrm{Tr}_{L \mid K}(x_2 y) \in R,$$

so $x_1 + x_2 \in M^*$. If $a \in R$, $x \in M^*$, $y \in M$, then

$$\mathrm{Tr}_{L \mid K}((ax)y) = a\,\mathrm{Tr}_{L \mid K}(xy) \in R$$

so $ax \in M^*$.

Now, let us assume that $T . M \subseteq M$. If $b \in T$, $x \in M^*$ and $y \in M$ then $\text{Tr}_{L|K}((bx)y) = \text{Tr}_{L|K}(x(by)) \in R$ since $by \in M$.

(2) It is obvious.

(3) Since T is integral over R, which is integrally closed, then

$$\text{Tr}_{L|K}(T) \subseteq R.$$

Now, if $x, y \in T$ then $\text{Tr}_{L|K}(xy) \in R$, hence $x \in T^*$; that is, $T \subseteq T^*$.

(4) We have $\text{Tr}_{L|K}(x_i^* x_j) = 0$ when $i \neq j$ and $\text{Tr}_{L|K}(x_i^* x_i) = 1$ for every index $i = 1, \ldots, n$. Hence each x_i^* belongs to M^* and therefore $\Sigma_{i=1}^n Rx_i^* \subseteq M^*$. Conversely, let $\Sigma_{i=1}^n a_i x_i^* \in M^*$, with $a_i \in K$. Then for every $j = 1, \ldots, n$ we have

$$a_j = \text{Tr}_{L|K}\left(\left(\sum_{i=1}^n a_i x_i^*\right) x_j\right) \in R, \quad \text{so } M^* \subseteq \sum_{i=1}^n Rx_i^*.$$

Therefore, $M^{**} = M$. ∎

The next result dates back essentially to Euler:

7C. *Let $L = K(t)$, where t is integral over R and*

$$g = X^n + c_1 X^{n-1} + \cdots + c_n \in R[X]$$

is the minimal polynomial of t over K. Then:

(1) $\text{Tr}_{L|K}\left(\dfrac{t^i}{g'(t)}\right) = 0$ *when $i = 0, 1, \ldots, n - 2$,*

$$\text{Tr}_{L|K}\left(\frac{t^{n-1}}{g'(t)}\right) = 1.$$

(2) $R[t]^* = \dfrac{1}{g'(t)} R[t]$.

Proof:

(1) Let $t = t_1, t_2, \ldots, t_n$ be the conjugates of t over K, which are necessarily distinct and belong to a Galois extension L' of finite degree over K. We shall compute

$$\text{Tr}_{L|K}\left(\frac{t^i}{g'(t)}\right) = \sum_{k=1}^n \frac{t_k^{\,i}}{g'(t_k)}$$

for $i = 0, 1, \ldots, n - 1$.

Since g is the minimal polynomial of t, we have $g = \Pi_{k=1}^n (X - t_k)$, hence $1/g = \Pi_{k=1}^n 1/(X - t_k)$ and we may express the above produce as a sum $\Sigma_{k=1}^n a_k/(X - t_k)$, where the elements a_k will now be determined: from

$1/g = \Sigma_{k=1}^n a_k/(X - t_k)$ we have

$$1 = \sum_{k=1}^n \frac{a_k g}{X - t_k} = \sum_{k=1}^n a_k \left(\prod_{i \neq k} (X - t_i) \right)$$

hence for every $j = 1, \ldots, n$:

$$1 = \sum_{k=1}^n a_k \left(\prod_{i \neq k} (t_j - t_i) \right) = a_j \prod_{i \neq j} (t_j - t_i),$$

thus

$$a_j = \frac{1}{\prod_{i \neq j} (t_j - t_i)} = \frac{1}{g'(t_j)},$$

and we have found that

$$\frac{1}{g} = \sum_{k=1}^n \frac{1}{g'(t_k)(X - t_k)}.$$

By long Euclidean division, we may write

$$\frac{1}{g} = \frac{1}{X^n} + c_1 \frac{1}{X^{n+1}} + c_2 \frac{1}{X^{n+2}} + \cdots.$$

while

$$\sum_{k=1}^n \frac{1}{g'(t_k)(X - t_k)} = \sum_{k=1}^n \frac{1}{g'(t_k)} \left[\frac{1}{X} + \frac{t_k}{X^2} + \frac{t_k^2}{X^3} + \cdots \right].$$

Comparing the two formal power series, we conclude that

$$\sum_{k=1}^n \frac{t_k^i}{g'(t_k)} = 0 \quad \text{for } i = 0, 1, \ldots, n - 2$$

while

$$\sum_{k=1}^n \frac{t_k^{n-1}}{g'(t_k)} = 1.$$

(2) First we show that $t^j/g'(t) \in R[t]^*$ for $j = 0, 1, \ldots, n - 1$. Indeed, if

$$y = \sum_{i=0}^{n-1} a_i t^i \in R[t]$$

then

$$\mathrm{Tr}_{L|K} \left(\frac{t^j}{g'(t)} y \right) = \sum_{i=0}^{n-1} a_i \mathrm{Tr}_{L|K} \left(\frac{t^{j+i}}{g'(t)} \right) = a_{n-1-j} \in R.$$

The elements $t^j/g'(t)$ (for $j = 0, 1, \ldots, n - 1$) are linearly independent over K. For if $y = \Sigma_{j=0}^{n-1} a_j(t^j/g'(t)) = 0$ (with $a_j \in K$) then $0 = \mathrm{Tr}_{L|K}(y) = a_{n-1}$; again $0 = yt = \Sigma_{j=0}^{n-2} a_j(t^{j+1}/g'(t))$, hence $a_{n-2} = 0$. In this way we establish successively that all coefficients a_j are equal to zero.

To prove the inclusion $R[t]^* \subseteq (1/g'(t))R[t]$ let $y \in R[t]^*$, hence we may write $y = \sum_{j=0}^{n-1} a_j(t^j/g'(t))$ because the elements $t^j/g'(t)$ $(j = 0, 1, \ldots, n-1)$ form a basis of L over K. Then

$$\text{Tr}_{L|K}(y) = \sum_{j=0}^{n-1} a_j \, \text{Tr}_{L|K}\left(\frac{t^j}{g'(t)}\right) = a_{n-1},$$

hence $a_{n-1} \in R$ because $y \in R[t]^*$. Similarly,

$$\text{Tr}_{L|K}(yt) = \sum_{j=0}^{n-1} a_j \, \text{Tr}_{L|K}\left(\frac{t^{j+1}}{g'(t)}\right) = a_{n-2} + a_{n-1} \, \text{Tr}_{L|K}\left(\frac{t^n}{g'(t)}\right)$$

$$= a_{n-2} - a_{n-1}\left(\sum_{i=1}^{n} c_i \, \text{Tr}_{L|K}\left(\frac{t^{n-i}}{g'(t)}\right)\right) = a_{n-2} - a_{n-1}c_1$$

because $t^n = -(c_1 t^{n-1} + c_2 t^{n-2} + \cdots + c_n)$ with $c_i \in R$. Since

$a_{n-2} - a_{n-1}c_1 \in R$ then $a_{n-2} \in R$. Proceeding in the same manner, we deduce

that $a_i \in R$ for every $i = 0, 1, \ldots, n-1$ and therefore $R[t]^* = \frac{1}{g'(t)} R[t]$. ∎

7D. *T^* is a fractional ideal of L (with respect to T).*
Proof: It is enough to show that the T-module T^* is finitely generated, because if $b \in T$ is a non-zero common denominator of the generators of T^* then $bT^* \subseteq T$.

Since $R[t]$ is a finitely generated free R-module, it follows from (7A) that $R[t]^*$ is a finitely generated R-module. From $R[t] \subseteq T$ it follows that $T^* \subseteq R[t]^*$; since R is a Dedekind domain, hence a Noetherian ring, we deduce that T^* is also a finitely generated R-module (see Chapter 7, (G) and (D)); a fortiori, T^* is a finitely generated T-module. ∎

Definition 6. The ideal of T equal to the inverse of the fractional ideal T^* is called the *different of T over R*, and denoted by $\Delta(T \mid R)$.

Since $T \subseteq T^*$ then $\Delta(T \mid R)$ is a non-zero integral ideal of T.

Since T is a Dedekind domain, the ideal $\Delta(T \mid R)$ may be written in unique way as $\Delta(T \mid R) = \Pi Q^{s_Q}$ where each Q is a non-zero prime ideal of T and $s_Q \geqslant 0$ is an integer. Moreover, $s_Q > 0$ only for a finite number of prime ideals Q.

The integer s_Q is called *the exponent at Q of the different* $\Delta(T \mid R)$.

There is a situation where the computation of the different offers no difficulty:

7E. *Let $L = K(t)$, where $t \in T$ and $g \in R[X]$ is the minimal polynomial of t over K. Then, $\Delta(T \mid R) = T \cdot g'(t)$ if and only if $T = R[t]$.*

Proof: If $T = R[t]$ we have seen in (7C) that $T^* = (1/g'(t))T$, hence $\Delta(T \mid R) = T \cdot g'(t)$.

Conversely, let $T^* = (1/g'(t))T$ and consider an arbitrary element $z \in T$. Then there exists a polynomial $h \in K[X]$, having degree less than $n = [L:K]$ and such that $z = h(t)$. In the proof of (7C) we have seen that

$$1 = \sum_{k=1}^{n} \frac{g}{g'(t_k) \cdot (X - t_k)} = \sum_{k=1}^{n} \prod_{i \neq k} \frac{X - t_i}{t_k - t_i}$$

where $t = t_1, \ldots, t_n$ are the conjugates of t over K.

The polynomial

$$h_1 = \sum_{k=1}^{n} h(t_k) \prod_{i \neq k} \frac{X - t_i}{t_k - t_i}$$

has degree less than n and

$$h_1(t_j) = \sum_{k=1}^{n} h(t_k) \prod_{i \neq k} \frac{t_j - t_i}{t_k - t_i} = h(t_j)$$

because if $k \neq j$ then $\prod_{i \neq k} (t_j - t_i)/(t_k - t_i) = 0$. So $h = h_1$ since the polynomial $h - h_1$ has already n roots. But

$$\mathrm{Tr}_{L \mid K}\left(\frac{z \cdot g}{g'(t) \cdot (X - t)}\right) = \mathrm{Tr}_{L \mid K}\left(z \prod_{i \neq 1} \frac{X - t_i}{t - t_i}\right) = \sum_{k=1}^{n} h(t_k) \prod_{j \neq k} \frac{X - t_j}{t_k - t_j} = h.$$

Every coefficient of $g/(X - t)$ is in L and integral over R, thus

$$\frac{g}{X - t} \in T[X].$$

Since $z/g'(t) \in T^*$ then

$$h = \mathrm{Tr}_{L \mid K}\left(\frac{z \cdot g}{(g'(t) \cdot (X - t))}\right) \in R[X]$$

and we conclude that $z = h(t) \in R[t]$ proving that $T = R[t]$. ∎

The different satisfies the following characteristic property:

7F. *Let J be a fractional ideal of T. Then $\mathrm{Tr}_{L \mid K}(J) \subseteq R$ if and only if $J \subseteq T^* = \Delta(T \mid R)^{-1}$.*
Proof: If $J \subseteq T^*$ then $\mathrm{Tr}_{L \mid K}(J) \subseteq R$. Conversely, from $J = T \cdot J$ and $\mathrm{Tr}_{L \mid K}(J) \subseteq R$ we deduce that $J \subseteq T^*$. ∎

Another useful property is the transitivity of the different. Let $L' \mid L$ be a separable extension of finite degree, and let T' be a Dedekind domain, having field of quotients equal to L' and equal to the integral closure of T in L'. With these notations:

7G. $\Delta(T' \mid R) = T' \Delta(T \mid R) \cdot \Delta(T' \mid T)$.

Proof: A fractional ideal J' of T' is such that $J' \subseteq \Delta(T' \mid T)^{-1}$ if and only if $T \supseteq \operatorname{Tr}_{L'\mid L}(J')$. This means that

$$\Delta(T \mid R)^{-1} \supseteq \Delta(T \mid R)^{-1} \cdot \operatorname{Tr}_{L'\mid L}(J') = \operatorname{Tr}_{L'\mid L}(T' \Delta(T \mid R)^{-1} \cdot J'),$$

so

$$R \supseteq \operatorname{Tr}_{L\mid K}(\Delta(T \mid R)^{-1})$$

$$\supseteq \operatorname{Tr}_{L\mid K}(\operatorname{Tr}_{L'\mid L}(T' \Delta(T \mid R)^{-1} \cdot J')) = \operatorname{Tr}_{L'\mid K}(T' \Delta(T \mid R)^{-1} \cdot J').$$

Again, this means that $T' \Delta(T \mid R)^{-1} \cdot J' \subseteq \Delta(T' \mid R)^{-1}$, that is,

$$J' \subseteq T' \Delta(T \mid R) \cdot \Delta(T' \mid R)^{-1}.$$

So, we have shown that $\Delta(T' \mid T)^{-1} = T' \Delta(T \mid R) \cdot \Delta(T' \mid R)^{-1}$; that is, $\Delta(T' \mid R) = T' \Delta(T \mid R) \cdot \Delta(T' \mid T)$. $\blacksquare$

We shall apply the theory just developed in two main instances.

Let K be an algebraic number field, $L \mid K$ an extension of degree n, let $R = A$, and $T = B$ be the rings of algebraic integers of K and L respectively.

Definition 7. The different $\Delta(B \mid A)$ is also denoted by $\Delta_{L\mid K}$ and it is called *the different of $L \mid K$*.

In the special case where the ground field is $K = \mathbb{Q}$ the different of $L \mid \mathbb{Q}$ is also called *the absolute different of L*, and sometimes denoted by Δ_L.

Now, let P be a non-zero prime ideal of A, S the set complement of P in A, let $A' = S^{-1}A$, $B' = S^{-1}B$, so B' is the integral closure of the principal ideal domain A' in L and we may take $R = A'$, $T = B'$.

Definition 8. The different $\Delta(B' \mid A')$ is called *the different of $L \mid K$ above P*. Sometimes we denote it by $\Delta_P(L \mid K)$ or simply Δ_P.

We wish to compare these differents.

7H. *With above notations $B' \cdot \Delta_{L\mid K} = \Delta(B' \mid A')$.*

Proof: Let $x \in B' \cdot \Delta_{L\mid K}$; it may be written in the form $x = y/s$ with $y \in \Delta_{L\mid K} = \Delta(B \mid A)$, $s \in S$. Let $z \in B'^*$ (the complementary module of the A'-module B'), so $\operatorname{Tr}_{L\mid K}(zB') \subseteq A'$. We know that B is a finitely generated A-module; let $\{t_1, \ldots, t_m\}$ be a system of generators, let $\operatorname{Tr}_{L\mid K}(zt_i) = a_i/s_i$ with $a_i \in A$, $s_i \in S$. If $s_0 = s_1 \ldots s_m \in S$ then

$$\operatorname{Tr}_{L\mid K}(zs_0t_i) = s_0 \operatorname{Tr}_{L\mid K}(zt_i) \in A$$

for every $i = 1, \ldots, m$. Thus $\operatorname{Tr}_{L\mid K}(zs_0B) \subseteq A$ so $zS_0 \in B^*$ (complementary module of the A-module B); that is, $yzs_0 \in B$ because $y \in \Delta(B \mid A)$. We deduce that $xz = yzs_0/ss_0 \in B'$ showing that $x \in \Delta(B' \mid A')$ and the inclusion $B' \cdot \Delta_{L\mid K} \subseteq \Delta(B' \mid A')$.

Conversely, let $x \in \Delta(B' \mid A')$. B^* is a fractional ideal of B, hence a finitely generated A-module. Let $\{z_1, \ldots, z_m\}$ be a system of generators of the A-module B^*. We have $\mathrm{Tr}_{L\mid K}(z_i B) \subseteq A$, and since $S \subseteq K$ then

$$\mathrm{Tr}_{L\mid K}(z_i B') \subseteq A',$$

so $z_i \in B'^*$ hence $xz_i \in B' = S^{-1}B$; so we may write $xz_i = b_i/s_i$. Let

$$s = s_1 \ldots s_m \in S,$$

then $sxz_i \in B$ for every $i = 1, \ldots, m$, hence also $sxB^* \subseteq B$. This proves that $sx \in \Delta(B \mid A)$ and $x \in B' . \Delta(B \mid A)$. ∎

We are able to compute the different $\Delta_{L\mid K} = \Pi\, Q^{s_Q}$. Let e_Q be the ramification index of Q in $L \mid K$.

7I. *For every non-zero prime ideal Q of B we have $s_Q \geqslant e_Q - 1$. Moreover, $s_Q = e_Q - 1$ if and only if the characteristic of B/Q does not divide the ramification index e_Q.*

Proof: Given the non-zero prime ideal Q_1 of B let $P = Q_1 \cap A$, and let S be the multiplicative set complement of P in A, $A' = S^{-1}A$, $B' = S^{-1}B$. We note $\Delta_{L\mid K} = \Pi\, Q^{s_Q}$, with $s_Q \geqslant 0$ integer, and $BP = \Pi_{i=1}^{g} Q_i^{e_i}$, hence $B'P = \Pi_{i=1}^{g} B'Q_i^{e_i}$. From (7H) we have

$$\Delta(B' \mid A') = B' . \Delta_{L\mid K} = \prod_{i=1}^{g} B'Q_i^{s_i},$$

where $s_i = s_{Q_i}$ (for $i = 1, \ldots, g$). Thus the complementary module of the A'-module B' is $B'^* = \Pi_{i=1}^{g} B'Q_i^{-s_i}$. The inequalities

$$s_i \geqslant e_i - 1 \qquad (i = 1, \ldots, g)$$

hold if and only if $\Pi_{i=1}^{g} B'Q_i^{1-e_i} \subseteq B'^*$.

Thus let $x \in \Pi_{i=1}^{g} B'Q_i^{1-e_i}$; we recall that $P' = A'P$ is a principal ideal, so there exists $t \in K$ such that $P' = A't$; since $B't = B'P = \Pi_{i=1}^{g} B'Q_i^{e_i}$, then $xt \in \Pi_{i=1}^{g} B'Q_i \subseteq \cap_{i=1}^{g} B'Q_i$. This implies that $\mathrm{Tr}_{L\mid K}(xt) \in A'P$. Indeed, we consider the smallest Galois extension $\tilde{L}$ of K containing L and its ring of integers $\tilde{B}$; it follows that $xt \in S^{-1}\tilde{B}\tilde{Q}$ for every prime ideal $\tilde{Q}$ of $\tilde{B}$ such that $\tilde{Q} \cap A = P$. So the same holds for all the conjugates of xt in $\tilde{L} \mid K$ and

$$\mathrm{Tr}_{\tilde{L}\mid K}(xt) = [\tilde{L}:L] . \mathrm{Tr}_{L\mid K}(xt) \in \left(\underset{\tilde{Q}}{\cap}\, S^{-1}\tilde{B}\tilde{Q} \right) \cap A' = A'P,$$

hence

$$\mathrm{Tr}_{L\mid K}(xt) = t . \mathrm{Tr}_{L\mid K}(x) \in A'P = A't.$$

We conclude that $\mathrm{Tr}_{L\mid K}(x) \in A'$.

Now, if $y \in B'$ then xy belongs again to $\Pi_{i=1}^{g} B'Q_i^{1-e_i}$ so $\mathrm{Tr}_{L\mid K}(xy) \in A'$. This shows that $x \in B'^*$.

Now we assume that the characteristic of $B/Q_1 = B'/B'Q_1$ divides the ramification index e_1; we wish to show that

$$J = B'Q_1^{-e_1} \cdot \prod_{i=2}^{g} B'Q_i^{1-e_i} \subseteq B'^*$$

hence $s_1 \geqslant e_1$. Let $x \in J$; by the previous argument $xt \in \cap_{i=2}^{g} B'Q_i$. It follows from **(5G)** that if $\psi: A' \to A'/P'$ and $\psi_i: B' \to B'/B'Q_i$ then

$$\psi(\mathrm{Tr}_{L|K}(xt)) = \sum_{i=1}^{g} e_i \cdot \mathrm{Tr}_{(B'/B'Q_i)|(A'/P')}(\psi_i(xt))$$

$$= e_1 \cdot \mathrm{Tr}_{(B'/B'Q_1)|(A'/P')}(\psi_1(xt)).$$

But e_1 is a multiple of the characteristic of $B'/B'Q_1$, hence $\psi(\mathrm{Tr}_{L|K}(xt)) = 0$, so $t \cdot \mathrm{Tr}_{L|K}(x) = \mathrm{Tr}_{L|K}(xt) \in P' = A't$ and therefore $\mathrm{Tr}_{L|K}(x) \in A'$. Now, if $y \in B'$, then $xy \in J$, so $\mathrm{Tr}_{L|K}(xy) \in A'$, and this shows that $x \in B'^*$.

Conversely, if the characteristic of $B/Q_1 = B'/B'Q_1$ does not divide the ramification index e_1 we proceed as follows. Let $x \in B'$ be an element such that its image $\psi_1(x) \in B'/B'Q_1$ has a non-zero trace. By Chapter 7, **(K)** (applied to the Dedekind domain B') there exists an element $y \in B'$ such that $y - x \in B'Q_1$, and $y \in B'Q_i^{e_i}$ for $i = 2, \ldots, g$. Then by **(5G)**

$$\psi(\mathrm{Tr}_{L|K}(y)) = \sum_{i=1}^{g} e_i \cdot \mathrm{Tr}_{(B'/B'Q_i)|(A'/P')}(\psi_i(y))$$

$$= e_1 \cdot \mathrm{Tr}_{(B'/B'Q_1)|(A'/P')}(\psi_1(x)) \neq 0$$

since e_1 is not a multiple of the characteristic of $B'/B'Q_1$. Therefore

$$\mathrm{Tr}_{L|K}(y) \notin P' = A't$$

and so

$$\mathrm{Tr}_{L|K}\left(\frac{y}{t}\right) = \frac{1}{t}, \qquad \mathrm{Tr}_{L|K}(y) \notin A'.$$

This shows that $y/t \notin B'^*$; since $B't = B'P$ and $y \in B'Q_i^{e_i}$ ($i = 2, \ldots, g$), then $y/t \in B'Q_1^{-e_i}$ hence $B'Q_1^{-e_i}$ is not contained in B'^*, it is not true that $-e_1 \geqslant -s_1$, so $e_1 > s_1 \geqslant e_1 - 1$; therefore, $s_1 = e_1 - 1$. ∎

Theorem 2. *An ideal Q of B is ramified in $L \mid K$ if and only if Q divides the different $\Delta_{L|K}$.*

Proof: We assume that Q is ramified in $L \mid K$, that is $e_Q \geqslant 2$, hence $s_Q \geqslant 1$ and so Q divides the different $\Delta_{L|K}$.

Conversely, if $e_Q = 1$, since the characteristic of B/Q cannot divide e_Q we conclude that $s_Q = e_Q - 1 = 0$, so Q does not divide the different. ∎

From Theorems 1 and 2 we expect a relationship between the relative different $\Delta_{L|K}$ and the relative discriminant $\delta_{L|K}$. In fact, we have:

7J. $N_{L|K}(\Delta_{L|K}) = \delta_{L|K}.$

Proof: Let P be any non-zero prime ideal of A, let S be the multiplicative set, complement of P in A, $A' = S^{-1}A$, $B' = S^{-1}B$. Then B' is a Dedekind domain, having only finitely many prime ideals (by **(4A)**), so B' is a principal ideal domain (by Chapter 7, **(L)**). Let B'^* be the complementary module of the free A'-module B'; it is a fractional ideal of B', hence there exists $y \in L$, $y \neq 0$ such that $B'^* = B'y$, and therefore $B'y^{-1} = \Delta(B' \mid A') = B' . \Delta_{L|K}$ (by **(7H)**). If $\{x_1', \ldots, x_n'\}$ is any basis of the A'-module B' then

$$\{yx_1', \ldots, yx_n'\}$$

is a basis of B'^*. By **(7B)** B'^* has also the complementary basis

$$\{x_1'^*, \ldots, x_n'^*\}$$

over A'. By **(5C)** and **(6C)**

$$\text{discr}_{L|K}(x_1'^*, \ldots, x_n'^*) = \text{discr}_{B'^*|A'}(x_1'^*, \ldots, x_n'^*)$$

and

$$\text{discr}_{L|K}(yx_1', \ldots, yx_n') = \text{discr}_{B'^*|A'}(yx_1', \ldots, yx_n')$$

are associated elements of A'. But

$$\text{discr}_{L|K}(x_1', \ldots, x_n') . \text{discr}_{L|K}(x_1'^*, \ldots, x_n'^*) = 1,$$

and

$$\text{discr}_{L|K}(yx_1', \ldots, yx_n') = [N_{L|K}(y)]^2 . \text{discr}_{L|K}(x_1', \ldots, x_n');$$

therefore, $[\text{discr}_{L|K}(x_1', \ldots, x_n')]^2 . [N_{L|K}(y)]^2$ is a unit of A'; so

$$A' . \text{discr}_{L|K}(x_1', \ldots, x_n') = A' . N_{L|K}(y^{-1}).$$

Now $y = z/a$ where $z \in B$, $a \in S$; since $B'y^{-1} = B'\Delta_{L|K}$ then $B'(Ba) = B'(Bz . \Delta_{L|K})$ hence by **(1A)** $Ba = Bz . \Delta_{L|K}$ and taking norms we have $Aa^n = A . N_{L|K}(z) . N_{L|K}(\Delta_{L|K})$, hence $A'a^n = A'N_{L|K}(z) . A'N_{L|K}(\Delta_{L|K})$ so $A'N_{L|K}(y^{-1}) = A'N_{L|K}(\Delta_{L|K})$.

We have shown that for every basis $\{x_1', \ldots, x_n'\}$ of the A'-module B' we have $A' . \text{discr}_{L|K}(x_1', \ldots, x_n') = A' . N_{L|K}(\Delta_{L|K})$.

This being seen, we show the inclusion $\delta_{L|K} \subseteq N_{L|K}(\Delta_{L|K})$. Let $\{x_1, \ldots, x_n\}$ be any K-basis of L such that each element x_i belongs to B. For every prime ideal P, let $\{x_1', \ldots, x_n'\}$ be any basis of the A'-module B'. We may write

$$x_j = \sum_{i=1}^n a_{ij}'x_i \quad (j = 1, \ldots, n) \text{ with } a_{ij}' \in A'.$$

Thus

$$\text{discr}_{L|K}(x_1, \ldots, x_n) \in A' . \text{discr}_{L|K}(x_1', \ldots, x_n') \subseteq A_P . N_{L|K}(\Delta_{L|K}).$$

This shows that $\delta_{L|K} \subseteq \cap_P A_P . N_{L|K}(\Delta_{L|K}) = N_{L|K}(\Delta_{L|K})$ (see **(4C)**).

Conversely, let P be any non-zero prime ideal of A, let $\delta_{L|K} = P^s \cdot J$, and $N_{L|K}(\Delta_{L|K}) = P^{s'} \cdot J'$, where J, J' are ideals of A, not multiples of P. We shall see that $s \leqslant s'$; since this holds for every prime ideal $P \neq 0$ of A, then $\delta_{L|K}$ divides, hence contains $N_{L|K}(\Delta_{L|K})$.

Now, for every prime ideal $P \neq 0$ of A we choose a basis $\{x_1', \ldots, x_n'\}$ of the A'-module B'; after multiplication with an element of S we may assume that each $x_i' \in B$. Let $A \cdot \mathrm{discr}_{L|K}(x_1', \ldots, x_n') = P^r I$, where I is an ideal of A not multiple of P. By **(4F)** $A' \cdot \mathrm{discr}_{L|K}(x_1', \ldots, x_n') = (A'P)^r$ and since $A' \cdot \mathrm{discr}_{L|K}(x_1', \ldots, x_n') = A' \cdot N_{L|K}(\Delta_{L|K}) = (A'P)^{s'}$ then $s' = r$. But $\delta_{L|K}$ contains, hence divides $A \cdot \mathrm{discr}_{L|K}(x_1', \ldots, x_n')$, so $s \leqslant r = s'$. This concludes the proof. ∎

Using this result and the transitivity of the different, we obtain the transitivity of the discriminant:

7K. *Let $K \subseteq L \subseteq L'$ be algebraic number fields. Then*

$$\delta_{L'|K} = (\delta_{L|K})^{[L':L]} \cdot N_{L|K}(\delta_{L'|L}).$$

Proof: From **(7G)** we have the following relation between the differents: $\Delta_{L'|K} = B'\Delta_{L|K} \cdot \Delta_{L'|L}$, where B' is the ring of integers of L'. Taking norms, it follows from **(1H)**, **(1J)**, and **(7J)** that

$$N_{L'|K}(\Delta_{L'|K}) = N_{L'|K}(B'\Delta_{L|K}) \cdot N_{L'|K}(\Delta_{L'|L})$$

so

$$\begin{aligned}
\delta_{L'|K} &= N_{L|K}[N_{L'|L}(B'\Delta_{L|K})] \cdot N_{L|K}[N_{L'|L}(\Delta_{L'|L})] \\
&= N_{L|K}(\Delta_{L|K}^{[L':L]}) \cdot N_{L|K}(\delta_{L'|L}) \\
&= (\delta_{L|K})^{[L':L]} \cdot N_{L|K}(\delta_{L'|L}).
\end{aligned}$$

As an application of these computations, we may establish the following general result, which is due to Ore:

7L. *Let K be an algebraic number field, let W be the multiplicative group of roots of unity in K, and w the number of elements in W. Then w divides $2\delta_K$.*
Proof: We have seen in Chapter 9, **(C)** that W is a cyclic group. Let $w = \prod_{i=1}^{s} p_i^{k_i}$, where $k_i \geqslant 1$ and $p_1, \ldots, p_s$ are distinct prime numbers.

Let η be a primitive root of unity of order w, let $w_i = \prod_{j \neq i} p_j^{k_j}$; then $\zeta_i = \eta^{w_i}$ is a primitive root of unity of order $p_i^{k_i}$ belonging to K (for every $i = 1, \ldots, s$).

It is enough to show that if K contains a primitive p^kth root of unity ζ then p^k divides $2\delta_K$; this will imply that $w = \prod_{i=1}^{s} p_i^{k_i}$ divides $2\delta_K$.

By **(7K)** the discriminant $\delta_{\mathbb{Q}(\zeta)}$ divides δ_K, so it sufficient to prove that p^k divides $2\delta_{\mathbb{Q}(\zeta)}$.

We have seen in **(3A)**, **(3B)** that p is the only ramified prime in $\mathbb{Q}(\zeta)$ and $Ap = (A\xi)^{\varphi(p^k)}$, thus the different of $\mathbb{Q}(\zeta) \mid \mathbb{Q}$ is $\Delta_{\mathbb{Q}(\zeta)} = (A\xi)^s$ and by **(7I)** $\varphi(p^k) - 1 \leqslant s$. Taking norms we have $|\delta_{\mathbb{Q}(\zeta)}| = p^s$ (by **(7J)** and **(3A)**.

But

$$s \geqslant \varphi(p^k) - 1 = p^{k-1}(p-1) - 1 \geqslant [1 + (k-1)(p-1)](p-1) - 1;$$

this last quantity is at least equal to k when $p > 2$ and equal to $k - 1$ when $p = 2$. In any case p^k divides $2p^s = 2\,|\delta_{\mathbb{Q}(\zeta)}|$. $\blacksquare$

We indicate now a system of generators for the different $\Delta_{L \mid K}$, which will allow its explicit computation in certain cases.

If $t \in L$ and g is the minimal polynomial of t over K then

$$g'(t) = \prod_{i=2}^{n} (t - \sigma_i(t))$$

where $\sigma_1 = \varepsilon, \sigma_2, \ldots, \sigma_n$ are the K-isomorphisms of L.

It is customary to call $g'(t)$ *the different of the element t in L $\mid$ K*. The reason for the name is quite simple: $g'(t) \neq 0$ exactly when t is different from all its conjugates, that is, t is a primitive element of $L \mid K$.

We want to compare the fractional ideals B^* and $A[t]^*$. As we know, $B^* \subseteq A[t]^* = A[t]/g'(t)$. Let $F_t = \{x \in B \mid x \,.\, A[t]^* \subseteq B^*\}$. F_t is an ideal of B called *the conductor of A[t] in B* and equal to $F_t = g'(t) \,.\, B^*$.

Indeed, $g'(t) \,.\, A[t]^* = A[t] \subseteq B$, hence

$$(g'(t) \,.\, B^*) \,.\, A[t]^* = g'(t) \,.\, A[t]^* \,.\, B^* \subseteq B^*.$$

Conversely, if $x \in B$, $x \,.\, A[t]^* \subseteq B^*$ then $x \in x \,.\, A[t] \subseteq g'(t) \,.\, B^*$.

The conductor has the following characteristic property:

7M. *F_t is the largest ideal of B contained in $A[t]$.*

Proof: Since $B^* \subseteq A[t]^* = A[t]/g'(t)$, then $F_t = g'(t) \,.\, B^* \subseteq A[t]$.

Now, let J be an ideal of B which is contained in $A[t]$ and $x \in J$. By **(7C)** $\mathrm{Tr}_{L \mid K}(x/g'(t)) \in A$, hence $x/g'(t) \in B^*$ by definition of B^*. Therefore, $x \in g'(t) \,.\, B^* = F_t$. $\blacksquare$

7N. *$\Delta_{L \mid K}$ is the ideal of B generated by the differents $g'(t)$ of all the elements $t \in B$.*

Proof: If $t \in B$ then $B^* \subseteq A[t]^* = A[t]/g'(t) \subseteq B \,.\, (1/g'(t))$ hence

$$g'(t) \,.\, B^* \subseteq B \quad \text{so} \quad g'(t) \in \Delta_{L \mid K}.$$

To show the converse, it is enough to establish that for every non-zero prime ideal Q of B there exists an element $t \in B$ such that Q does not contain

F_t. If this has been shown, the ideal $\Sigma_{t\in B} F_t$ generated by $\bigcup_{t\in B} F_t$ is equal to B. Hence

$$\Delta_{L|K} = \Delta_{L|K} . B = \Delta_{L|K} \left(\sum_{t\in B} F_t\right) = \sum_{t\in B} \Delta_{L|K} . F_t$$

$$= \sum_{t\in B} \Delta_{L|K} . g'(t) . B^* = \sum_{t\in B} B . g'(t).$$

So, let us prove the above assertion.

Let $Q \cap A = P$ and $B . P = Q^e . J$, where $e \geqslant 1$, Q does not divide J. First, we show: It is possible to choose an element $t \in J$, $t \notin Q$ such that:

(1) The image $\bar{t}$ of t in B/Q is a generator of the multiplicative cyclic group of non-zero elements of the finite field B/Q.

(2) For every integer $l > 0$ and for every $b \in B$ there exists $c \in A[t]$ such that $b - c \in Q^l$.

Indeed, let $u \in B$, $u \notin Q$ be such that the image $\bar{u}$ of u in B/Q is a generator of the multiplicative cyclic group of non-zero elements of B/Q. If $v = N(Q) = \#(B/Q)$ then $u^{v-1} \equiv 1 \pmod{Q}$, hence $u^v \equiv u \pmod{Q}$. If it happens that $u^v \equiv u \pmod{Q^2}$, let $v \in Q$, $v \notin Q^2$; then $\overline{u + v} = \bar{u}$ and $(u + v)^v = u^v + vu^{v-1}v + \cdots + v^v \in u^v + Q^2$ because $N(Q) \in Q$; it follows that $(u + v)^v \equiv (u + v) \pmod{Q}$, and $(u + v)^v - (u + v) \in (u^v - u) - v + z$ with $z \in Q^2$ (since $v \geqslant 2$), hence taking $u_1 = u + v$, it satisfies the conditions: $u_1{}^v - u_1 \in Q$, $u_1{}^v - u_1 \notin Q^2$. Now, since Q does not divide J then $B = Q^2 + J$, hence we may write $u_1 = v_1 + t$, with $v_1 \in Q^2$, $t \in J$; so the image of t in B/Q is $\bar{t} = \bar{u}_1 = \bar{u}$, therefore t is a generator of the multiplicative cyclic group of non-zero elements of B/Q, so $t \notin Q$. Moreover, by the same arguments as before,

$$t^v - t = (u_1{}^v - u_1) - v_1 + z_1,$$

with $z_1 \in Q^2$; from $v_1 \in Q^2$ we conclude that $t^v - t \notin Q^2$.

Now, let S be a system of representatives of B modulo Q. If $w = t^v - t$ then for every $l > 0$,

$$R = \{s_0 + s_1 w + s_2 w^2 + \cdots + s_{l-1} w^{l-1} \mid s_0, s_1, \ldots, s_{l-1} \in S\}$$

is a system of representatives of B modulo Q (see Chapter 8, (**B**)). In particular, we may take $S = \{1, t, t^2, \ldots, t^{v-1}\}$. Then for every element $b \in B$ there exists a unique element $c = s_0 + s_1 w + \cdots s_{l-1} w^{l-1} \in A[t]$, with each $s_i \in \{1, t, \ldots, t^{v-1}\}$, such that $b - c \in Q^l$.

Once the choice of t, satisfying conditions (1) and (2) has been made, we proceed in the following way to prove that Q does not contain the conductor F_t.

Let $a \in B . g'(t) \cap A$, $a \neq 0$; let $Aa = P^r . I$, where $r \geqslant 0$ and P does not divide the ideal I. If h is the class number of K, we obtain principal ideals by considering hth powers of ideals; let $Aa_1 = Aa^h$, $P^{rh} = Aa_2$, hence $a_1 = a_2a_3$, where $a_3 \in I^h$. Moreover, $a_3 \notin P$, since the exact power of P which divides a_1 is rh.

Let us show that the principal ideal $B(a_3t^{rh})$ is contained in $A[t]$. Given $b \in B$, let $c \in A[t]$ be an element such that $b - c \in Q^{erh}$. Then $ba_3t^{rh} = (b - c)a_3r^{rh} + ca_3t^{rh}$. As $ca_3t^{rh} \in A[t]$ it is enough to show that

$$(b - c)a_3t^{rh} \in A[t].$$

Now, we have

$$B(b - c)a_3t^{rh} = \frac{Ba_2a_3(b - c)t^{rh}}{B \cdot P^{rh}} \subseteq \frac{Ba_1 \cdot Q^{erh} \cdot Bt^{rh}}{Q^{erh} \cdot J^{rh}}$$

$$\subseteq Ba^h \cdot B \subseteq g'(t) \cdot B \subseteq A[t]$$

because $t \in J$ and $g'(t)B^* \subseteq A[t]$. Therefore, $B(a_3t^{rh}) \in A[t]$. It follows from (7M) that $B(a_3t^{rh}) \subseteq F_t$.

To conclude, we note that $a_3 \notin P$, so $a_3 \notin Q$ and $t \notin Q$, hence $a_3t^{rh} \notin Q$, thus F_t is not contained in Q. ∎

One might ask whether a similar result holds for the relative discriminant. More precisely, for every primitive integral element x of $L \mid K$ we may consider the discriminant $\delta(x) = \text{discr}_{L|K}(1, x, x^2, \ldots, x^{n-1})$.

By definition $\delta(x) \in \delta_{L|K}$ and we wish to compare the ideal $\delta_{L|K}$ of A with the ideal $\delta^*_{L|K}$ which is generated by all the above elements $\delta(x)$. We quote without proof the following results due to Hensel:

For every x we have an integral ideal I_x such that $\delta(x) = I_x^2 \cdot \delta_{L|K}$. A non-zero prime ideal P of A which divides I_x for every primitive integral element x is called an *unessential factor of the discriminant*. In order that $\delta_{L|K} = \delta(x)$ for some primitive integral element x it is necessary and sufficient that there exists no unessential factor for the discriminant.

A necessary and sufficient condition for a prime ideal P of A not to be an unessential factor of the discriminant $\delta_{L|K}$ is the following:

$$g(f) \leqslant \frac{1}{f} \sum_{d \mid f} \mu\left(\frac{f}{d}\right) N(P)^d$$

(for any natural number $f \geqslant 1$) where $g(f)$ denotes the number of prime ideals Q of B such that $Q \cap A = P$ and Q has inertial degree f over P.

It follows that if P is an unessential factor of $\delta_{L|K}$ then $N(P) < [L:K] = n$. The converse is also true when there exist n distinct prime ideals Q_i of B such that $Q_i \cap A = P$.

So, if $[L:K] = 2$, then there are no unessential factors for $\delta_{L|K}$. However, Dedekind has shown the existence of unessential prime factors for the discriminant of a cubic field (see Chapter 13, Example 3).

We conclude this section with results relating the degree, rings, integers and discriminants of fields and their compositum.

7O. *Let K_1, K_2 be two algebraic number fields, extensions of the field K, let $L = K_1K_2$ be their compositum; let P be a non-zero prime ideal of the ring of integers A of K. Then:*

(1) *P is unramified in $L \mid K$ if and only if it is unramified in $K_1 \mid K$ and in $K_2 \mid K$.*

(2) *$\Delta_{L \mid K_2}$ divides $B\Delta_{K_1 \mid K}$, $\Delta_{L \mid K_1}$ divides $B\Delta_{K_2 \mid K}$ (where B is the ring of integers of L).*

(3) *$N_{K_2 \mid K}(\delta_{L \mid K_2})$ divides $\delta_{K_1 \mid K}^{[L:K_1]}$ and $N_{K_1 \mid K}(\delta_{L \mid K_1})$ divides $\delta_{K_2 \mid K}^{[L:K_2]}$.*

Proof: Clearly, if P is unramified in $L \mid K$ then it is also unramified in $K_1 \mid K$ and in $K_2 \mid K$.

Conversely, we assume that P is ramified in $L \mid K$ and unramified in $K_2 \mid K$. By Theorem 1, P divides $\delta_{L \mid K}$ and P does not divide $\delta_{K_2 \mid K}$.

By (7K) $\delta_{L \mid K} = (\delta_{K_2 \mid K})^{[L:K_2]} \cdot N_{K_2 \mid K}(\delta_{L \mid K_2})$, hence P divides

$$N_{K_2 \mid K}(\delta_{L \mid K_2}) = N_{K_2 \mid K}(N_{L \mid K_2}(\Delta_{L \mid K_2})) = N_{L \mid K}(\Delta_{L \mid K_2}).$$

Now we show that $\Delta_{L \mid K_2}$ divides $B\Delta_{K_1 \mid K}$; that is, $\Delta_{K_1 \mid K} \subseteq \Delta_{L \mid K_2}$ (where B is the ring of integers of L). This will follow from (7N). Let $t \in A_1 \subseteq B$ (where A_i is the ring of integers of K_i, $i = 1, 2$), let $g \in K[X]$ be the minimal polynomial of t over K, let $h \in K_2[X]$ be the minimal polynomial of t over K_2, so h divides g in $K_2[X]$, that is there exists $k \in K_2[X]$ such that $g = hk$; but t is an integer so $g \in A[X]$, $h \in A_2[X]$ and h is monic, so $k \in A_2[X]$. It follows that $g'(t) = h'(t) \cdot k(t)$ with $h'(t) \in \Delta_{L \mid K_2}$, $k(t) \in B$; by (7N) $\Delta_{K_1 \mid K} \subseteq \Delta_{L \mid K_2}$.

So P divides

$$N_{L \mid K}(B\Delta_{K_1 \mid K}) = N_{K_1 \mid K}(N_{L \mid K_1}(B\Delta_{K_1 \mid K})) = N_{K_1 \mid K}(\Delta_{K_1 \mid K})^{[L:K_1]} = \delta_{K_1 \mid K}^{[L:K_1]},$$

hence P divides $\delta_{K_1 \mid K}$. By Theorem 1, P is ramified in $K_1 \mid K$.

In the course of the proof we have also established (2).

Finally, (3) follows from (2) and (7J) by taking norms: $N_{K_2 \mid K}(\delta_{L \mid K_2})$ divides

$$N_{K_2 \mid K}(N_{L \mid K_2}(B\Delta_{K_1 \mid K})) = N_{L \mid K}(B\Delta_{K_1 \mid K}) = N_{K_1 \mid K}(N_{L \mid K_1}(B\Delta_{K_1 \mid K}))$$
$$= N_{K_1 \mid K}(\Delta_{K_1 \mid K})^{[L:K_1]} = \delta_{K_1 \mid K}^{[L:K_1]}.$$

As a corollary, we have:

7P. *Let $K \subseteq L \subseteq L'$ be algebraic number fields such that L' is the smallest field containing L for which $L' \mid K$ is a Galois extension. Let P be a non-zero prime ideal of the ring of integers of K. Then P is unramified in $L \mid K$ if and only if P is unramified in $L' \mid K$.*

Proof: We apply the preceding result for the fields $L = L_1, L_2, \ldots, L_m$, where each L_i is a conjugate of L over K, noting that L' is the compositum of these fields.   ∎

For algebraic number fields with relatively prime absolute discriminants, we have:

7Q. *Let K_1, K_2 be algebraic number fields of degree n_1, n_2 respectively and such that δ_{K_1}, δ_{K_2} are relatively prime. Let $L = K_1 K_2$. Then:*
(1) $[L:\mathbb{Q}] = n_1 n_2$
(2) $\delta_L = \delta_{K_1}^{\ n_2}\delta_{K_2}^{\ n_1}$
(3) *If A_1, A_2, B are the rings of integers of K_1, K_2, L respectively, then $B = A_1 A_2$; if $\{x_1, \ldots, x_{n_1}\}$ is an integral basis of K_1, if $\{y_1, \ldots, y_{n_2}\}$ is an integral basis of K_2 then $\{x_1 y_1, \ldots, x_{n_1}y_{n_2}\}$ is an integral basis of L.*

Proof:
(1) We have $[L:\mathbb{Q}] = [K_2:\mathbb{Q}] \cdot [L:K_2] = n_2[L:K_2]$. If $[L:\mathbb{Q}] < n_1 n_2$ then $[L:K_2] < n_1$. Let $K_1 = \mathbb{Q}(t)$, let $g \in \mathbb{Q}[X]$ be the minimal polynomial of t over $\mathbb{Q}$, so $\deg(g) = n_1$. Since $[L:K_2] < n_1$, the minimal polynomial h of t over K_2 has degree smaller than n_1 and its divides g; let K' be the subfield of K_2 generated by the coefficients of h, so K' is not equal to $\mathbb{Q}$. From $K' \subseteq K_2$ we deduce that $\delta_{K'}$ divides δ_{K_2}.

On the other hand, the coefficients of h are elementary symmetric functions of the roots of h, which are among the conjugates of t; thus $h \in K_1'[X]$, where K_1' is the smallest Galois extension of $\mathbb{Q}$ containing K_1. Thus $K' \subseteq K_1'$ and again $\delta_{K'}$ divides $\delta_{K_1'}$.

If p is a prime number dividing $\delta_{K'}$, it divides δ_{K_2} and $\delta_{K_1'}$; by Theorem 1 and (**7P**) p divides δ_{K_1}, which is against the hypothesis.

Thus $|\delta_{K'}| = 1$ and by Chapter 9, (**K**) we conclude that $K' = \mathbb{Q}$, a contradiction. This proves (1).

(2) We have $\delta_L = N_{K_1|\mathbb{Q}}(\delta_{L|K_1}) \cdot \delta_{K_1}^{\ n_2} = N_{K|\mathbb{Q}}(\delta_{L|K_2}) \cdot \delta_{K_2}^{\ n_1}$ by (**7K**). Hence $\delta_{K_1}^{\ n_2}$ and $\delta_{K_2}^{\ n_1}$ divide δ_L, and by hypothesis $\delta_{K_1}^{\ n_2}\delta_{K_2}^{\ n_1}$ divides δ_L.

On the other hand, from (**7O**) we know that $N_{K_2|\mathbb{Q}}(\delta_{L|K_2})$ divides $\delta_{K_1}^{[L:K_1]} = \delta_{K_1}^{\ n_2}$ by (1); hence $\delta_L = N_{K_2|\mathbb{Q}}(\delta_{L|K_2}) \cdot \delta_{K_2}^{\ n_1}$ divides $\delta_{K_1}^{\ n_2}\delta_{K_2}^{\ n_1}$ and this establishes the equality.

(3) Let $A_1 A_2$ denote the smallest subring of L containing A_1 and A_2; so $A_1 A_2 \subseteq B$.

$A_1 A_2$ has the $\mathbb{Z}$-basis $\{x_1 y_1, \ldots, x_{n_1}y_{n_2}\}$. We shall compute the discriminant of this basis and show that it is equal to $\delta_{K_1}^{\ n_2} \cdot \delta_{K_2}^{\ n_1} = \delta_L$. Expressing the elements of the integral basis of $A_1 A_2$ in terms of an integral basis $\{z_1, \ldots, z_{n_1 n_2}\}$ of B as linear combinations with coefficients in $\mathbb{Z}$ and taking the discriminants, it follows from (**6A**) that the matrix of the linear transformation has determinant with absolute value equal to 1. Therefore $z_1, \ldots, z_{n_1 n_2}$

are expressible in terms of $x_i y_j$ with coefficients in $\mathbb{Z}$; this will establish that $A_1 A_2 = B$ and (3).

For the computation of $\operatorname{discr}_{L|\mathbb{Q}}(x_1 y_1, \ldots, x_{n_1} y_{n_2})$ we observe that if σ is any isomorphism of L, if σ_{K_1}, σ_{K_2} denote the restrictions of σ to K_1 and K_2, then the mapping $\sigma \to (\sigma_{K_1}, \sigma_{K_2})$ is injective, because $L = K_1 K_2$, and also surjective since $[L:\mathbb{Q}] = n_1 n_2$. Hence,

$$\operatorname{discr}_{L|\mathbb{Q}}(x_1 y_1, \ldots, x_{n_1} y_{n_2}) = [\det(\sigma_i \tau_j(x_k y_l))]^2,$$

where $\sigma_1, \ldots, \sigma_{n_1}$ are the isomorphisms of K_1 and $\tau_1, \ldots, \tau_{n_2}$ are the isomorphisms of K_2.

We have to compute the determinant of a matrix which is the Kronecker product of the matrices

$$(\sigma_i(x_k))_{i,k=1,\ldots,n_1} \quad \text{and} \quad (\tau_j(y_l))_{j,l=1,\ldots,n_2}.$$

As is well known, we obtain the result

$$[\det(\sigma_i(x_k))]^{2n_2} \times [\det(\tau_j(y_l))]^{2n_1} = \delta_{K_1}{}^{n_2} \cdot \delta_{K_2}{}^{n_1}. \qquad \blacksquare$$

EXERCISES

1. Let $K = \mathbb{Q}(x)$, where $x^3 - x^2 - 2x + 8 = 0$. If $y = (x^2 - x)/2$, compute its characteristic polynomial in $K \mid \mathbb{Q}$ and its minimal polynomial over $\mathbb{Q}$.

2. Give an independent proof of Theorem 1 for the case of a quadratic field $K = \mathbb{Q}(\sqrt{d})$.

3. Let p, q be distinct prime numbers, ζ a primitive pth root of unity, η a primitive qth root of unity.
   (a) Find an integral basis for $K = \mathbb{Q}(\zeta, \eta)$ and the discriminant of this field.
   (b) Let K' be the maximal real subfield of K. Show that the relative different $\Delta_{K|K'}$ is the unit ideal.

4. Let p, q be distinct prime numbers, $p \equiv 1 \pmod 4$, $q \equiv 1 \pmod 4$. Let $K = \mathbb{Q}(\sqrt{p}, \sqrt{q})$.
   (a) Find an integral basis and the discriminant of K.
   (b) Let $K' = \mathbb{Q}(\sqrt{pq})$; show that the relative different $\Delta_{K|K'}$ is the unit ideal.

5. Applying Hensel's criterion for the existence of an unessential factor of the discriminant, show the assertions of the text:
   (a) If P is an unessential factor of $\delta_{L|K}$ then $N(P) < [L:K] = n$.
   (b) If $N(P) < n$ and BP is decomposed into the product of n distinct prime ideals of B then P is an unessential factor.

6. Let K be an algebraic number field, A the ring of integers, and R a subring of A. Show that there exists an ideal F of A such that:

(a) $F \subseteq R$.

(b) If I is an ideal of A such that $I \subseteq R$ then $I \subseteq F$.

F is called the *conductor* of R in A.

7. Let K be an algebraic number field, A the ring of integers, R a subring of A, F the conductor of R in A. If I is an ideal of R let AI denote the ideal of A generated by I. I is said to be a *regular ideal* of R when $gcd(AI,F) = A$. Show:

(a) If J is an ideal of A such that $gcd(J,F) = A$ there exists an ideal I of R such that $J = AI$.

(b) If I, I' are regular ideals of R then $A(I \cdot I') = AI \cdot AI'$.

(c) The number of congruence classes modulo F of elements $a \in R$ such that $gcd(Aa,F) = A$ divides $\varphi(F)$ (see Chapter 8, Exercise 2).

CHAPTER 11

The Ramification of Prime Ideals in Galois Extensions

Let K be an algebraic number field, $L \mid K$ a finite extension of degree n, and as before, let A,B be respectively the rings of integers of K,L. Let $\underline{P}$ be a prime ideal of A, and let $B\underline{P} = \Pi_{i=1}^{g} P_i^{e_i}$ be the decomposition of $B\underline{P}$ into product of prime ideals, with $f_i = [B/P_i : A/\underline{P}]$. We shall study in more detail how this decomposition takes place. This has been done by Hilbert, assuming that $L \mid K$ is a Galois extension.

Accordingly, let $\mathscr{K} = G(L \mid K)$ be the Galois group of $L \mid K$, so $\mathscr{K}$ has n elements, the K-automorphisms of L. We shall make appeal to the discussion in Chapter 10, preceding and including (1C), (1D), and (1L).

We shall also adopt the following notation: $\bar{K} = A/\underline{P}$, $\bar{L}_i = B/P_i$, each field $\bar{L}_i$ is isomorphic with the extension of degree f_i of the finite field $\bar{K}$. Since $f_1 = \cdots = f_g$, all the fields $\bar{L}_i$ are actually ismorphic; however, since there is no natural isomorphism between these fields, we shall not identify them.

Definition 1. With preceding notations, the subgroup $\mathscr{Z}_i$ of $\mathscr{K}$, defined by $\mathscr{Z}_i = \{\sigma \in \mathscr{K} \mid \sigma(P_i) = P_i\}$ is called the *decomposition group of P_i in the extension $L \mid K$.* The field of invariants of $\mathscr{Z}_i$ is denoted by Z_i and called *the decomposition field of P_i in the extension $L \mid K$.*

If necessary, we may also use the following notations:

$$\mathscr{Z}_i = \mathscr{Z}(P_i \mid \underline{P}) = \mathscr{Z}_{P_i}(L \mid K),$$
$$Z_i = Z(P_i \mid \underline{P}) = Z_{P_i}(L \mid K).$$

A. *The subgroups $\mathscr{Z}_1, \ldots, \mathscr{Z}_g$ of $\mathscr{K}$ are conjugate (by inner automorphisms of $\mathscr{K}$). In particular, if $\mathscr{K}$ is an Abelian group, then $\mathscr{Z}_1 = \cdots = \mathscr{Z}_g$.*
Proof: Let P_i,P_j be distinct prime ideals of B such that $P_i \cap A = P_j \cap A = \underline{P}$. Since $\mathscr{K}$ acts transitively on the set of prime ideals $\{P_1, \ldots, P_g\}$, there exists $\sigma \in \mathscr{K}$ such that $\sigma(P_i) = P_j$. Then $\mathscr{Z}_i = \sigma^{-1}\mathscr{Z}_j\sigma$. ∎

B. *For every integer $i = 1, \ldots, g$ we have: $[Z_i : K] = (\mathscr{K} : \mathscr{Z}_i) = g$.*
Proof: This proof has already been made in Chapter 10, (3B), while discussing the cyclotomic field. We repeat it for the convenience of the reader. We have $\sigma\mathscr{Z}_i = \tau\mathscr{Z}_i$ (with $\sigma,\tau \in \mathscr{K}$) if and only if $\sigma(P_i) = \tau(P_i)$. morphism of $\bar{L}$; moreover $\overline{\sigma . \tau} = \bar{\sigma} . \bar{\tau}$ for any $\sigma,\tau \in \mathscr{Z}$, and therefore we have a group-homomorphism.

221

In fact, if $\sigma \mathcal{L}_i = \tau \mathcal{L}_i$, then $\sigma^{-1}\tau \in \mathcal{L}_i$, so $\sigma^{-1}\tau(P_i) = P_i$, hence $\tau(P_i) = \sigma(P_i)$. Conversely, if $\sigma(P_i) = \tau(P_i)$, then $\sigma^{-1}\tau \in \mathcal{L}_i$ hence $\tau \mathcal{L}_i = \sigma \mathcal{L}_i$.

Thus, the number g of distinct prime ideals P_i is the same as the number of distinct cosets modulo $\mathcal{L}_i$; that is, the index $(\mathcal{K} : \mathcal{L}_i)$. From Galois theory, we have $(\mathcal{K} : \mathcal{L}_i) = [Z_i : K]$. ∎

The decomposition field has the following minimality property:

C. *If $Q_i = P_i \cap Z_i$ (prime ideal of the ring of integers $B \cap Z_i$ of Z_i), then P_i is the only ideal of the ring of integers of L which extends Q_i. Conversely, if Z_i' is a field, $K \subseteq Z_i' \subseteq L$, if $Q_i' = P_i \cap Z_i'$ and P_i is the only extension of Q_i' to L, then $Z_i \subseteq Z_i'$.*

Proof: We have $\mathcal{L}_i = G(L \mid Z_i)$; by Chapter 12, **(1C)**, $\mathcal{L}_i$ acts transitively on the set of prime ideals of B extending Q_i; but, by definition $\sigma(P_i) = P_i$ for every $\sigma \in \mathcal{L}_i$, thus P_i is the only extension of Q_i.

Next, if P_i is the only extension of $Q_i' = P_i \cap Z_i'$, then every element of the Galois group $\mathcal{L}_i' = G(L \mid Z_i')$ fixes P_i, hence belongs to $\mathcal{L}_i$; thus, we have the opposite inclusion for the fixed fields: $Z_i' \supseteq Z_i$. ∎

We fix our attention on one of the prime ideals P_i, which we shall denote by P for simplicity.

Let $Z = Z_P(L \mid K)$, $\mathcal{L} = \mathcal{L}_P(L \mid K)$, $\bar{L} = B/P$, $\mathcal{K} = G(\bar{L} \mid \bar{K})$. We denote also by $B_Z = B \cap Z$ the ring of integers of Z, by $P_Z = P \cap Z$ the prime ideal of B_Z defined by P and by $\bar{Z} = B_Z/P_Z$, the corresponding residue class field; accordingly, let e be the ramification index and f the inertial degree of P over $\underline{P}$.

D. (1) *$\bar{Z} = \bar{K}$, so the inertial degrees are $f(P_Z \mid \underline{P}) = 1, f(P \mid P_Z) = f$; the ramification indices are $e(P_Z \mid \underline{P}) = 1, e(P \mid P_Z) = e$.*

(2) *The mapping $\sigma \in \mathcal{L} \to \bar{\sigma} \in G(\bar{L} \mid \bar{K})$ is a group-homomorphism onto $G(\bar{L} \mid \bar{K})$, having kernel equal to the normal subgroup*

$$\mathcal{T} = \{\sigma \in \mathcal{L} \mid \sigma(x) \equiv x \pmod{P} \text{ for every element } x \in B\}.$$

Proof: (1) By the fundamental relation and **(C)**, we have

$$[L : Z] = e(P \mid P_Z) \cdot f(P \mid P_Z).$$

On the other hand, $[L : K] = efg$ and by **(B)**, $[Z : K] = g$. Therefore, $ef = e(P \mid P_Z) \cdot f(P \mid P_Z)$. By the transitivity of the ramification index and inertial degree, we must have $e(P \mid P_Z) = e$, $f(P \mid P_Z) = f$ and $e(P_Z \mid \underline{P}) = 1$, $f(P_Z \mid \underline{P}) = 1$. Therefore,

$$[\bar{L} : \bar{Z}] = f(P \mid P_Z) = f = [\bar{L} : \bar{K}] \text{ implies that } \bar{Z} = \bar{K}.$$

(2) If $\sigma \in \mathcal{L}$ then $\sigma(P) = P$ hence σ induces the mapping $\bar{\sigma} : \bar{L} \to \bar{L}$, defined by $\bar{\sigma}(\bar{x}) = \overline{\sigma(x)}$ for every $x \in B$. It is immediate that $\bar{\sigma}$ is a $\bar{Z}$-auto-

Now, we shall prove that the image is equal to $G(\bar{L} \mid \bar{K})$. Since $\bar{K}$ is a finite field, there exists $b \in B$ such that $\bar{L} = \bar{K}(\bar{b})$. If $\xi \in G(\bar{L} \mid \bar{K})$ then $\xi(\bar{b})$ is a conjugate of $\bar{b}$ over $\bar{K}$.

Let h be the minimal polynomial of b over Z; since $L \mid Z$ is a Galois extension, all the conjugates of b over Z are still in L, and in fact in B; thus h decomposes as $h = \Pi_{\sigma \in \mathscr{L}}(X - \sigma(b))$; considering the images of the co-efficients by the canonical mapping $B \to \bar{L}$ (which extends $B_Z \to \bar{Z} = \bar{K}$), we have $\bar{h} = \Pi(X - \overline{\sigma(b)}) \in \bar{K}[X]$; of course, $\bar{b}$ is among the roots of $\bar{h}$; the conjugates of $\bar{b}$ over $\bar{K}$ are the roots of its minimal polynomial, which divides $\bar{h}$, thus the conjugates of $\bar{b}$ are among the elements $\overline{\sigma(b)} \in \bar{L}$. In particular $\xi(\bar{b}) = \overline{\sigma(b)} = \bar{\sigma}(\bar{b})$, for some $\sigma \in \mathscr{L}$, and therefore ξ and $\bar{\sigma}$ must coincide on every element of $\bar{L}$.

The kernel of the group-homomorphism is obviously the set of all $\sigma \in \mathscr{L}$ such that $\bar{\sigma}(\bar{x}) = \bar{x}$ for every $\bar{x} \in \bar{L}$, that is, $\sigma(x) \equiv x \pmod{P}$ for every $x \in B$. ∎

Thus, we have the group-isomorphism $\mathscr{L}/\mathscr{T} \cong G(\bar{L} \mid \bar{K})$, for every prime ideal P.

It is convenient to remark that $\mathscr{T}$ is also equal to the set of all $\sigma \in \mathscr{L}$ such that $\sigma(x) \equiv x \pmod{B_P P}$ for every $x \in B_P$. For if σ satisfies this latter condition, if $x \in B$ then $\sigma(x) - x \in B \cap B_P P = P$, Conversely, if $\sigma \in \mathscr{T}$, if $x \in B_P$, we may write $x = b/s$, with $b,s \in B$, $s \notin P$; let $a = N_{L|Z}(s)$, product of the conjugates of s over Z, so $a = ss' \in B \cap Z$, $a \notin P$ (because if $\sigma \in \mathscr{L}$ then $\sigma(P) = P$, so $\sigma(s) \notin P$) and $x = bs'/a$, with $bs' \in B$, $a \in B \cap Z$, $a \notin P$; thus $\sigma(x) - x = (\sigma(bs') - bs')/a \in B_P P$ for $\sigma \in \mathscr{T}$.

Definition 2.   For every prime ideal P_i of B, $\mathscr{T}_i$ is called the *inertial group of* P_i in the extension $L \mid K$. The field of invariants of $\mathscr{T}_i$ is denoted by T_i and called the *inertial field of* P_i in $L \mid K$.

We may also adopt the following notations:

$$\mathscr{T}_i = \mathscr{T}(P_i \mid \underline{P}) = \mathscr{T}_{P_i}(L \mid K),$$
$$T_i = T(P_i \mid \underline{P}) = T_{P_i}(L \mid K).$$

When we fix our attention on one of the prime ideals P_i, which we denote by P for simplicity, then we write $T = T_P(L \mid K)$ and $\mathscr{T} = \mathscr{T}_P(L \mid K)$. We denote also by $B_T = B \cap T$ the ring of integers of T, P_T the prime ideal $P \cap T = P_T$, and by $\bar{T}$, the corresponding residue class field.

With these notations, we have the first important result:

Theorem 1.   (1) $T \mid Z$ is a Galois extension and $G(T \mid Z) = \mathscr{L}/\mathscr{T} \cong \bar{\mathscr{K}}$.
   (2) $[T:Z] = f$, $[L:T] = e$.
   (3) $\bar{L} = \bar{T}$

so the inertial degrees are

$$f(P_T \mid P_Z) = f, \qquad f(P \mid P_T) = 1.$$

(4) *The ramification indices of the ideals in question are*

$$e(P_T \mid P_Z) = 1, \qquad e(P \mid P_T) = e.$$

Groups	Fields	Ideals	Degrees	Ramification Indices	Residue Class Fields	Inertial Degrees

Proof: The assertion (1) is now obvious.

By (1), we have $[T:Z] = \#(\mathscr{Z}/\mathscr{T}) = \#(\overline{\mathscr{K}}) = [\bar{L}:\bar{K}] = f$.

Since $n = efg$, $[Z:K] = g$ (by (**B**)) and $[T:Z] = f$, then $[L:T] = e$.

To show that $\bar{L} = \bar{T}$ we consider the extension $L \mid T$. Then $\mathscr{Z}_P(L \mid T) = \mathscr{T}$ and $\mathscr{T}_P(L \mid T) = \mathscr{T}$, as it is obvious. Hence, by (1) $G(\bar{L} \mid \bar{T}) = \mathscr{T}/\mathscr{T}$ so $\bar{L} = \bar{T}$. Therefore, $[\bar{T}:\bar{Z}] = [\bar{L}:\bar{K}] = f$.

Considering the Galois extension $L \mid T$, we have

$$[L:T] = e = e(P \mid P_T) \cdot f(P \mid P_T);$$

but $f(P \mid P_T) = 1$, so $e(P \mid P_T) = e$ and by transitivity of ramification, $e(P_T \mid P_Z) = 1$. ∎

It is also useful to know the relative behavior of these groups:

E. *If $K \subseteq F \subseteq L$ are algebraic number fields, such that $L \mid K$ and $F \mid K$ are Galois extensions, if P is a prime ideal of the ring of integers of L, then:*

$$\mathscr{Z}_{P_F}(F \mid K) \cong \mathscr{Z}_P(L \mid K)/\mathscr{Z}_P(L \mid F),$$
$$\mathscr{T}_{P_F}(F \mid K) \simeq \mathscr{T}_P(L \mid K)/\mathscr{T}_P(L \mid F).$$

Proof: If $\sigma \in \mathscr{Z}_P(L \mid K)$ let $\sigma \mid F$ denote the restriction of σ to the field F; thus $\sigma \mid F \in G(F \mid K)$ and actually $\sigma \mid F \in \mathscr{Z}_{P_F}(F \mid K)$. The mapping $\sigma \to \sigma \mid F$ is obviously a group-homomorphism and its kernel is $G(L \mid F) \cap \mathscr{Z}_P(L \mid K) = \mathscr{Z}_P(L \mid F)$. It remains to show that the image of the mapping is $\mathscr{Z}_{P_F}(F \mid K)$. Given $\tau \in \mathscr{Z}_{P_F}(F \mid K)$ there exists an extension σ of τ to a K-automorphism of L, so $\sigma \mid F = \tau$; then $\sigma(P)$ is such that $\sigma(P) \cap F = P_F$, because $\sigma \mid F = \tau$ leaves P_F fixed. By Chapter 10, (**1C**) there exists $\sigma' \in G(L \mid F)$ such that $\sigma'(\sigma(P)) = P$. It follows that $\sigma'\sigma \in \mathscr{Z}_P(L \mid K)$ and $\tau = (\sigma'\sigma) \mid F \in \mathscr{Z}_{P_F}(F \mid K)$.

The second assertion is proved in the same way. If $\sigma \in \mathscr{T}_P(L \mid K)$ then $\sigma \mid F \in \mathscr{T}_{P_F}(F \mid K)$, the kernel of the homomorphism in question is

$$\mathscr{T}_P(L \mid F) = \mathscr{T}_P(L \mid K) \cap G(L \mid F).$$

Finally, the mapping has image equal to $\mathscr{T}_{P_F}(K \mid F)$, as we may easily see: $\mathscr{T}_P(L \mid K)/\mathscr{T}_P(L \mid F) \subseteq \mathscr{T}_{P_F}(F \mid K)$ (up to isomorphism), hence, considering the orders of these groups, which are ramification indices (by Theorem 1, (2)), we have $e_P(L \mid K)/e_P(L \mid F) \leqslant e_{P_F}(F \mid K)$; from Chapter 10, (**1B**) we must have equality, proving thereby the second assertion. ∎

Now we shall study the ramification, which by Theorem 1, (2) occurs in the extension $L \mid T$. Since $[L:T] = e_P(L \mid T)$, we say that P_T is *totally ramified* in $L \mid T$.

We have to consider the following situation:

$L \mid T$ *is a Galois extension of degree* e, *with Galois group* $\mathscr{T}$, *the prime ideal* $P_T = P \cap T$ *of the ring* B_T *of integers of* T *has only one extension* P *to the ring* B *of* L, $BP_T = P^e$, *and the residue class fields are* $\bar{L} = \bar{T}$. *Thus* $\mathscr{Z}_P(L \mid T) = \mathscr{T}_P(L \mid T) = \mathscr{T}$.

F. *Under above hypothesis, let* $t \in B_P$ *be a generator of the prime ideal* $B_P P$ *of* B_P, *that is* $B_P P = B_P t$. *Then* $\{1, t, \ldots, t^{e-1}\}$ *is a basis of the free module* B_P *over* $(B_T)_{P_T}$; *in particular* $B_P = (B_T)_{P_T}[t]$. *Moreover,* t *is the root of an Eisenstein polynomial with coefficients in* $(B_T)_{P_T}$.

Proof: First we show that if $a \in T$, $a \neq 0$, then $B_P a = B_P P^{se}$ (for some $s \in \mathbb{Z}$). Indeed, since $(B_T)_{P_T}$ is a principal ideal domain, by Chapter 10, (**4G**) there exists $s \in \mathbb{Z}$ such that $(B_T)_{P_T} a = (B_T)_{P_T} P_T^s$, then $B_P a = B_P (B_T)_{P_T} a = B_P (B_T)_{P_T} P_T^s = B_P P_T^s = B_P B P_T^s = B_P P^{se}$.

Next, we prove that if $x = \Sigma_{i=1}^{e-1} a_i t^i$ with $a_i \in T$, if $a_i \neq 0$, $a_j \neq 0$ for $i \neq j$, $0 \leqslant i, j \leqslant e - 1$ then $s_i e + i \neq s_j e + j$ (where $B_P a_i = B_P P^{s_i e}$); this is clear since otherwise we would have $0 \neq i - j = (s_j - s_i)e$ and $|i - j| < e$, which is impossible.

So, if $x = \Sigma_{i=0}^{e-1} a_i t^i$, with $a_i \in T$, and some $a_i \neq 0$, if

$$m = \min\{s_i e + i \mid a_i \neq 0, B_P a_i = B_P P^{s_i e}\} \text{ then } B_P x = B_P P^m.$$

Indeed, let i be such that $m = s_i e + i$; then for all j such that $a_j \neq 0$, we have $m < s_j e + j$, $x \in B_P P^m$ and if $x \in B_P P^{m+1}$ then $a_i t^i = x - \Sigma_{j \neq i} a_j t^j \in B_P P^{m+1}$, hence $B_P a_i t^i = B_P P^m \subseteq B_P P^{m+1}$, which is not true.

This implies that $\{1, t, \ldots, t^{e-1}\}$ are linearly independent over T. Because if $\Sigma_{i=0}^{e-1} a_i t^i = 0$ with $a_i \in T$ and some $a_i \neq 0$, then $0 = B_P P^m$, with $m \in \mathbb{Z}$, which is not possible.

Since $[L:T] = e$, the set $\{1, t, \ldots, t^{e-1}\}$ is a T-basis of L. These elements generate the $(B_T)_{P_T}$-module B_P, hence form a basis. Indeed, if $x \in B_P$ we may already write $x = \sum_{i=0}^{e-1} a_i t^i$ with coefficients $a_i \in T$. We have to show that $a_i \in (B_T)_{P_T}$, and we assume that $x \neq 0$, so some $a_i \neq 0$. From $x \in B_P$, if $B_P x = B_P P^m$ then $m \geqslant 0$. However, as we have seen,

$$m = \min\{s_i e + i \mid a_i \neq 0, \ B_P a_i = B_P P^{s_i e}\}$$

so $s_i e + i \geqslant 0$, $s_i \geqslant -i/e > -1$; therefore $s_i \geqslant 0$ since it is an integer; thus, $a_i \in B_P \cap T = (B_T)_{P_T}$.

Let $g = X^e + a_1 X^{e-1} + \cdots + a_e \in T[X]$ be the minimal polynomial of t over T; we shall show that $a_i \in (B_T)_{P_T} P_T$ for $i = 1, \ldots, e$, but $a_e \notin (B_T)_{P_T} P_T^2$; in such a case, g is called an *Eisenstein polynomial*. From $g(t) = 0$, we deduce that $-t^e = a_1 t^{e-1} + \cdots + a_e$; therefore,

$$e = \min\{s_i e + (e - i) \mid a_i \neq 0, \ B_P a_i = B_P P^{s_i e}\};$$

so $e \leqslant s_i e + (e - i)$, $0 < i/e \leqslant s_i$; therefore $a_i \in (B_T)_{P_T} P_T$ for $i = 1, \ldots, e$. On the other hand, if $i \neq e$ then $s_i e + e - i \geq e + e - i > e$; thus the above minimum is attained when $i = e$, that is $s_e e = e$, hence $(B_T)_{P_T} P_T = (B_T)_{P_T} a_e$, and $a_e \notin (B_T)_{P_T} P_T^2$. ∎

We note at this point that $B_P P$ has a generator $t' \in B$, for if $B_P P = B_P t$, with $t = t'/s$, $t' \in B$, $s \in B$, $s \notin P$ then $B_P t = B_P t'$.

G. *For every* $i = 0, 1, 2, \ldots$ *let* $\mathscr{V}_i = \{\sigma \in \mathscr{T} \mid \sigma(x) \equiv x (P^{i+1})$ *for every* $x \in B\}$. *Then:*

(1) *Each* $\mathscr{V}_i$ *is a normal subgroup of* $\mathscr{L} = \mathscr{T}$ *and*
$$\mathscr{T} = \mathscr{V}_0 \supseteq \mathscr{V}_1 \supseteq \mathscr{V}_2 \supseteq \cdots.$$

(2) *There exists an index r such that $\mathscr{V}_r$ is the trivial group.*

(3) *If $t \in B$ is an element such that $B_P P = B_P t$ then*
$$\mathscr{V}_i = \{\sigma \in \mathscr{T} \mid \sigma(t). \, t^{-1} \equiv 1 \ (\mathrm{mod}\ B_P P^i)\} \ \textit{for every } i = 0, 1, 2, \ldots.$$

Proof:

(1) We consider the ring B/P^{i+1}; if $\sigma \in \mathscr{L}$ then $\sigma(P) = P$, hence $\sigma(P^{i+1}) = P^{i+1}$; then each σ acts on B/P^{i+1} in natural way: $\bar{\sigma}(\bar{x}) = \overline{\sigma(x)}$. Thus, $\sigma \in \mathscr{V}_i$ if and only if σ acts trivially on B/P^{i+1}; that is, $\bar{\sigma}$ is the identity mapping; $\mathscr{V}_i$ is therefore the kernel of the group-homomorphism $\sigma \to \bar{\sigma}$, so $\mathscr{V}_i$ is a normal subgroup of $\mathscr{T}$. Obviously, $\mathscr{T} = \mathscr{V}_0 \supseteq \mathscr{V}_1 \supseteq \mathscr{V}_2 \supseteq \cdots$.

(2) $\bigcap_{i=0}^{\infty} \mathscr{V}_i$ is the trivial group, for if σ belongs to this intersection then $\sigma(x) - x \in \bigcap_{i=0}^{\infty} P^{i+1}$, that is $\sigma(x) = x$, for every $x \in B$.

Since $\mathscr{T}$ is a finite group, then there exists r such that $\mathscr{V}_r$ consists only of the trivial automorphism.

(3) If $\sigma \in \mathscr{V}_i$ then $\sigma(t) - t \in B_P P^{i+1}$, hence $\sigma(t)/t - 1 \in B_P P^i$. Conversely, this implies that $\sigma(t) - t \in B_P P^{i+1}$. Next, if $x \in B \subseteq B_P$ we may write $x = \sum_{i=0}^{e-1} a_i t^i$ with $a_i \in (B_T)_{P_T} \subseteq T$, hence

$$\sigma(x) - x = \sum_{i=0}^{e-1} a_i(\sigma(t)^i - t^i);$$

but

$$\sigma(t)^i - t^i = [\sigma(t) - t] \cdot [\sigma(t)^{i-1} + \sigma(t)^{i-2}t + \cdots + t^{i-1}] \in B_P P^{i+1},$$

hence $\sigma(x) - x \in B \cap B_P P^{i+1} = P^{i+1}$. ∎

Definition 3. For every $i = 0,1,2,\ldots$, $\mathscr{V}_i$ is called the ith *ramification group of P* in $L \mid K$. The field of invariants of $\mathscr{V}_i$ is denoted by V_i and called the ith *ramification field* of P in $L \mid K$.

If necessary, we may use the following more precise notations: $\mathscr{V}_i = \mathscr{V}_i(P \mid \underline{P}) = \mathscr{V}_{i_P}(L \mid K)$ and $V_i = V_i(P \mid \underline{P}) = V_{i_P}(L \mid K)$.

Thus, in our case:

$$K \subseteq Z \subseteq T = V_0 \subseteq V_1 \subseteq V_2 \subseteq \cdots \subseteq V_r = L$$

and each extension $V_i \mid V_0$ is Galoisian, with Galois group $G(V_i \mid V_0) = \mathscr{T}/\mathscr{V}_i$.

As for the inertial group, let us note that $\mathscr{V}_i$ is also equal to $\{\sigma \in \mathscr{T} \mid \sigma(x) \equiv x(B_P P^{i+1})$ for every $x \in B_P\}$; the proof is the same, therefore will be omitted.

It is our purpose now to study the structure of the group $\mathscr{T} = \mathscr{V}_0$.

Theorem 2.

(1) *there exists a natural group-isomorphism* θ *from* $\mathscr{T}/\mathscr{V}_1$ *into* $\bar{L}^\cdot$ (*multiplicative group of non-zero elements of* $\bar{L}$), *hence* $\mathscr{T}/\mathscr{V}_1$ *is a cyclic group whose order is prime to* p, *where* $\mathbb{Z}p = P \cap \mathbb{Z}$.

(2) *for every* $i = 1, 2, \ldots$ *there exists an isomorphism* θ_i *from the group* $\mathscr{V}_i/\mathscr{V}_{i+1}$ *into the additive group of* $\bar{L}$, *hence* $\mathscr{V}_i/\mathscr{V}_{i+1}$ *is an elementary Abelian p-group* (*that is, a finite dimensional vector space over the field* $\mathbb{F}_p$).

(3) $\mathscr{V}_1$ *is a p-group.* $\mathscr{T}$ *is a solvable group.*

(4) *If* $m = [V_1:T] = \#(\mathscr{T}/\mathscr{V}_1)$ *then* p *does not divide* m *and* $e = mp^s$, *for some* $s \geqslant 0$, $p^s = [L:V_1] = \#(\mathscr{V}_1)$.

Proof:

(1) Let $t \in B$ be a generator of the principal ideal $B_P P$ so $t \in B \cap B_P P = P$. If $\sigma \in \mathscr{T}$, since $\sigma(P) = P$, then $\sigma(t) \in P \subseteq B_P P = B_P t$, hence there exists $c_\sigma \in B_P$ such that $\sigma(t) = c_\sigma t$.

We show that $c_\sigma \notin B_P P$, because considering $\sigma^{-1} \in \mathscr{T}$, $\sigma^{-1}(t) = c_{\sigma^{-1}}t$ with $c_{\sigma^{-1}} \in B_P$, $t = \sigma(\sigma^{-1}(t)) = \sigma(c_{\sigma^{-1}})\sigma(t) = [\sigma(c_{\sigma^{-1}}) \cdot c_\sigma]t$ and therefore

$$\sigma(c_{\sigma^{-1}}) \cdot c_\sigma = 1,$$

c_σ is invertible in B_P, thus $c_\sigma \notin B_P P$. By Chapter 12, **(4H)** $B_P/B_P P \cong B/P = \bar{L}$ and the image of c_σ in $\bar{L}$ is $\bar{c}_\sigma \neq \bar{0}$.

We define the mapping $\tilde{\theta}:\mathscr{T} \to \bar{L}^\cdot$ by $\tilde{\theta}(\sigma) = \bar{c}_\sigma$. To be convinced that $\tilde{\theta}$ is natural, we have to show that $\tilde{\theta}$ is independent of the element $t \in B$. If $t' \in B$ is also such that $B_P P = B_P t'$, then $t' = ut$ where $u \in B_P$ is invertible in B_P, hence $u \notin B_P P$; let $\sigma(t') = c_\sigma' t'$, hence $\sigma(ut) = c_\sigma' ut$. But $\sigma \in \mathscr{T}$, therefore $\sigma(u) \equiv u \pmod{B_P P}$, so $\sigma(u) = u + vt$ with $v \in B_P$, and therefore $(u + vt)c_\sigma t = c_\sigma' ut$, $uc_\sigma + vc_\sigma t = c_\sigma' u$ and considering the images in $\bar{L}$, we have $\bar{u} \cdot \bar{c}_\sigma = \bar{c_\sigma}' \cdot \bar{u}$, so $\bar{c_\sigma} = \bar{c_\sigma}'$. This shows that $\tilde{\theta}$ is independent of the choice of t.

$\tilde{\theta}$ is a group-homomorphism: $\tilde{\theta}(\sigma\tau) = \tilde{\theta}(\sigma)\tilde{\theta}(\tau)$. In fact, if $\sigma(t) = c_\sigma t$, $\tau(t) = c_\tau t$ then $\sigma\tau(t) = \sigma(c_\tau)\sigma(t) = (c_\tau + vt)c_\sigma t = (c_\tau c_\sigma + vc_\sigma t)t$ where $v \in B_P$; thus $\tilde{\theta}(\sigma\tau) = \overline{c_\tau c_\sigma + vc_\sigma t} = \overline{c_\tau c_\sigma} = \tilde{\theta}(\sigma)\tilde{\theta}(\tau)$.

The kernel of $\tilde{\theta}$ is $\mathscr{V}_1$, hence $\tilde{\theta}$ induces an isomorphism θ from $\mathscr{T}/\mathscr{V}_1$ into $\bar{L}^\cdot$. In fact, if $\sigma \in \mathscr{V}_1$ then $\sigma(t) \equiv t(P^2)$ hence $\sigma(t) = (1 + bt)t$ with $b \in B$, thus $\tilde{\theta}(\sigma) = \overline{1 + bt} = \bar{1}$. Conversely, if $\sigma \in \mathscr{T}$ is such that $\tilde{\theta}(\sigma) = \bar{1}$, that is $\bar{c}_\sigma = \bar{1}$ then $\sigma(t) - t = (c_\sigma - 1)t \in B_P t^2$, hence $\sigma(t)/t \equiv 1(B_P P)$ and $\sigma \in \mathscr{V}_1$ (by **(G)**).

Thus $\mathscr{T}/\mathscr{V}_1$ is isomorphic to a subgroup of $\bar{L}^\cdot$. If $\mathbb{Z}p = \mathbb{Z} \cap P$, then $\bar{L}$ is a finite field containing $\mathbb{F}_p$, so its non-zero elements form a cyclic group of order $\#(\bar{L}) - 1$, which is not a multiple of p; therefore $\mathscr{T}/\mathscr{V}_1$ is also cyclic of order not multiple of p.

(2) Let $i \geqslant 1$. If $\sigma \in \mathscr{V}_i$ then $\sigma(t) = t + d_\sigma t^{i+1}$ with $d_\sigma \in B$. Let $\tilde{\theta}_i(\sigma) = \bar{d}_\sigma \in \bar{L}$. We shall show that the mapping $\tilde{\theta}_i : \mathscr{V}_i \to \bar{L}$ is a homomorphism into $\bar{L}$. In fact, if $\sigma, \tau \in \mathscr{V}_i$ then

$$\sigma\tau(t) = \sigma(t + d_\tau t^{i+1}) = \sigma(t) + \sigma(d_\tau) \cdot \sigma(t)^{i+1}$$

$$= t + d_\sigma t^{i+1} + (d_\tau + d't^{i+1})(t + d_\sigma t^{i+1})^{i+1}$$

$$= t + (d_\sigma + d_\tau)t^{i+1} + ct^{i+2},$$

where $d', c \in B$; hence $\tilde{\theta}_i(\sigma\tau) = \overline{d_\sigma + d_\tau} = \tilde{\theta}_i(\sigma) + \tilde{\theta}_i(\tau)$.

The kernel of $\tilde{\theta}_i$ is $\mathscr{V}_{i+1}$, hence $\tilde{\theta}_i$ induces an isomorphism θ from $\mathscr{V}_i/\mathscr{V}_{i+1}$ into $\bar{L}$. In fact, if $\sigma \in \mathscr{V}_{i+1}$ then $\sigma(t) = t + ct^{i+2}$, so $d_\sigma = ct$ and $\bar{d}_\sigma = \bar{0}$. Conversely, if $\bar{d}_\sigma = \bar{0}$, then $\sigma(t) \equiv t(P^{i+2})$ so $\sigma(t)/t \equiv 1(P^{i+1})$ and therefore $\sigma \in \mathscr{V}_{i+1}$.

Noting that $\bar{L}$ is a finite dimensional vector space over $\mathbb{F}_p$, the same holds for $\mathscr{V}_i/\mathscr{V}_{i+1}$.

(3) The group $\mathscr{T}$ has the following sequence of subgroups, each being actually a normal subgroup:

$$\mathscr{T} \supseteq \mathscr{V}_1 \supseteq \mathscr{V}_2 \supseteq \cdots \supseteq \mathscr{V}_r = \{\varepsilon\}$$

(for some r, by **(G)**). Since $\mathscr{T}/\mathscr{V}_1$ is a cyclic group and each group $\mathscr{V}_i/\mathscr{V}_{i+1}$ is an elementary Abelian p-group, then $\mathscr{T}$ is a solvable group and $\mathscr{V}_1$ is a p-group.

(4) We have $e = [L:V_1] \cdot [V_1:T]$ (with $T = K$), $[L:V_1] = \#\mathscr{V}_1$ which is a power of p, $[V_1:T] = \#(\mathscr{T}/\mathscr{V}_1)$ which is relatively prime to p. ∎

In particular, if $\bar{K}$ has characteristic p not dividing the degree $e = [L:T]$, then $\#(\mathscr{V}_1) = 1$, so $V_1 = V_2 = \cdots = L$. Thus, in this case there is no higher ramification present, and $\underline{P}$ is said to be *tamely ramified in* $L \mid K$. If p divides e, then $\underline{P}$ is said to be *wildly ramified in* $L \mid K$.

Combining Theorem 1, (1), and Theorem 2, (3), we have:

H. *If P is any prime ideal of B such that $P \cap A = \underline{P}$, the decomposition group $\mathscr{L}$ of $\underline{P}$ in $L \mid K$ is a solvable group.*

Proof: We have $\mathscr{L} \supseteq \mathscr{T} \supseteq \{\varepsilon\}$, where $\mathscr{T}$ is solvable and $\mathscr{L}/\mathscr{T}$ is a cyclic group; thus $\mathscr{L}$ itself is a solvable group. ∎

An interesting application is the following:

I. *Let K be an algebraic number field, let $f \in K[X]$ be an irreducible polynomial, L the splitting field of f over K. If there exists a prime ideal $\underline{P}$ of the ring of integers A of K which divides only one prime ideal P of the ring of integers of L, then the polynomial f is solvable by radicals.*

Proof: Let $\mathscr{K} = G(L \mid K)$, the Galois group of the polynomial f over K. We have $\mathscr{L}_P(L \mid K) = \chi$, so χ is a solvable group (by (**H**)). By the theorem of Galois, the polynomial f is solvable by radicals. ∎

Later, we shall need the result which follows. Since $\mathscr{T}, \mathscr{V}_1$ are normal subgroups of $\mathscr{L}$, then $\sigma\tau\sigma^{-1} \in \mathscr{T}$ for every $\tau \in \mathscr{T}$, $\sigma\mathscr{V}_1\sigma^{-1} = \mathscr{V}_1$ thus considering the cosets of $\mathscr{T}$ by $\mathscr{V}_1$, we have $\sigma(\tau\mathscr{V}_1)\sigma^{-1} = (\sigma\tau\sigma^{-1})\mathscr{V}_1$. Therefore σ acts on the quotient group $\mathscr{T}/\mathscr{V}_1$ by conjugation, defining $\bar{\sigma}(\tau\mathscr{V}_1) = (\sigma\tau\sigma^{-1})\mathscr{V}_1$. The following proposition due to Speiser describes this action:

J. *Let $\sigma \in \mathscr{L}$ be such that its image by the homomorphism $\mathscr{L} \to \mathscr{L}/\mathscr{T} \cong G(\bar{L} \mid \bar{K})$ corresponds to the Frobenius automorphism of $\bar{L} \mid \bar{K}$. Then:*
(1) *If $\tau \in \mathscr{T}$ then $\bar{\sigma}(\tau\mathscr{V}_1) = \tau^q\mathscr{V}_1$ where $q = \#(\bar{K})$.*
(2) *If $\mathscr{L}$ is an Abelian group then $\tau^{q-1} \in \mathscr{V}_1$ for every $\tau \in \mathscr{T}$, and $\mathscr{T}/\mathscr{V}_1$ has order dividing $q - 1$.*
Proof:
(1) Let $B_P P = B_P t$, with $t \in B$ and let us compute $\sigma\tau\sigma^{-1}(t)$. We write $\sigma(t) = c_\sigma t$, $\sigma^{-1}(t) = c_{\sigma^{-1}}t$ with $c_\sigma, c_{\sigma^{-1}} \in B_P$ and as in the proof of Theorem 2, $\sigma(c_{\sigma^{-1}}) \cdot c_\sigma = 1$. If $\tau \in \mathscr{T}$ then $\sigma\tau\sigma^{-1}(t) = \sigma\tau(c_{\sigma^{-1}}t) = \sigma(\tau(c_{\sigma^{-1}}) \cdot \tau(t)) = \sigma(c_{\sigma^{-1}} + vt) \cdot \sigma(c_\tau) \cdot \sigma(t) = (\sigma(c_{\sigma^{-1}}) + \sigma(v) \cdot c_\sigma t)\sigma(c_\tau)c_\sigma t \equiv \sigma(c_\tau)t \equiv c_\tau^q t \pmod{B_P P^2}$ (where $v \in B_P$), recalling that the Frobenius automorphism $\bar{\sigma}$ is defined as being the raising to the power $q = \#(\bar{K})$.

On the other hand, $\tau(t) = c_\tau t$, $\tau^2(t) = \tau(c_\tau) \cdot \tau(t) = (c_\tau + ut)c_\tau t \equiv c_\tau^2 t \pmod{B_P P^2}$ (where $u \in B_P$); similarly $\tau^q(t) \equiv c_\tau^q t \pmod{B_P P^2}$. Thus

$\sigma\tau\sigma^{-1}(t) \equiv \tau^q(t)(B_P P^2)$ hence $\tau^{-q}\sigma\tau\sigma^{-1}(t) \equiv t(B_P P^2)$ and
$(\tau^{-q}\sigma\tau\sigma^{-1}(t))/t \equiv 1(B_P P)$, so $\tau^{-q}\sigma\tau\sigma^{-1} \in \mathscr{V}_1$ and $\sigma(\tau\mathscr{V}_1) = \sigma\tau\sigma^{-1}\mathscr{V}_1 = \tau^q\mathscr{V}_1$.

(2) If $\mathscr{L}$ is an Abelian group, then $(\tau^{-q}\sigma\tau\sigma^{-1})^{-1} = \tau^{q-1} \in \mathscr{V}_1$ for every $\tau \in \mathscr{T}$.

Since $\mathscr{T}/\mathscr{V}_1$ is a cyclic group and $\tau^{q-1}\mathscr{V}_1 = \mathscr{V}_1$ (unit of $\mathscr{T}/\mathscr{V}_1$), then the order of $\mathscr{T}/\mathscr{V}_1$ divides $q - 1$. ∎

In order to have a better insight into the higher ramification, we shall now consider the different above a prime ideal $\underline{P}$ of A.

We assume as before that $L \mid K$ is a Galois extension. Moreover, $B \cdot \underline{P} = P^e$, $e \geqslant 1$. This implies that P is the only prime ideal of B such that $P \cap A = \underline{P}$. If S is the multiplicative set complement of $\underline{P}$ in A, if $A' = S^{-1} \cdot A$, $B' = S^{-1} \cdot B$ then $B' \subseteq B_P$. But under the above hypotheses, we have actually $B' = B_P$.

Indeed, let $b/s \in B_P$ with $b,s \in B$, $s \notin P$; the conjugates of b/s are the elements $\sigma(b)/\sigma(s)$ (where σ is any K-isomorphism of L), which belong to B_P, since $L \mid K$ is a Galois extension and P is the only extension of $\underline{P}$ to an ideal of B. It follows that the coefficients of the minimal polynomial of b/s over K, which are the elementary symmetric polynomials on the conjugates of this element, belong necessarily to $K \cap B_P$. Now $A_{\underline{P}} \subseteq K \cap B_P \neq K$, hence by Chapter 10, (4G) we have $K \cap B_P = A_{\underline{P}}$. This proves that b/s is integral over $A_{\underline{P}}$. By Chapter 10, (4E), $B_P \subseteq B'$, hence we have the equality $B' = B_P$.

We note $\Delta_{\underline{P}}(L \mid K) = \Delta(B_P \mid A_{\underline{P}})$ the different of $L \mid K$ above $\underline{P}$. By Chapter 10, (7H) we have $\Delta_{\underline{P}}(L \mid K) = B' \cdot \Delta_{L|K}$, hence by Chapter 10, (7H) we have $\Delta_{\underline{P}}(L \mid K) = B_P P^s$ where $s \geqslant 0$ is an integer; it is the exponent at P of the different $\Delta_{L|K}$. Sometimes we denote it also by $s_P(L \mid K)$.

We shall now compute the exponent of the different; it turns out that this expression will involve the orders of the various ramifications groups.

Theorem 3. *Let $L \mid K$ be a Galois extension and P the only extension of $\underline{P}$ to L. We assume that there exists an element t such that $B_P P = B_P t$ and $B_P = A_{\underline{P}}[t]$ (for example, by* (F)*, this holds when P is totally ramified over $\underline{P}$ that is, the inertial field T is equal to K). Then the exponent of the different of P in $L \mid K$ is*

$$s_P(L \mid K) = \sum_{i=0}^{r-1} [\#(\mathscr{V}_i) - 1],$$

where $\mathscr{T} = \mathscr{V}_0 \supseteq \mathscr{V}_1 \supseteq \cdots \supseteq \mathscr{V}_r = \{\varepsilon\}$ are the ramification groups of P in $L \mid K$.

Proof: We have seen in Chapter 12, (7N) that $\Delta_P = B_P \cdot g'(t)$, where g is the minimal polynomial of t over K. We write $g = \Pi_{\sigma \in \mathscr{K}}(X - \sigma(t))$, where $\mathscr{K} = G(L \mid K)$. Then $g'(t) = \Pi_{\sigma \neq \varepsilon}(t - \sigma(t))$.

Since P is the only extension of $\underline{P}$, then the decomposition group of P is $\mathscr{Z} = \chi$. If $\sigma \in \mathscr{Z}$, $\sigma \notin \mathscr{T}$ then $\sigma(t) - t \in B$, but $\sigma(t) - t \notin P$, and similarly, if $\sigma \in \mathscr{V}_i$ but $\sigma \notin \mathscr{V}_{i+1}$ then $\sigma(t) - t \in P^{i+1}$, but $\sigma(t) - t \notin P^{i+2}$ (by **(G)**). If $s = s_P(L \mid K)$ then $B_P P^s = B_P g'(t) = \Pi_{\sigma \neq \varepsilon} B_P(t - \sigma(t))$ and writing $B_P(t - \sigma(t)) = B_P P^{s(\sigma)}$ then

$$
\begin{aligned}
s &= \sum_{\sigma \neq \varepsilon} s(\sigma) = \sum_{i=0}^{r-1} \sum_{\sigma \in \mathscr{V}_i, \sigma \notin \mathscr{V}_{i+1}} s(\sigma) = \sum_{i=0}^{r-1} [\#(\mathscr{V}_i) - \#(\mathscr{V}_{i+1})](i + 1) \\
&= [\#(\mathscr{V}_0) - \#(\mathscr{V}_1)] + 2[\#(\mathscr{V}_1) - \#(\mathscr{V}_2)] \\
&\quad + 3[\#(\mathscr{V}_2) - \#(\mathscr{V}_3)] + \cdots + r[\#(\mathscr{V}_{r-1}) - 1] \\
&= \#(\mathscr{V}_0) + \#(\mathscr{V}_1) + \#(\mathscr{V}_2) + \cdots + \#(\mathscr{V}_{r-1}) - r \\
&= \sum_{i=0}^{r-1} [\#(\mathscr{V}_i) - 1]. \quad \blacksquare
\end{aligned}
$$

EXERCISES

1. Let K be an algebraic number field, $L \mid K$ a finite Galois extension, and P a prime ideal of the ring B of integers of L. Show that if K' is a field, $K \subseteq K' \subseteq L$, A' the ring of integers of K' and $P' = P \cap A'$ is unramified and inert over P, then $K' \subseteq Z$ (decomposition field of P in $L \mid K$).

2. Let K be an algebraic number, $L \mid K$ a finite Galois extension, and P a prime ideal of the ring B of integers of L. Show that if K' is a field, $K \subseteq K' \subseteq L$, A' the ring of integers of K', and $P' = P \cap A'$, if P' is unramified over P, then $K' \subseteq T$ (inertial field of P in $L \mid K$).

3. Let K be an algebraic number field, $L \mid K$ a finite extension, L' the smallest field containing L and such that $L' \mid K$ is a Galois extension. Show that if a prime ideal of the ring of integers of K decomposes completely in $L \mid K$ then it also decomposes completely in $L' \mid K$.

4. Let K be an algebraic number field, P a prime ideal of its ring of integers. Let $L_1 \mid K$ and $L_2 \mid K$ be finite extensions. Show that if P decomposes completely in $L_1 \mid K$ and $L_2 \mid K$ then it decomposes completely in $L_1 L_2 \mid K$.

5. Let $K \mid \mathbb{Q}$ be a Galois extension of degree n, let I be an ideal of the ring A of integers of K, such that $\sigma(I) = I$ for every $\sigma \in G(K \mid \mathbb{Q})$. Show that $I^{n!}$ is generated by some rational integer.

6. Use the previous exercise to show that given the ideal $I \neq 0$ of A there exists an ideal $J \neq 0$ of A such that IJ is a principal ideal (see Chapter 7, **(H)**).

7. Let K be an algebraic number field, let $L \mid K$ be a Galois extension of degree n. Show that the relative different $\Delta_{L \mid K}$ is invariant by the Galois group. Hence $\Delta_{L \mid K}^n$ is the ideal generated by the relative discriminant $\delta_{L \mid K}$.

8. Let $n \geqslant 1$ be an integer. Show that there exists an integer $l(n)$ (depending only on n) such that: If $K \mid \mathbb{Q}$ is a Galois extension of degree n, if p is a prime number and p^s divides the discriminant δ_K then $s \leqslant l(n)$ (compare this statement with Chapter 9, (**O**)).

   Hint: Apply Theorem 3.

9. Let $K \mid \mathbb{Q}$ be a quadratic extension, let A be the ring of algebraic integers of K and P a non-zero prime ideal of A. Discuss in all cases the decomposition, inertia and ramification groups and fields of P.

10. Do the previous exercise in case $K \mid \mathbb{Q}$ is a Galois extension of degree p, an odd prime number.

11. Do the previous exercise in case $K \mid \mathbb{Q}$ is a Galois extension with Galois group equal to the Klein group $\mathscr{K} = \{\varepsilon, \sigma, \tau, \sigma\tau\}$, $\sigma^2 = \tau^2 = \varepsilon$, $\sigma\tau = \tau\sigma = \varepsilon$.

CHAPTER 12

The Fundamental Theorem of Abelian Extensions

In Chapter 2, (8), we have stated that every cyclotomic field $\mathbb{Q}(\zeta)$ (where ζ is a primitive nth root of $1, n > 2$) is an Abelian extension of $\mathbb{Q}$.

From the arithmetical point of view, the cyclotomic fields have been fairly well studied. In order to investigate other algebraic number fields, it is reasonable to consider first the case of Galois extensions of $\mathbb{Q}$, for which we have already indicated in the preceding section a rather elaborate theory of decomposition of ideals. Assuming moreover that the Galois group is Abelian proves to be a workable hypothesis; in fact, a whole branch of the theory of algebraic numbers, called *class field theory* is devoted to the study of Abelian extensions.

Therefore the natural question to ask is the following: which are the possible Abelian extensions of $\mathbb{Q}$? We certainly cannot attempt to solve this problem in such an elementary text. However, we shall prove a rather interesting theorem about Abelian extensions.

Already in Chapter 4, (**P**), we have shown that every quadratic extension of $\mathbb{Q}$ (which is certainly an Abelian extension) is contained in a cyclotomic field. We have also indicated its generalization, by Kronecker and Weber. It is our intention now to prove this theorem; this will provide an excellent ground for the application of the concepts and techniques developed in the preceding chapters.

Theorem 1. *If L is an algebraic number field, Abelian extension of $\mathbb{Q}$, then there exists a root of unity ζ, such that $L \subseteq \mathbb{Q}(\zeta)$.*
Proof: We shall consider two particular crucial cases first and then show how to reduce the general case to these two special ones. We shall require several steps.

Case 1: $[L:\mathbb{Q}] = p^m$, $\delta_L = p^k$ where p is an odd prime, $m, k \geqslant 1$.

A. *There exists only one prime ideal P in L such that $P \cap \mathbb{Z} = \mathbb{Z}p$; moreover, P is totally ramified in $L \mid \mathbb{Q}$ and the inertial field T and first ramification fields of P in $L \mid \mathbb{Q}$ are $V_1 = T = \mathbb{Q}$.*

Proof: Let P, P' be prime ideals such that $P \cap \mathbb{Z} = P' \cap \mathbb{Z} = \mathbb{Z}p$. Since $\mathscr{K} = G(L \mid \mathbb{Q})$ is an Abelian group, then the decomposition groups $\mathscr{L}, \mathscr{L}'$ of P, P' respectively, must coincide (Chapter 11, (**A**)).

By Chapter 10, (**1C**), there exists $\sigma \in \mathscr{K}$ such that $\sigma(P) = P'$, hence considering the inertial groups of P, P', we have $\mathscr{T}' = \sigma \mathscr{T} \sigma^{-1}$ and therefore $\mathscr{T}' = \mathscr{T}$, because $\mathscr{K}$ is Abelian.

Let T be the inertial field of P and of P'. We show that $\delta_T = 1$. By Chapter 11, Theorem 1, (4), p is unramified in $T \mid \mathbb{Q}$, since $e(P_T \mid p) = 1$, so every other prime ideal of T extending p (if it exists) must also be unramified in $T \mid \mathbb{Q}$, because $T \mid \mathbb{Q}$ is a Galois extension (by Chapter 10, (**1D**)). By Chapter 10, Theorem 1, p does not divide the discriminant δ_T.

Now, if q is a prime, $q \neq p$, and q divides δ_T, then again by the same theorem, q is ramified in $T \mid \mathbb{Q}$, hence in $L \mid \mathbb{Q}$, so q divides $\delta_L = p^k$, a contradiction. Thus, δ_T has no prime factor; therefore, $|\delta_T| = 1$. In view of Chapter 9, (**K**), $T = \mathbb{Q}$. If Z is the decomposition field of any prime ideal P containing p, since $T \supseteq Z \supseteq \mathbb{Q}$, then $Z = \mathbb{Q}$. By Chapter 11, (**C**), there exists only one prime ideal P in L, containing p.

Since $[T : \mathbb{Q}] = f(P \mid p)$, then $f(P \mid p) = 1$. Finally, from $[L : \mathbb{Q}] = efg$, with $f = g = 1$, we deduce that p is totally ramified in $L \mid \mathbb{Q}$.

Also, since $[V_1 : T]$ has degree prime to p (characteristic of the residue class field), by Chapter 11, Theorem 2, (4), $V_1 = T$. ∎

B. *Let H be a field, $\mathbb{Q} \subseteq H \subseteq L$, such that $[H : \mathbb{Q}] = p$. If C is the ring of integers of H, $Q = P \cap C$, then the exponent of the different of Q in $H \mid \mathbb{Q}$ is equal to $2(p - 1)$; this value is therefore independent of the field H, provided $[H : \mathbb{Q}] = p$.*

Proof: Since p is totally ramified in $L \mid \mathbb{Q}$ by transitivity of the ramification index, p is totally ramified in $H \mid \mathbb{Q}$. If z is a generator of the ideal $C_Q Q$, that is $C_Q Q = C_Q z$, then by Chapter 11, (**F**), $C_Q = \mathbb{Z}_p[z]$, $\{1, z, \ldots, z^{p-1}\}$ are linearly independent over $\mathbb{Q}$ and z is a root of an Eisenstein polynomial $g = X^p + a_1 X^{p-1} + \cdots + a_p$, with $a_i \in \mathbb{Z}$, p dividing each a_i but p^2 not dividing a_p. Then $g = \prod_{\sigma \in \mathscr{G}} (X - \sigma(z))$ where $\mathscr{G}$ is the Galois group of $H \mid \mathbb{Q}$ and $g'(z) = \prod_{\sigma \neq \varepsilon} (z - \sigma(z))$; by Chapter 10, (**7N**) $\Delta_Q(H \mid \mathbb{Q}) = C_Q g'(z)$. We have to compute the largest power of $C_Q Q$ dividing the principal ideal generated by

$$g'(z) = pz^{p-1} + (p-1)a_1 z^{p-2} + \cdots + a_{p-1}.$$

Since p is totally ramified in $H \mid \mathbb{Q}$, then $C_Q p = C_Q z^p$ (p being the degree of $H \mid \mathbb{Q}$); on the other hand, if p' is any prime different from p, then $p' \notin Q$ so $C_Q p' = C_Q z^0 = C_Q$; hence, for every integer $a \in \mathbb{Z}$, if p^m divides a, but p^{m+1} does not divide a, then $C_Q a = C_Q z^{pm}$, with $m \geqslant 0$. In particular, let $C_Q(p - i)a_i = C_Q z^{ps_i}$ so $s_i \geqslant 1$ for every i and $a_0 = 1$. Hence,

$$C_Q[(p - i)a_i z^{p-i-1}] = C_Q z^{ps_i+(p-i-1)}.$$

By the argument in the proof of Chapter 11, (**F**), if i, j are distinct indices, $0 \leqslant i, j \leqslant p - 1$ then $ps_i + (p - i - 1) \neq ps_j + (p - j - 1)$; hence, if $C_Q g'(z) = C_Q z^s$ then

$$s = \min_{0 \leqslant i \leqslant p-1} \{ps_i + (p - i - 1)\}.$$

It follows that $p \leqslant s$ since $s_i \geqslant 1$, so $p \leqslant ps_i + (p - i - 1)$ for $i = 0, 1, \ldots, p - 1$. On the other hand, $C_Q p z^{p-1} = C_Q z^{p+(p-1)} = C_Q z^{2p-1}$; therefore, we have the inequalities $p \leqslant s \leqslant 2p - 1$.

But, by Chapter 11, Theorem 3, the exponent of the different is

$$s = \sum_{i=0}^{r-1} [\#(\mathscr{V}_i') - 1]$$

where $\mathscr{V}_i'$ denotes the ith ramification group of Q in $H \mid \mathbb{Q}$. Since $[H : \mathbb{Q}] = p$ then $\#(\mathscr{V}_i')$ is either 1 or p, thus $p - 1$ divides s, and therefore

$$1 < \frac{p}{p - 1} \leqslant \frac{s}{p - 1} \leqslant \frac{2p - 1}{p - 1} = 2 + \frac{1}{p - 1} < 3,$$

(because $p \neq 2$). We conclude that $s = 2(p - 1)$. ∎

C. *Let i be the smallest index such that $\mathscr{V}_i \neq \mathscr{K} = G(L \mid \mathbb{Q})$ (hence $i > 1$ by (**A**)). Then $[V_i : \mathbb{Q}] = p$ and V_i is the only field of degree p over $\mathbb{Q}$, contained in L.*

Proof: We have $[V_i : \mathbb{Q}] = (\mathscr{V}_{i-1} : \mathscr{V}_i) = \#(\mathscr{V}_{i-1}/\mathscr{V}_i)$. By Chapter 11, Theorem 2, (2), $\mathscr{V}_{i-1}/\mathscr{V}_i$ is isomorphic to a subgroup of the additive group $\bar{L}$. Since $f(L \mid \mathbb{Q}) = 1$ then $\bar{L} = \mathbb{F}_p$, so from $\mathscr{V}_{i-1} \neq \mathscr{V}_i$ it follows that $\#(\mathscr{V}_{i-1}/\mathscr{V}_i) = p$, and therefore $[V_i : \mathbb{Q}] = p$.

Now, let H be any field such that $\mathbb{Q} \subseteq H \subseteq L$, $[H : \mathbb{Q}] = p$ and assume that $H \neq V_i$; we shall arrive at a contradiction. For this purpose we compute the differents $\Delta_p(L \mid V_i)$ and $\Delta_p(L \mid H)$, using Theorem 3 of Chapter 11.

Let $\mathscr{H} = G(L \mid H)$, then $\mathscr{V}_j(L \mid H) = \mathscr{V}_j(L \mid \mathbb{Q}) \cap \mathscr{H}$ and similarly $\mathscr{V}_j(L \mid V_i) = \mathscr{V}_j(L \mid \mathbb{Q}) \cap \mathscr{V}_i$ for every $j \geqslant 0$.

Thus, $\mathscr{V}_0(L \mid V_i) = \cdots = \mathscr{V}_i(L \mid V_i) = \mathscr{V}_i$ while $\mathscr{V}_j(L \mid V_i) = \mathscr{V}_j$ for $j \geqslant i + 1$ (as before $\mathscr{V}_j = \mathscr{V}_j(L \mid \mathbb{Q})$ for every $j \geqslant 0$). Similarly, $\mathscr{V}_0(L \mid H) = \cdots = \mathscr{V}_{i-1}(L \mid H) = \mathscr{H}$ (since $\mathscr{V}_{i-1} = \mathscr{K}$) while $\mathscr{V}_i(L \mid H)$ is properly contained in $\mathscr{V}_i$ (otherwise $\mathscr{V}_i(L \mid H) = \mathscr{H}$ hence $V_i = H$ against the hypothesis) and $\mathscr{V}_j(L \mid H) \subseteq \mathscr{V}_j$ for $j \geqslant i + 1$. Therefore

$$s_P(L \mid V_i) = \sum_{j=0}^{r-1} [\#\mathscr{V}_j(L \mid V_i) - 1] > \sum_{j=0}^{r-1} [\#\mathscr{V}_j(L \mid H) - 1] = s_P(L \mid H).$$

However, from the transitivity of the different we have $\Delta_p(L \mid \mathbb{Q}) = B_P \Delta_{P \cap V_i}(V_i \mid \mathbb{Q}) \cdot \Delta_P(L \mid V_i)$ and also $\Delta_P(L \mid \mathbb{Q}) = B_P \Delta_{P \cap H}(H \mid \mathbb{Q}) \cdot \Delta_P(L \mid H)$; since P is totally ramified in $L \mid \mathbb{Q}$, and $[H:\mathbb{Q}] = [V_i:\mathbb{Q}] = p$ by (**B**), we deduce that the exponents of the differents $\Delta_P(L \mid V_i)$ and $\Delta_P(L \mid H)$ must coincide, and this is a contradiction. ∎

D. $\mathscr{K} = G(L \mid \mathbb{Q})$ *is a cyclic group.*
Proof: By hypothesis, $\mathscr{K}$ is an Abelian group of order p^m. By (**C**), $\mathscr{K}$ has only one subgroup of order p^{m-1}. This implies necessarily that $\mathscr{K}$ is a cyclic group: it is a well-known fact in the theory of finite Abelian groups which may be proved either directly or else by means of the structure theorem of finite Abelian groups (see Lemma 1 below, or Chapter 3, Theorem 3). ∎

For the convenience of the reader, we shall establish this and another easy fact about finite Abelian groups.

First we recall that if p is a prime number dividing the order n of the finite Abelian group G, then G has an element of order p (see Chapter 3, (**M**)).

Lemma 1. *Let G be a finite Abelian group.*
(1) *If G has order p^m, where $m \geqslant 1$ and p is a prime, if H is a subgroup of G of order p^h, if $h \leqslant h' \leqslant m$, then there exists a subgroup H' of G, having order $p^{h'}$ and containing H.*
(2) *If G has order p^m, $m \geqslant 2$, p a prime, if G has only one subgroup of order p^{m-1}, then G is a cyclic group.*
Proof:
(1) It is enough to assume that $h' = h + 1 \leqslant m$ and then repeat the argument. Let $\bar{G} = G/H$ be the quotient group so $\#\bar{G} = p^{m-h}$; thus there exists an element $\bar{x} \in G/H$ of order p. Let H' be the subgroup generated by H and x, so H' contains H properly (since $x \notin H$); but $H' = H \cup Hx \cup \cdots \cup Hx^{p-1}$, because $x^p \in H$; thus $\#H' = p^{h+1}$.
(2) Let H be the only subgroup of order p^{m-1} of G, let $x \in G$, $x \notin H$ and assume that x has order less than p^m. By (1), the cyclic group generated by x is contained in a subgroup of order p^{m-1} of G, which must be equal to H, by hypothesis, so $x \in H$, a contradiction. This means that x has order p^m and G is a cyclic group. ∎

E. *In Case 1, $L \subseteq \mathbb{Q}(\zeta)$, where ζ is a root of unity.*
Proof: Let $R = \mathbb{Q}(\zeta)$ where ζ is a primitive root of unity of order p^{m+1}. Thus $R \mid \mathbb{Q}$ is an extension of degree $\varphi(p^{m+1}) = p^m(p - 1)$, with Galois group isomorphic to the group $P(p^{m+1})$ of prime residue classes modulo p^{m+1} (see Chapter 2, (8)); by Chapter 3, (**L**), $R \mid \mathbb{Q}$ is a cyclic extension (since $p \neq 2$). By Chapter 13, (**4A**) the discriminant $\delta_{R|\mathbb{Q}} = \delta_R$ is a power of p.

The cyclic group $G(R \mid \mathbb{Q})$ has a subgroup of order $p - 1$ (if σ is a generator then σ^{p^m} has order $p - 1$), whose field of invariants we denote by R', so $[R':\mathbb{Q}] = p^m$. Thus, $R' \mid \mathbb{Q}$ is a cyclic extension and the discriminant $\delta_{R'}$ is again a power of p (if q is a prime, dividing $\delta_{R'}$ then q is ramified in $R' \mid \mathbb{Q}$ hence also ramified in $R \mid \mathbb{Q}$, hence q divides δ_R and so $q = p$, using Theorem 1 of Chapter 10.

Let LR' be the compositum field of L and R'; by Chapter 2, (7), $LR' \mid \mathbb{Q}$

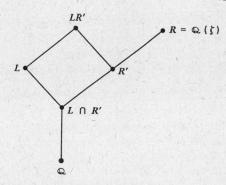

is also an Abelian extension, with degree $[LR':\mathbb{Q}] = [LR':R'] \cdot [R':\mathbb{Q}] = [L: L \cap R'] \cdot [R':\mathbb{Q}]$ which is a power of p. Now we show that the discriminant $\delta_{LR'}$ is also a power of p. In fact, if q is a prime dividing $\delta_{LR'}$ then by Chapter 10, Theorem 1, q is ramified in LR'. By Chapter 10, (7O) or Lemma 3 below, q is ramified in $L \mid \mathbb{Q}$ or q is ramified in $R' \mid \mathbb{Q}$. Hence q divides δ_L or q divides $\delta_{R'}$. In both cases $q = p$ and therefore $\delta_{LR'}$ is a power of p.

We may now apply (**D**) to the Abelian extension $LR' \mid \mathbb{Q}$ with degree and discriminant powers of p; it follows that $LR' \mid \mathbb{Q}$ is a cyclic extension, and by Galois theory, $G(LR' \mid L \cap R') \cong G(L \mid L \cap R') \times G(R' \mid L \cap R')$. Now, it is quite obvious that such a decomposition as a Cartesian product of cyclic groups of orders powers of p has to be trivial, namely, one of the groups $G(L \mid L \cap R')$ or $G(R' \mid L \cap R')$ has to be trivial (see Lemma 2 below). If $L = L \cap R'$ then $L \subseteq R'$, if $R' = L \cap R'$ then $R' \subseteq L$ and since L, R' have the same degree over $\mathbb{Q}$, then in both cases $R' = L$ and so $L = R' \subseteq \mathbb{Q}(\zeta)$. ∎

This lemma is included again for the convenience of the reader:

Lemma 2. *Let G be a cyclic group of order p^m, where p is any prime number. If $G \cong H \times H'$, then H, H' are cyclic groups of orders $p^h, p^{h'}$ respectively with either $h = 1$, or $h' = 1$ (hence $G \cong H'$ or $G \cong H$ respectively).*
Proof: This follows at once from the uniqueness asserted in Theorem 3 of Chapter 1. ∎

The following lemma has already been proved in Chapter 10, (7O). For the case of Galois extensions we have however a simpler proof, independent of the theory of the different:

Lemma 3. *Let K, K' be algebraic number fields, Galois extensions of $\mathbb{Q}$, let $L = K . K'$ be the compositum of these fields. If q is a prime number, unramified in $K \mid \mathbb{Q}$ and in $K' \mid \mathbb{Q}$ then q is also unramified in $L \mid \mathbb{Q}$.*
Proof: We recall that $L \mid (K \cap K')$ is a Galois extension and $G(L \mid K \cap K') \cong G(K \mid K \cap K') \times G(K' \mid K \cap K')$; this isomorphism associates with every $\sigma \in G(L \mid K \cap K')$ the couple $(\sigma_K, \sigma_{K'})$, where σ_K denotes the restriction of σ to K and $\sigma_{K'}$ the restriction of σ to K'.

Let Q be any prime ideal of the ring C of integers of L such that $Q \cap \mathbb{Z} = \mathbb{Z}q$; let $\mathcal{T}_Q(L \mid \mathbb{Q})$ be the corresponding inertial group. If

$$\sigma \in \mathcal{T}_Q(L \mid \mathbb{Q}) \cap G(L \mid K \cap K')$$

then $\sigma_K \in \mathcal{T}_{Q \cap K}(K \mid K \cap K')$, $\sigma_{K'} \in \mathcal{T}_{Q \cap K'}(K' \mid K \cap K')$ (as one sees immediately from the definition of the inertial groups). By hypothesis the inertial groups of the prime ideals of K, K' which extend q are necessarily trivial; a fortiori, $\sigma_K, \sigma_{K'}$ are the identity automorphism, hence σ is the identity automorphism. This proves that Q is unramified in $L \mid (K \cap K')$. But $Q \cap (K \cap K')$ is also unramified in $(K \cap K') \mid \mathbb{Q}$ because of the hypothesis. Therefore Q is unramified in $L \mid \mathbb{Q}$, showing that q is unramified in $L \mid \mathbb{Q}$. ∎

The proof of the Case **1** of the theorem is now complete. We shall continue, considering the case where $p = 2$.

Case **2**: $[L:\mathbb{Q}] = 2^m$, $\delta_L = 2^k$.

F. *Given $m \geqslant 1$ there exists a real field K such that $[K:\mathbb{Q}] = 2^m$, δ_K is a power of 2 and $K \subseteq \mathbb{Q}(\xi)$ for some root of unity ξ.*
Proof: Let ξ be a primitive root of unity, of order 2^{m+2}, and let $K' = \mathbb{Q}(\xi)$. Then $[K':\mathbb{Q}] = \varphi(2^{m+2}) = 2^{m+1}$ and hence $i = \sqrt{-1} \in K'$.

Let $K = K' \cap \mathbb{R}$, so $K' = K(i)$; in fact the conjugates of ξ belong to K', are either real or appear in pairs of complex conjugates $a + bi$, $a - bi$ with $a, b \in \mathbb{R}$; then $2a, 2b \in K' \cap \mathbb{R} = K$, so $a, b \in K$ and $K' = K(i)$. It follows that $[K':K] = 2$, hence $[K:\mathbb{Q}] = 2^m$.

We shall show that the discriminant δ_K is a power of 2. If q is a prime dividing δ_K then q is ramified in $K \mid \mathbb{Q}$, hence also ramified in $K' \mid \mathbb{Q}$; therefore q divides $\delta_{K'}$ (by Chapter 10, Theorem 1). But, if $K' = \mathbb{Q}(\xi)$, then by Chapter 13, (4A), $\delta_{K'}$ is a power of 2, so $q = 2$ and δ_K is a power of 2. ∎

G. *Given $m \geqslant 1$, there exists one real field K such that $K \mid \mathbb{Q}$ is an Abelian extension, $[K:\mathbb{Q}] = 2^m$ and δ_K is a power of 2.*

Proof: If $m = 1$ and $[F:\mathbb{Q}] = 2$, $F \subseteq \mathbb{R}$ then $F = \mathbb{Q}(\sqrt{d})$ with $d > 0$, d square-free. But $\delta_F = d$ when $d \equiv 1 \pmod 4$, $\delta_F = 4d$ when $d \equiv 2$ or 3 $\pmod 4$, so necessarily $d = 2$, $F = \mathbb{Q}(\sqrt{2})$.

Thus, we may assume that $m \geqslant 2$. If F is an Abelian extension of $\mathbb{Q}$, $F \subseteq \mathbb{R}$, $[F:\mathbb{Q}] = 2^m$, and δ_F power of 2, then the Galois group $G(F \mid \mathbb{Q})$ contains a subgroup of order 2^{m-1} (by Lemma 1), hence F contains a subfield H such that $[H:\mathbb{Q}] = 2$ and the discriminant of H must be a power of 2 (by the same argument); so $H = \mathbb{Q}(\sqrt{2})$. Thus $G(F \mid \mathbb{Q})$ contains only one subgroup of order 2^{m-1} and by Lemma 1, it must be cyclic.

If F is different from the field K, obtained in (**F**), we consider the compositum FK; thus $FK \subseteq \mathbb{R}$, $FK \mid \mathbb{Q}$ is again an Abelian extension of degree power of 2 and with discriminant power of 2 (see the argument in (**E**) and Lemma 3); hence by our proof just above (considering FK in place of F), $G(FK \mid \mathbb{Q})$ is a cyclic group, and

$$G(FK \mid F \cap K) \cong G(F \mid F \cap K) \times G(K \mid F \cap K).$$

By Lemma 2, either $F \subseteq K$ or $K \subseteq F$ and since both fields have the same degree 2^m then $F = K$. ∎

H. *If $L \mid \mathbb{Q}$ is an Abelian extension of degree 2^m and the discriminant δ_L is a power of 2, then there exists a root of unity ζ such that $L \subseteq \mathbb{Q}(\zeta)$.*

Proof: Since $\mathbb{Q}(i)$, L are Abelian extensions of $\mathbb{Q}$ then the compositum $L(i)$ is also an Abelian extension of $\mathbb{Q}$. By previous arguments $L(i) \mid \mathbb{Q}$ has degree and discriminant powers of 2.

Let $K = L(i) \cap \mathbb{R}$, hence K is a real Abelian extension of $\mathbb{Q}$, with degree and discriminant powers of 2. By (**G**) and (**F**) there exists a root of unity ξ such that $K \subseteq \mathbb{Q}(\xi)$.

Let $L(i) = K(a + bi)$, where $a, b \in \mathbb{R}$. The complex conjugate $a - bi$, which is a conjugate of $a + bi$ over $\mathbb{Q}$, still belongs to $L(i)$; hence $a \in L(i) \cap \mathbb{R} = K$, and $bi \in L(i)$, thus $b^2 \in L(i) \cap \mathbb{R} = K$; it follows that $a + bi$ is a root of the polynomial $X^2 - 2aX + (a^2 + b^2)$ with coefficients in K, so $[L(i):K] = 2$; since $i \notin K$ then $L \subseteq L(i) = K(i) \subseteq \mathbb{Q}(\xi, i) \subseteq \mathbb{Q}(\zeta)$, where ζ is a root of unity. ∎

It remains now to show how it is possible to reduce the general case of the theorem to the preceding ones.

Reduction to cases **1** *and* **2**:

I. *If the theorem is true for Abelian extensions having degree power of a prime, then it is true for any finite Abelian extension of $\mathbb{Q}$.*

Proof: Let L be an algebraic number field, Abelian extension of degree n over $\mathbb{Q}$. We prove now that L is the compositum of finitely many fields $L = L_1 \ldots L_s$, where each L_i is an extension having degree power of a prime.

By Chapter 3, (**O**), the Abelian Galois group $G(L \mid \mathbb{Q})$ is isomorphic to the Cartesian product of p_i-groups: $G(L \mid \mathbb{Q}) \cong \Pi_{i=1}^s \mathscr{H}_i$, $\#(\mathscr{H}_i) = p_i^{h_i}$, $[L:\mathbb{Q}] = n = \Pi_{i=1}^s p_i^{h_i}$. Let $\mathscr{L}_i = \Pi_{j \neq i} \mathscr{H}_j$ for every $i = 1, \ldots, s$, and let L_i denote the fixed field of the subgroup $\mathscr{L}_i$ of $G(L \mid \mathbb{Q})$; then $[L_i:\mathbb{Q}]$ is a power of p_i, since $G(L_i \mid \mathbb{Q}) \cong \mathscr{H}_i$ $(i = 1, \ldots, s)$. Moreover, if $L_1 \ldots L_s$ denotes the compositum of the fields $L_1, \ldots, L_s$ then the Galois group $G(L \mid L_1 \ldots L_s) \subseteq \cap_{i=1}^s \mathscr{L}_i = \{\varepsilon\}$ thus $L = L_1 \ldots L_s$.

Assuming the theorem true for each of the Abelian extensions L_i of $\mathbb{Q}$, we may write $L_i \subseteq \mathbb{Q}(\xi_i)$ where ξ_i is a primitive root of unity; let ζ be a primitive root of unity of order equal to the least common multiple of the orders of $\xi_1, \ldots, \xi_s$; then

$$L = L_1 L_2 \ldots L_s \subseteq \mathbb{Q}(\xi_1, \ldots, \xi_s) \subseteq \mathbb{Q}(\zeta). \quad \blacksquare$$

J. *If the theorem is true for Abelian extensions having degree and discriminant which are powers of the same prime p, then it is also true for Abelian extensions of degree power of p.*

Proof: In order to establish (**J**), we shall need to prove the following reduction step:

K. *Let $L \mid \mathbb{Q}$ be an Abelian extension of degree n. For every prime q dividing δ_L but not dividing n, there exists an Abelian extension $L' \mid \mathbb{Q}$ such that $[L':\mathbb{Q}]$ divides n, $L \subseteq L'(\xi)$ where ξ is a qth root of unity, q does not divide $\delta_{L'}$ and if q' is a prime dividing $\delta_{L'}$ then q' divides δ_L too.*

Assuming (**K**), we may proceed as follows: If $L \mid \mathbb{Q}$ is an Abelian extension of degree p^m (where p is a prime number), if δ_L is also a power of p, then we are already in the first or second case and the theorem is true.

If there exists a prime q, different from p, such that q divides δ_L, by (**K**) there exists an Abelian extension $L_1 \mid \mathbb{Q}$ and a qth root of unity ξ_1 such that $L \subseteq L_1(\xi_1)$, $[L_1:\mathbb{Q}]$ is still a power of p, q does not divide the discriminant δ_{L_1} and if q' is any prime dividing δ_{L_1} then already q' divides δ_L; thus, δ_{L_1} has *less* prime factors than δ_L.

If δ_{L_1} is not a power of p, we repeat the same argument; hence there exists an Abelian extension $L_2 \mid \mathbb{Q}$ and a root of unity ξ_2 such that $L_1 \subseteq L_2(\xi_2)$, $[L_2:\mathbb{Q}]$ is a power of p and δ_{L_2} has less prime factors than δ_{L_1}.

After a finite number of steps, we arrive at an Abelian extension $L_r \mid \mathbb{Q}$ such that $[L_r:\mathbb{Q}]$ is a power of p, $L_{r-1} \subseteq L_r(\xi_r)$, where ξ_r is a root of unity, and finally δ_{L_r} is now a power of p (perhaps equal to 1, in which case $L_r = \mathbb{Q}$). At worst, by the first or second case, $L_r \subseteq \mathbb{Q}(\xi_{r+1})$, where ξ_{r+1} is a root of unity.

Then $L \subseteq L_1(\xi_1)$, $L_1 \subseteq L_2(\xi_2), \ldots, L_{r-1} \subseteq L_r(\xi_1)$, $L_r \subseteq \mathbb{Q}(\xi_{r+1})$ and so $L \subseteq \mathbb{Q}(\xi_1, \ldots, \xi_{r+1}) \subseteq \mathbb{Q}(\zeta)$ where ζ is a root of unity of order equal to the least common multiple of the orders of the roots $\xi_1, \ldots, \xi_{r+1}$. This proves the theorem, except for the need of establishing **(K)**. ∎

Proof of **(K)**:

Case 1: L contains a primitive qth root of unity ξ. Then $L \supseteq \mathbb{Q}(\xi) \supseteq \mathbb{Q}$. Let Q be a prime ideal of the ring of L such that $Q \cap \mathbb{Z} = \mathbb{Z}q$. Since q does not divide $n = [L:\mathbb{Q}]$ then q does not divide $e = e_Q(L \mid \mathbb{Q})$. By Chapter 11, Theorem 2, if $V_1 = V_{1Q}(L \mid \mathbb{Q})$ then $[L:V_1]$ is a power of q and divides n, so $L = V_1$. By Chapter 11, **(J)**, the ramification index $e = \#(\mathcal{T}/\mathcal{V}_1) = \#(\mathcal{T})$ divides $q - 1$.

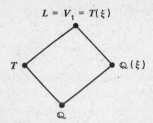

$$L = V_1 = T(\xi)$$
$$T \qquad\qquad \mathbb{Q}(\xi)$$
$$\mathbb{Q}$$

On the other hand, $e = e_Q(L \mid \mathbb{Q}) = e_Q(L \mid \mathbb{Q}(\xi)) \cdot e_{Q'}(\mathbb{Q}(\xi) \mid \mathbb{Q})$, where $Q' = Q \cap \mathbb{Q}(\xi)$. By Chapter 10, **(3A)**, $e_{Q'}(\mathbb{Q}(\xi) \mid \mathbb{Q}) = q - 1$, thus $q - 1$ divides e, therefore $e = q - 1$, thus $e_Q(L \mid \mathbb{Q}(\xi)) = 1$.

Let $L' = T$, the inertial field of Q in $L \mid \mathbb{Q}$ and we shall prove that it satisfies the required conditions.

Of course, $T \mid \mathbb{Q}$ is an Abelian extension, and its degree divides n. The inertial group of Q in $L \mid \mathbb{Q}(\xi)$ is $\mathcal{T} \cap G(L \mid \mathbb{Q}(\xi))$, hence the inertial field is $T(\xi)$; similarly $V_{1Q}(L \mid \mathbb{Q}(\xi)) = V_{1Q}(L \mid \mathbb{Q}) \cdot \mathbb{Q}(\xi) = V_1 = L$. Thus $[L:T(\xi)] = e_Q(L \mid \mathbb{Q}(\xi)) = 1$, hence $L = T(\xi) = L'(\xi)$.

Next, we note that q does not divide δ_T because q is unramified in T (this being the inertial field). If q' is any prime, $q' \neq q$, and q' divides δ_T then q' is ramified in $T \mid \mathbb{Q}$ hence also in $L \mid \mathbb{Q}$, thus q' divides δ_L (by Chapter 10, Theorem 1).

Case 2:   General.

We adjoin a primitive qth root of unity ξ to L, obtaining the Abelian extension $L(\xi) \mid \mathbb{Q}$. Let $F = L \cap \mathbb{Q}(\xi)$, so

$$G(L(\xi) \mid F) \cong G(L \mid F) \times G(\mathbb{Q}(\xi) \mid F).$$

Then $[L(\xi):\mathbb{Q}] = [L(\xi):F] \cdot [F:\mathbb{Q}] = [L:F] \cdot [\mathbb{Q}(\xi):F] \cdot [F:\mathbb{Q}]$ divides $n(q - 1)$ since $[\mathbb{Q}(\xi):\mathbb{Q}] = q - 1$.

We may apply Case 1 to the Abelian extension $L(\xi) \mid \mathbb{Q}$. Let q be a prime dividing δ_L but not dividing n; then q is ramified in $L \mid \mathbb{Q}$, hence also in

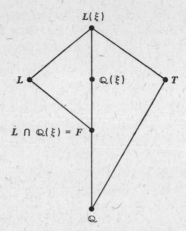

$L(\xi) \mid \mathbb{Q}$, thus q divides $\delta_{L(\xi)}$ (Chapter 10, Theorem 1). Also q does not divide $[L(\xi):\mathbb{Q}]$, because this degree divides $n(q-1)$. By Case 1, if Q is a prime ideal of the ring of integers of $L(\xi)$, $Q \cap \mathbb{Z} = \mathbb{Z}q$, if T is the inertial field of Q in $L(\xi) \mid \mathbb{Q}$, then $T(\xi) = L(\xi)$ and $[L(\xi):T] = e_Q(L(\xi) \mid \mathbb{Q}) = q - 1$, $[L(\xi):\mathbb{Q}] = [L(\xi):T] \cdot [T:\mathbb{Q}] = (q-1) \cdot [T:\mathbb{Q}]$ hence $[T:\mathbb{Q}]$ divides n. Since q is unramified in $T \mid \mathbb{Q}$ then q does not divide δ_T.

Now, if q' is a prime different from q, and dividing δ_T, then q' is ramified in $T \mid \mathbb{Q}$, hence also in $L(\xi) \mid \mathbb{Q}$. By Lemma 3, either q' is ramified in $L \mid \mathbb{Q}$ or in $\mathbb{Q}(\xi) \mid \mathbb{Q}$. Since $q' \neq q$, it is not ramified in $\mathbb{Q}(\xi) \mid \mathbb{Q}$, and q' divides the discriminant of $L \mid \mathbb{Q}$. Thus, we only need to take $L' = T$. This concludes the proof of the theorem. ∎

EXERCISES

1. Let I be a set of indices, let $\leqslant$ be an order relation on I and $(S_i)_{i \in I}$ a family of sets. For every couple of indices $(i,j) \in I \times I$, such that $i \leqslant j$, let $\pi_{ji}: S_j \to S_i$ be a mapping such that:

(1) π_{ii} is the identity map.

(2) if $i \leqslant j \leqslant k$ then $\pi_{ki} = \pi_{ji} \circ \pi_{kj}$.

Show:

(a) There exists a set S and a family of maps $\pi_i: S \to S_i$ with the following properties:

   (1) If $i \leqslant j$ then $\pi_i = \pi_{ji} \circ \pi_j$.

   (2) If S' is a set, if for every $i \in I$, $\pi_i': S' \to S_i$ is a map such that $\pi_i' = \pi_{ji} \circ \pi_j'$ when $i \leqslant j$ then there exists a unique map $\theta: S' \to S$ such that $\pi_i \circ \theta = \pi_i'$ for every $i \in I$.

(b) S, $(\pi_i)_{i \in I}$ are uniquely defined, in the following sense: if $\tilde{S}$, $(\tilde{\pi}_i)_{i \in I}$ satisfy the properties of (a) then there exists a bijection $\sigma: S \to \tilde{S}$ such that $\tilde{\pi}_i \circ \sigma = \pi_i$ for every $i \in I$.

$(S, (\pi_i)_{i \in I})$ is called the *inverse limit* or *projective limit* of the family $(S_i)_{i \in I}$, with respect to the maps π_{ji} (for $i \leqslant j$). It is denoted by $S = \varprojlim S_i$.

2. In the previous exercise, assume that each set is a group and that each map is a group-homomorphism. Show that S is a group and the maps π_i are group-homomorphisms. Explicitly, S is the subgroup of the product $\Pi_{i \in I} S_i$, consisting of the families $(s_i)_{i \in I}$ such that if $i \leqslant j$ then $\pi_{ji}(s_j) = s_i$.

3. Let $K \mid K_0$ be a Galois extension of infinite degree, let $\mathscr{G}$ be the family of fields F, $K_0 \subseteq F \subseteq K$ such that $F \mid K_0$ is a Galois extension of finite degree. Show:
   (a) If $F, F' \in \mathscr{G}$ then $FF' \in \mathscr{G}$.
   If $F \mid K_0$, $F' \mid K_0$ are Galois extensions, $K_0 \subseteq F' \subseteq F \subseteq K$, let

$$\rho_{FF'} : G(F \mid K_0) \to G(F' \mid K_0)$$

be the group-homomorphism which associates with every K_0-automorphism of F its restriction to F'.
   (b) If $F'' \subseteq F' \subseteq F$ belong to $\mathscr{G}$ then $\rho_{FF''} = \rho_{F'F''} \circ \rho_{FF'}$ and ρ_{FF} is the identity.
   (c) If $F' \subseteq F$ belong to $\mathscr{G}$ then $\rho_{KF'} = \rho_{FF'} \circ \rho_{KF}$.
   (d) Let H be a group, for every $F \in \mathscr{G}$ let $\theta_F : H \to G(F \mid K_0)$ be a group homomorphism such that if $F' \subseteq F$ then $\rho_{FF'} \circ \theta_F = \theta_{F'}$. Show that there is a unique group-homomorphism $\theta : H \to G(K \mid K_0)$ such that $\rho_{KF} \circ \theta = \theta_F$ for every $F \in \mathscr{G}$. Conclude that $G(K \mid K_0)$ is the inverse limit of the family of finite groups $G(F \mid K_0)$ (for $F \in \mathscr{G}$) with respect to the group-homomorphisms $\rho_{FF'}$.

4. Let $\xi_1, \ldots, \xi_k$ be roots of unity, m_i the order of ξ_i. Let $m = lcm(m_1, \ldots, m_k)$ and let ζ be a primitive mth root of unity. Show that $\mathbb{Q}(\xi_1, \ldots, \xi_k) = \mathbb{Q}(\zeta)$.

5. Let p be a prime number and T_p the field generated over $\mathbb{Q}$ by all p^mth roots of unity (for $m = 1, 2, \ldots$). Show:
   (a) $T_p \mid \mathbb{Q}$ is a Galois extension of infinite degree.
   (b) If T_p' denotes the field generated by $\cup_{q \neq p} T_q$, then $T_p \cap T_p' = \mathbb{Q}$ (for every prime number p).
   (c) the Abelian closure of $\mathbb{Q}$, denoted by $\mathbb{A}b$, is the field generated by $\cup T_p$ (for all prime numbers p).
   Hint: Apply the theorem of Kronecker and Weber.

6. Let $K \mid K_0$ be a Galois extension. For every $i = 1, 2, \ldots$ let $L_i \mid K_0$ be a Galois extension, $L_i \subseteq K$, such that:
   (a) If L_i' is the subfield of K generated by $\cup_{j \neq i} L_j$ then $L_i \cap L_i' = K_0$.
   (b) K is generated by $\cup_{i=1}^{\infty} L_i$.

Show:

$$G(K \mid K_0) \cong \prod_{i=1}^{\infty} G(L_i \mid K_0).$$

7. Show that $G(\bigwedge b \mid \mathbb{Q}) \cong \Pi_p G(T_p \mid \mathbb{Q})$.
   Hint: Apply the two previous exercises.

8. Let p be an odd prime and $G_p = G(T_p \mid \mathbb{Q})$. Prove that

$$G_p \cong \mathbb{Z}/(p-1) \times \mathbb{Z}_p,$$

where $\mathbb{Z}_p = \lim_{\leftarrow} \mathbb{Z}/\mathbb{Z}p^m$ (for $m \geqslant 1$), with respect to the canonical group-homomorphisms $\mathbb{Z}/\mathbb{Z}p^m \rightarrow \mathbb{Z}/\mathbb{Z}p^n \cong (\mathbb{Z}/\mathbb{Z}p^n)/(\mathbb{Z}p^m/\mathbb{Z}p^n)$ (when $m \geqslant n$).
   Hint: Apply the previous exercises and Chapter 3, (**J**).

9. Show that $G_2 = G(T_2 \mid \mathbb{Q})$ is isomorphic to $\mathbb{Z}/2 \times \mathbb{Z}_2$, where $\mathbb{Z}_2 = \lim_{\leftarrow} \mathbb{Z}/\mathbb{Z}2^m$ (for $m \geqslant 1$).
   Hint: Apply Chapter 3, (**K**).

CHAPTER 13

Miscellaneous Numerical Examples

In this chapter we shall exhibit some numerical examples to illustrate the theories developed so far.

The handling of a definite example may offer many difficulties. These may be due to the non-effectiveness of the methods of proof of the theorems; that is, no algorithm in a finite number of steps may be seen from the proof indicated. Sometimes, even though there is a theoretically finite algorithm to solve the problem, it may be too long to perform and therefore shortcuts have to be found.

We should stress that much more is known about computational methods than we are able to explain in this text. We suggest therefore that the reader consult more specialized books about these problems.

We are mainly interested in the following questions: the determination of the ring of integers and of an integral basis; the computation of the different and the discriminant; the decomposition of a natural number into a product of prime ideals of the ring of integers of the field; the determination of a fundamental system of units and the computation of the regulator. Concerning the class-number, there exist important analytical methods for its determination; however, these will not be dealt with in this book.

1. SOME THEORETICAL CONSIDERATIONS

1A. Let $K = \mathbb{Q}(t)$, where t is a root of the monic polynomial $f \in \mathbb{Z}[X]$.

The degree $n = [K:\mathbb{Q}]$ is equal to the degree of the minimal polynomial of t. This is an irreducible factor of f and may be determined in a finite number of steps, as it is well known.

So, we may already assume that $f \in \mathbb{Z}[X]$ is irreducible, $f(t) = 0$.

1B. The discriminant d of the polynomial f may be computed explicitly, as was indicated in Chapter 2, **(11)**. For this purpose, we make use of Newton formulas or of the norm in $K \mid \mathbb{Q}$ of the derivative of f. By definition, d is equal to the discriminant of the $\mathbb{Q}$-basis $\{1, t, \ldots, t^{n-1}\}$ of K.

Let $\{x_1, \ldots, x_n\}$ be an integral basis of K. From $\mathbb{Z}[t] \subseteq A$, we may write

$$t^j = \sum_{i=1}^{n} a_{ij} x_i \quad \text{(for } j = 0, 1, \ldots, n-1) \quad \text{with } a_{ij} \in \mathbb{Z}.$$

Thus,

$$d = \text{discr}_{K|\mathbb{Q}}(1, t, \ldots, t^{n-1})$$

$$= [\det(a_{ij})]^2 \cdot \text{discr}_{K|\mathbb{Q}}(x_1, \ldots, x_n) = m^2 \delta_K,$$

where $m \in \mathbb{Z}$, and δ_K is the discriminant of K. We know also from Chapter 6, (O) that δ_K is congruent to 0 or to 1 modulo 4.

The decomposition of d into a product of prime factors is obtainable in a finite number of steps, so we know whether d has a square factor $m^2 \neq 1$. If no such factor exists then $\delta_K = d$ and $A = \mathbb{Z}[t]$ because we may express the elements $x_1, \ldots, x_n$ in terms of the $\mathbb{Q}$-basis $\{1, t, \ldots, t^{n-1}\}$, as linear combinations with coefficients in $\mathbb{Z}$.

However, if $d = m^2 \delta_K$, with $m^2 > 1$, we may obtain further information in the following way. Suppose for example that $d = p^2 \delta_K$, where p is a prime number. Suppose also that we may find a subring A_1 of A such that $\mathbb{Z}[t]$ is properly contained in A_1. Since A_1 is also a free $\mathbb{Z}$-module of rank n, let $\{y_1, \ldots, y_n\}$ be a basis, hence $d = s^2 \cdot \text{discr}_{K|\mathbb{Q}}(y_1, \ldots, y_n)$. But $s^2 \neq 1$ (since $\mathbb{Z}[t] \neq A_1$). On the other hand,

$$\text{discr}_{K|\mathbb{Q}}(y_1, \ldots, y_n) = s_1^2 \cdot \text{discr}_{K|\mathbb{Q}}(x_1, \ldots, x_n) = s_1^2 \delta_K.$$

Comparing these relations we have $s_1^2 = 1$ hence $\delta_K = \text{discr}_{K|\mathbb{Q}}(y_1, \ldots, y_n)$, so $A_1 = A$ and $\{y_1, \ldots, y_n\}$ is an integral basis. Thus, in this case the discriminant would be found once the ring A_1 is found and a $\mathbb{Z}$-basis of A_1 is computed.

If $d = p_1^2 p_2^2 \ldots p_r^2 \delta_K$, where $p_1, p_2, \ldots, p_r$ are distinct primes, we may happen to find a strictly increasing chain of r distinct subrings containing properly $\mathbb{Z}[t]$, and then determine δ_K and an integral basis of the field K. However, it is important to note that it may well happen that the discriminant has a square factor $m^2 > 1$ which divides $\text{discr}_{K|\mathbb{Q}}(1, u, \ldots, u^{n-1})$ for every integer u of K such that $K = \mathbb{Q}(u)$. This possibility was indicated at the end of Chapter 10.

Let us mention here that at any rate the determination of the ring of integers may be effectively performed in a finite number of steps. This is seen as follows: Let t be an algebraic integer which is a primitive element of the extension $K \mid \mathbb{Q}$. Thus $\mathbb{Z}[t]$ is a free Abelian group of rank n. By Chapter 10, (7B) the complementary $\mathbb{Z}$-module $\mathbb{Z}[t]^*$ is a free Abelian group of rank n and its basis is explicitly computable (see Chapter 10, (7C)). We have $A \subseteq \mathbb{Z}[t]^* \subset K$ and $G = \mathbb{Z}[t]^*/\mathbb{Z}[t]$ is a finitely generated torsion Abelian group, thus a finite Abelian group. It is therefore possible to describe explicitly all the subgroups of G. Each subgroup H corresponds to a submodule M of $\mathbb{Z}[t]^*$ containing $\mathbb{Z}[t]$. Among the subgroups H of G some correspond to subrings M, a fact which may be verified from the multiplication table of the basis elements of M. The largest subring M so obtained is necessarily equal to A. Indeed, if M is not contained in A, let MA be the set of finite sums of elements of the form xy, where $x \in M$, $y \in A$. Then MA is a subring of K and A is properly contained in MA (if $x \in M$, $x \notin A$, then $x = x \cdot 1 \in MA$). Moreover, MA is a finitely generated $\mathbb{Z}$-module. Since every element of K is of the form a/m, where $a \in A$, $m \in \mathbb{Z}$, then there exists $r \in \mathbb{Z}$ such that $MA \subseteq (1/r)A$. By Chapter 7, Exercise **31**, (f) there exists a prime ideal P of A such that the subring MA contains the set

$$\{x \in K \mid \text{there exists } m \geqslant 0 \text{ such that } P^m Ax \subseteq A\} = \cup_{m \geqslant 0} P^{-m}.$$

So $P^{-m} \subseteq 1/rA$ and P^m divides Ar for every $m \geqslant 0$. This is impossible showing that every subring M, $\mathbb{Z}[t] \subseteq M \subseteq \mathbb{Z}[t]^*$ must be contained in A. Therefore A is determined as the largest of the subrings M which correspond to subgroups H of $G = \mathbb{Z}[t]^*/\mathbb{Z}[t]$.

1C.   Let K be an algebraic number field, A the ring of integers of K, P a prime ideal of A and $\varphi \colon A \to A/P = \bar{K}$, the canonical homomorphism with kernel equal to P. Let $L \mid K$ be an extension of degree n, let B' be a subring of the ring B of integers of L, such that $A \subseteq B' \subseteq B$. We consider homomorphisms $\psi \colon B' \to \bar{L}$ such that:

(1) $\bar{L}$ is a field extension of $\bar{K}$ of degree at most equal to n.
(2) $\psi(B') = \bar{L}$.
(3) ψ extends φ.

We say that two such homomorphisms ψ, ψ' are *equivalent* when there exists a $\bar{K}$-isomorphism $\tau \colon \bar{L} \to \bar{L'}$ such that $\tau \circ \psi = \psi'$. Since $\bar{K}$ is a finite field, then $\bar{L}$ is a Galois extension, hence necessarily $\bar{L'} = \bar{L}$ and τ belongs to the Galois group of $\bar{L} \mid \bar{K}$. We denote by $[\psi]$ the equivalence class of ψ and we note that if ψ, ψ' are equivalent then they have the same kernel.

Taking $B' = B$ we show that there is a one-to-one correspondence between the set of prime ideals Q of B such that $Q \cap A = P$ and the set of equivalence classes $[\psi]$. In fact, given Q, let $\psi \colon B \to B/Q$ be the canonical homomorphism. It satisfies properties (1), (2), and (3) and has kernel equal to Q. Conversely, if ψ is a homomorphism satisfying (1), (2), and (3), and if Q is the kernel of ψ then Q is a prime ideal of B and $Q \cap A = P$, because ψ extends φ. Moreover, $B/Q \cong \psi(B) = \bar{L}$, so the canonical homomorphism $B \to B/Q$ is equivalent to the homomorphism ψ.

Now let $\{y_1, \ldots, y_m\}$ be a system of generators of the A-module B and let

$$y_i y_j = \sum_{k=1}^{m} a_k^{ij} y_k$$

(for all $i, j = 1, \ldots, m$) where $a_k^{ij} \in A$. Then there is a one-to-one correspondence between the set of homomorphisms $\psi \colon B \to \bar{L}$ satisfying (1), (2), and (3) and the set of systems $\{\bar{y}_1, \ldots, \bar{y}_m\}$ of elements of L such that:

(1') $\bar{L} = \sum_{i=1}^{m} \bar{K} \bar{y}_i.$

(2') if $\sum_{i=1}^{m} a_i y_i = 0,$ then $\sum_{i=1}^{m} \varphi(a_i) \bar{y}_i = 0.$

(3') $\bar{y}_i \bar{y}_j = \sum_{k=1}^{m} \varphi(a_k^{ij}) \bar{y}_k$ (for all $i, j = 1, \ldots, m$).

In fact, if ψ is given and $\bar{y}_i = \psi(y_i)$, then $\{\bar{y}_1, \ldots, \bar{y}_m\}$ satisfy (1'), (2'), and (3'). Conversely, if the elements $\bar{y}_1, \ldots, \bar{y}_m$ are given with properties (1'), (2'), and (3'), let $\psi: B \to L$ be the mapping which is so defined:

$$\psi\left(\sum_{i=1}^{m} c_i y_i\right) = \sum_{i=1}^{m} \varphi(c_i) \bar{y}_i.$$

ψ is well-defined, as follows from (2'). Moreover, ψ is a ring-homomorphism onto I, as results from (1') and (3'), and properties (1), (2), and (3) are satisfied. It is also clear that the above correspondence is one-to-one.

Summarizing, the determination of the prime ideals Q of B such that $Q \cap A = P$ amounts to the possible assignments of images $\bar{y}_1, \ldots, \bar{y}_m$ satisfying the conditions indicated. Moreover, the fundamental relation $\Sigma_{i=1}^{g} e_i f_i = n$ of Chapter 10, (1L) and Theorem 1 in its sharper form allow to determine the inertial degrees and the ramification indices.

The above considerations required the knowledge of a system of generators of B as well as the relations which they satisfy. Thus B had to be completely known.

1D.   Even when B is not fully known, it is possible to find the decomposition of all but finitely many ideals BP (where P is a non-zero prime ideal of A).

We assume that A is a principal ideal domain, so B is a free A-module. Let $\{x_1, \ldots, x_n\}$ be an integral basis, thus $\delta_{L|K} = A \cdot \operatorname{discr}_{L|K}(x_1, \ldots, x_n)$. Let L' be the smallest Galois extension of K containing L and $\sigma_1, \ldots, \sigma_n$ the K-isomorphisms of L. For any elements $z_1, \ldots, z_n \in B$ let $D(z_1, \ldots, z_n) = [\det(\sigma_i(z_j))]^2 = \operatorname{discr}_{L|K}(z_1, \ldots, z_n)$. Let $t \in B$ be a primitive element of $L \mid K$, so $\{1, t, \ldots, t^{n-1}\}$ is a K-basis of L. If $y \in B$ we may write $y = \Sigma_{j=0}^{n-1} y_j t^j$ with $y_j = a_j/a_j' \in K$ and $a_j, a_j' \in A$. We may also assume that a_j, a_j' are relatively prime elements of the principal ideal domain A.

Applying the K-isomorphisms σ_i to the above relation, we obtain

$$\sigma_i(y) = \sum_{j=0}^{n-1} y_j \sigma_i(t^j)   (i = 1, \ldots, n).$$

This means that $\{y_0, \ldots, y_{n-1}\}$ is a solution of the system of linear equations in L', having coefficients $\sigma_i(t^j)$. By Cramer's rule, we have

$$y_j = \sqrt{\frac{D(1, t, \ldots, y, \ldots, t^{n-1})}{D(1, t, \ldots, t^j, \ldots, t^{n-1})}}.$$

But $A + At + \cdots + Ay + \cdots + At^{n-1} \subseteq B = \Sigma_{i=1}^n Ax_i$ and $\Sigma_{j=0}^{n-1} At^j \subseteq B = \Sigma_{i=1}^n Ax_i$, so, expressing the generators of the smaller module in terms of the integral basis with coefficients in A, we deduce that

$$D(1, t, \ldots, y, \ldots, t^{n-1}) = [\det(c_{ij}^{(j)})]^2 \cdot D(x_1, \ldots, x_n),$$
$$D(1, t, \ldots, t^j, \ldots, t^{n-1}) = [\det(c_{ik})]^2 \cdot D(x_1, \ldots, x_n),$$

with $c_{ik}, c_{ik}^{(j)} \in A$. We conclude that

$$\frac{a_j}{a_j'} = y_j = \frac{\det(c_{ik}^{(j)})}{\det(c_{ik})}.$$

So $a_j' \cdot \det(c_{ik}^{(j)}) = a_j \cdot \det(c_{ik})$ and a_j' divides

$$\det(c_{ik}) = \sqrt{\frac{D(1, t, \ldots, t^{n-1})}{D(x_1, x_2, \ldots, x_n)}} = \alpha.$$

Thus if P is a prime ideal of A and P does not divide α then P does not divide a_j' (for $j = 0, 1, \ldots, n - 1$). Hence, $y = \Sigma_{j=0}^{n-1} (a_j/a_j') \cdot t^j = z/a$ with $z \in A[t]$, $a \in A, a \notin P$.

We shall show how to determine the decomposition into prime ideals of the ideal BP, where P does not contain α. This result will therefore be valid for all but a finite number of prime ideals of A. In particular, if $B = A[t]$ then $\alpha = 1$ and it yields the decomposition of every prime ideal P of A.

As we have seen, the prime ideals Q of B such that $Q \cap A = P$ are in one-to-one correspondence with the equivalence classes of homomorphisms $\psi: B \to L$ which extend the homomorphism $\varphi: A \to A/P$ and have been described in (1C).

Each homomorphism ψ induces a homomorphism ψ_0 of the subring $A[t]$ into L. The image of ψ_0 is equal to L. Indeed, $\psi_0(A[t])$ is a subring of B containing $\varphi(A)$; since L is algebraic over A/P then $\psi_0(A[t])$ is itself a field. On the other hand, since B is contained in the field of quotients of $A[t]$ then $\psi(B) = L$ is also contained in the field of quotients of $\psi_0(A[t])$, thus $\psi_0(A[t]) = L$.

Conversely, if ψ_0 is a homomorphism from $A[t]$ onto L then it induces a unique extension to a homomorphism ψ from B to L. In fact, as we have shown every element $y \in B$ may be written in the form $y = z/a$ where $z \in A[t]$ and $a \in A, a \notin P$. Therefore, we may define $\psi(y) = \psi_0(z)/\varphi(a)$. $\psi: B \to L$ is a well-defined mapping and clearly the unique homomorphism of B onto L

extending ψ_0. It is also obvious that equivalent homomorphisms ψ, ψ' have restrictions ψ_0, ψ_0' which are equivalent homomorphisms from $A[t]$ (that is, ψ_0, ψ_0' have the same kernel).

Therefore, our problem becomes the determination of the equivalence classes of the homomorphisms $\psi_0: A[t] \to L$.

Let $f \in A[X]$ be the minimal polynomial of t over K, and let $\varphi(f) = \bar{f} \in \bar{K}[X]$ be the image of f. We consider its decomposition as a product of irreducible polynomials:

$$\bar{f} = \bar{g}_1^{l_1} \ldots \bar{g}_r^{l_r} \quad (l_i \geqslant 1)$$

where $g_i \in A[X]$ and $\bar{g}_1, \ldots, \bar{g}_r$ are distinct irreducible polynomials over A/P.

There is a one-to-one correspondence between the set of equivalence classes of homomorphisms $\psi_0: A[t] \to \bar{L}$ extending φ and the set of irreducible polynomials $\{\bar{g}_1, \ldots, \bar{g}_r\}$. In fact, given ψ_0, let $\bar{t} = \psi_0(t)$ so $\bar{t}$ is a root of $\bar{f}$, thus there exists a unique polynomial $\bar{g}_i$ such that $\bar{g}_i(\bar{t}) = 0$; $\bar{g}_i$ depends only on the equivalence class of ψ_0. This defines the correspondence $[\psi_0] \to \bar{g}_i$. Conversely, given the irreducible factor $\bar{g}_i$ of $\bar{f}$, we consider any root θ of $\bar{g}_i$ and let $\bar{L} = \bar{K}(\theta)$; then we define the homomorphism $\psi_0: A[t] \to \bar{L}$ extending φ, by $\psi_0(t) = \theta$. If we take another root θ' of $\bar{g}_i$ and define ψ_0' in similar way then ψ_0, ψ_0' are equivalent.

This establishes a one-to-one correspondence between the set of equivalence classes of extensions ψ_0 of φ and the set $\{\bar{g}_1, \ldots, \bar{g}_r\}$. Therefore the number of distinct prime ideals Q of B such that $Q \cap A = P$ is equal to the number r of irreducible factors $\bar{g}_i$ of $\bar{f}$ over A/P. If Q_i is the kernel of the extension $\psi^{(i)}$ of $\psi_0^{(i)}$ then the inertial degree of Q_i in $L \mid K$ is equal to

$$[B/Q_i : A/P] = [\psi^{(i)}(B) : A/P] = [\psi_0^{(i)}(B) : A/P] = \deg(\bar{g}_i)$$

where $\psi_0^{(i)}(t)$ is a root of $\bar{g}_i$.

Finally, we want to show that the ramification indices coincide with the exponents $l_1, \ldots, l_r$ which appeared in the decomposition of $\bar{f}$ into a product of irreducible polynomials.

If $BP = \Pi_{i=1}^r Q_i^{e_i}$, by the fundamental relation we have $n = \Sigma_{i=1}^n e_i f_i$, where $f_i = [B/Q_i : A/P] = \deg(\bar{g}_i)$ (Q_i corresponds to $\bar{g}_i$). On the other hand, $n = \deg(f) = \deg(\bar{f}) = \Sigma_{i=1}^r l_i \deg(\bar{g}_i)$. To show that $l_i = e_i$ it suffices to prove that $e_i \leqslant l_i$ (for $i = 1, \ldots, r$). Since $f - g_1^{l_1} \ldots g_r^{l_r} \in P[X]$ and $f(t) = 0$, then $g_1^{l_1}(t) \ldots g_r^{l_r}(t) \in A[t]P \subseteq BP = \Pi_{i=1}^r Q_i^{e_i}$. We shall determine the decomposition into a product of prime ideals of $Bg_1^{l_1}(t) \ldots g_r^{l_r}(t)$. We have $g_j(t) \notin Q_i$ when $i \neq j$ because $\bar{g}_j(\psi_i(t)) \neq 0$, and $g_i(t) \in Q_i$. We want to show that Q_i^2 does not divide $g_i(t)$. This will imply that

$$Bg_1^{l_i}(t) \ldots g_r^{l_r}(t) = Q_1^{l_i} \ldots Q_r^{l_r} \cdot J,$$

where J is an ideal not a multiple of $Q_1, \ldots, Q_r$. Since $\Pi_{i=1}^r Q_i^{e_i}$ divides $\Pi_{i=1}^r Q_i^{l_i}J$ then $e_i \leqslant l_i$ (for $i = 1, \ldots, r$) as we have to show.

For this purpose, we show that $Q_i \cap A[t]$ is the ideal generated by $A[t]P$ and $g_i(t)$. Indeed, given $z = h(t)$, with $h \in A[X]$, if $z \in Q_i$ then $0 = \psi_i(z) = \bar{h}().\bar{t}$ Hence $\bar{g}_i$ divides $\bar{h}$ so $\bar{h} = \bar{g}_i \bar{k}$ with $k \in A[X]$. Thus we may write $h = g_i k + m$, where $m \in P[X]$ and $z = h(t) = g_i(t)k(t) + m(t)$ showing our assertion.

Now we show that $g_i(t) \notin Q_i^2$. In fact, choosing $y \in Q_i$, $y \notin Q_i^2$, we may write $y = z/a$ with $z \in A[t]$, $a \in A$, $a \notin P$. Then $z = ay \in Q_i \cap A[t]$ but $z = ay \notin Q_i^2 \cap A[t]$. From the expression $z = g_i(t)k(t) + m(t)$ it follows that either $g_i(t) \notin Q_i^2$ or $m(t) \notin Q_i^2$. In the case where $g_i(t) \in Q_i^2$ and $m(t) \notin Q_i^2$, we just replace g_i by $g_i + m$, noting that $\bar{g}_i = \overline{g_i + m}$ and that the determination of the polynomials g_i was made modulo $P[X]$. Then $z = (g_i(t) + m(t))k(t) - m(t)(k(t) - 1)$ with $g_i(t) + m(t) \notin Q_i^2$.

2. SOME CUBIC FIELDS

Example 1. Let $K = \mathbb{Q}(t)$, where t is a root of $f = X^3 - X - 1$. This is an irreducible polynomial, because by Gauss' lemma the only roots in $\mathbb{Q}$ could be $1, -1$.

The discriminant of f is equal to $-(4(-1)^3 + 27(-1)^2) = -23$ (see Chapter 2, Exercise **48**).

Let δ_K be the discriminant of K. We have $-23 = m^2 \delta_K$, where $m \in \mathbb{Z}$; hence $m^2 = 1$ and $\delta_K = -23$. It follows also that $\{1, t, t^2\}$ is an integral basis. The only ramified prime is $p = 23$.

To see, for example, how the primes $2, 3, 5, 7, 23$ are decomposed, we consider the decomposition into irreducible polynomials of the images of f in the fields $\mathbb{F}_2, \mathbb{F}_3, \mathbb{F}_5, \mathbb{F}_7, \mathbb{F}_{23}$.

Over $\mathbb{F}_2, \bar{f}$ is irreducible, hence $A . 2 = Q_2$ where Q_2 has inertial degree 3.

Over $\mathbb{F}_3, \bar{f}$ is irreducible, hence $A . 3 = Q_3$ where Q_3 has inertial degree 3.

Over $\mathbb{F}_5$, $\bar{f} = (X - \bar{2})(X^2 + \bar{2}X + \bar{3})$, hence $A . 5 = Q_5 . Q_5'$, Q_5 has inertial degree 1, Q_5' has inertial degree 2. We have homomorphisms ψ_5, ψ_5' from A onto $\mathbb{F}_5$ with kernels respectively equal to Q_5, Q_5'. $\psi_5(t) = \bar{2}$, $\psi_5(t^2) = \bar{4}$, hence $Q_5 \supseteq \mathbb{Z} . 5 \oplus \mathbb{Z}(t - 2) \oplus \mathbb{Z}(t^2 - 4)$. On the other hand, if $a, b, c \in \mathbb{Z}$ and $\psi_5(a + bt + ct^2) = 0$ then $\bar{a} + \bar{2}\bar{b} + \bar{4}\bar{c} = \bar{0}$, so there exists $m \in \mathbb{Z}$ such that $a = 5m - 2b - 4c$, thus

$$a + bt + ct^2 = 5m + b(t - 2) + c(t^2 - 4).$$

This shows that $Q_5 = \mathbb{Z} . 5 \oplus \mathbb{Z}(t - 2) \oplus \mathbb{Z}(t^2 - 4)$. From

$$t^2 - 4 = (t + 2)(t - 2)$$

and

$$N_{K|\mathbb{Q}}(t - 2) = (-1)^3 f(2) = -5$$

it follows that 5, $t^2 - 4$ belong to the principal ideal $A(t - 2)$, hence $Q_5 = A(t - 2)$.

From $A \cdot 5 = Q_5 \cdot Q_5'$, it follows also that Q_5' is the principal ideal generated by $5/(t - 2) = -(t' - 2)(t'' - 2)$ where t', t'' are the conjugates of t. But $tt't'' = 1$, $t + t' + t'' = 0$, hence

$$(t' - 2)(t'' - 2) = t't'' - 2(t' + t'') + 4 = 1/t + 2t + 4 = (2t^2 + 4t + 1)/t.$$

From $N_{K|\mathbb{Q}}(t) = 1$, t is a unit, thus $Q_5' = A(2t^2 + 4t + 1)$.

Over $\mathbb{F}_7$, $\bar{f} = (X - \bar{5})(X^2 + \bar{5}X + \bar{3})$, hence $A \cdot 7 = Q_7 \cdot Q_7'$, Q_7 has inertial degree 1, Q_7' has inertial degree 2. We have homomorphisms ψ_7, ψ_7' from A onto $\mathbb{F}_7$ with kernels Q_7, Q_7'. $\psi_7(t) = \bar{5}$ $\psi_7(t^2) = \bar{4}$, hence by a calculation already explained $Q_7 = \mathbb{Z} \cdot 7 \oplus \mathbb{Z}(t - 5) \oplus \mathbb{Z}(t^2 - 4)$. But $t^2 - 4 = (t + 5)(t - 5) + 3 \times 7$ and $N_{K|\mathbb{Q}}(t - 5) = (-1)^3 f(5) = -119 = -7 \times 17$, hence $7 \notin A(t - 5)$; otherwise, $7 = x(t - 5)$, $x \in A$, and taking norms $343 = -N_{K|\mathbb{Q}}(x) \times 7 \times 17$, which is impossible because $N_{K|\mathbb{Q}}(x) \in \mathbb{Z}$. Therefore Q_7 is the ideal generated by 7, $t - 5$ (since the decomposition $A \cdot 7 = Q_7 \cdot Q_7'$ implies that $Q_7 \neq A \cdot 7$).

Next, $\psi_7'(7t) = 0$, $\psi_7'(t^2 + 5t + 3) = 0$ then $Q_7' = \mathbb{Z} \cdot 7 \oplus \mathbb{Z} \cdot 7t \oplus \mathbb{Z}(t^2 + 5t + 3)$. In fact, the generators are linearly independent over $\mathbb{Q}$ and if $\psi_7'(a + bt + ct^2) = 0$ then $\bar{a} + \bar{b}\bar{t} + \bar{c}(-\bar{5}\bar{t} - \bar{3}) = \bar{0}$ so

$$a = 7m + 3c, \quad b = 7l + 5c, \quad \text{with} \quad l, m \in \mathbb{Z},$$

and

$$a + bt + ct^2 = 7m + 7lt + c(t^2 + 5t + 3).$$

We conclude that Q_7' is the ideal generated by 7 and $t^2 + 5t + 3$.

Over $\mathbb{F}_{23}$, we know already that 23 is ramified, hence either $A \cdot 23 = Q_{23}^2 \cdot Q_{23}'$ or $A \cdot 23 = Q_{23}^3$. To decide what actually happens, we factorize $X^3 - X - \bar{1}$ into irreducible polynomials modulo 23. Since there will be a root of multiplicity at least 2, this will be a common root to $\bar{f} = X^3 - X - \bar{1}$, and to its derivative $\bar{f}' = \bar{3}X^2 - \bar{1}$. Multiplying $\bar{f}'$ by X and subtracting $\bar{3}\bar{f}$ we have $2X + 3 \equiv 0 \pmod{23}$ hence $t \equiv 10 \pmod{23}$ is a double root and this yields the congruence $X^3 - X - 1 \equiv (X - 10)^2(X - 3) \pmod{23}$. Therefore the decomposition is $A \cdot 23 = Q_{23}^2 \cdot Q_{23}'$ where Q_{23}, Q_{23}' have inertial degree equal to 1.

By a similar argument, we show that $Q_{23} = \mathbb{Z} \cdot 23 \oplus \mathbb{Z}(t - 10) \oplus \mathbb{Z}(t^2 - 8)$, $t^2 - 8 = (t + 10)(t - 10) + 4 \times 23$. Q_{23} is the ideal generated by 23 and $t - 10$, because $N_{K|\mathbb{Q}}(t - 10) = (-1)^3 f(10) = -989 = -23 \times 43$, $N_{K|\mathbb{Q}}(23) = 23^3$, hence $23 \notin A(t - 10)$ and $t - 10 \notin A \cdot 23$.

In the same way, $Q_{23}' = \mathbb{Z} \cdot 23 \oplus \mathbb{Z}(t - 3) \oplus \mathbb{Z}(t^2 - 9)$, $t^2 - 9 = (t + 3)(t - 3)$ and $N_{K|\mathbb{Q}}(t - 3) = (-1)^3 f(3) = -23$, thus Q_{23}' is the principal ideal generated by $t - 3$. In particular, Q_{23}^2 is the principal ideal generated by $3t^2 + 9t + 1$, because $-23 = (t - 3)(t' - 3)(t'' - 3)$, $t\, t' t'' = 1$, $t + t' + t'' = 0$, $N_{K|\mathbb{Q}}(t) = 1$, so t is a unit and

$$\frac{23}{t - 3} = \frac{1}{t}\,(3t^2 + 9t + 1).$$

Example 2. Let $K = \mathbb{Q}(t)$, where t is a root of $f = X^3 - 3X + 9$.

This is an irreducible polynomial (by Gauss' lemma the only roots in $\mathbb{Q}$ could be $\pm 1, \pm 3, \pm 9$ and none of these numbers is a root).

The discriminant of f is $d = -(4 \times (-3)^3 + 27 \times 9^2) = -27 \times 7 \times 11$.

We shall determine a subring A_1 of A which contains properly $\mathbb{Z}[t]$. From

$$t^3 - 3t + 9 = 0 \tag{1}$$

we have $1 - 3/t^2 + 9/t^3 = 0$ and multiplying by 3, $(3/t)^3 - (3/t)^2 + 3 = 0$. Let $u = 3/t$, so

$$u^3 - u^2 + 3 = 0 \tag{2}$$

and $u \in A$. The $\mathbb{Z}$-module A_1 generated by $\{1, t, u\}$ is actually a subring of A. In fact, dividing (1) by t and (2) by u, we have

$$\begin{cases} t^2 = 3 - 3u \\ u^2 = u - t \\ tu = 3 \end{cases}$$

and this provides the multiplication table of A_1. Also $u = 1 - (1/3)t^2$ and since this expression is unique (because $\{1, t, t^2\}$ is a $\mathbb{Q}$-basis of K) then $u \notin \mathbb{Z}[t]$. Therefore, $\mathbb{Z}[t]$ is properly contained in A_1. Considering the discriminants, we have $d = m^2 d_1$, where $d_1 = \mathrm{discr}_{K|\mathbb{Q}}(1, t, u)$, with $1 < m^2$. Hence $m^2 = 9$ (this is the only square dividing d) and $d_1 = -3 \times 7 \times 11$. Now, from $A_1 \subseteq A$ follows $d_1 = r^2 \delta_K$. But d_1 has no square-factors, thus $r^2 = 1$, $d_1 = \delta_K$, $A_1 = A$, and $\{1, t, u\}$ is an integral basis. The only ramified primes are therefore 3, 7, 11.

We shall describe in detail the decomposition of some primes p in A. Referring to our discussion in **(D)**, we see that

$$\alpha = \sqrt{\frac{\mathrm{discr}_{K|\mathbb{Q}}(1, t, t^2)}{\mathrm{discr}_{K|\mathbb{Q}}(1, t, u)}} = 3.$$

Over $\mathbb{F}_2$, $X^3 - 3X + 9 \equiv X^2 + X + 1 \pmod 2$ and this polynomial is irreducible over $\mathbb{F}_2$. Hence $A \cdot 2 = Q_2$ where the inertial degree of Q_2 in $K \mid \mathbb{Q}$ is equal to 3.

Over $\mathbb{F}_{17}$, $X^3 - 3X + 9$ has the root 5 mod 17 and we have $X^3 - 3X + 9 \equiv (X - 5)(X^2 + 5X + 5) \pmod{17}$ where $X^2 + 5X + 5$ is irreducible modulo 17. Thus $A \cdot 17 = Q_{17} \cdot Q_{17}'$, where Q_{17} has inertial degree 1, Q_{17}' has inertial degree 2.

Over $\mathbb{F}_7$ we know that $X^3 - 3X + 9$ must have at least a double root. So, we look for the roots common to $X^3 - 3X + 9$ and its derivative $3X^2 - 3$. 1 mod 7 is such a root and we have the decomposition $X^3 - 3X + 9 \equiv (X - 1)^2(X + 2) \pmod 7$. Thus $A \cdot 7 = Q_7^2 \cdot Q_7'$, where Q_7, Q_7' have inertial degree equal to 1.

The prime ideal Q_7 is the kernel of the homomorphism $\psi_7 \colon A \to \mathbb{F}_7$ such that $\psi_7(t) = \bar{1}$, while Q_7' is the kernel of ψ_7', with $\psi_7'(t) = \bar{2}$. From the relations between u,t we deduce that $3\psi_7(u) = \bar{3} - \psi_7(t^2) = \bar{2}$, hence $\psi_7(u) = \bar{3}$. Similarly $\psi_7'(u) = \bar{2}$. By the computation explained in Example 1, $Q_7 = \mathbb{Z} \cdot 7 \oplus \mathbb{Z}(t - 1) \oplus \mathbb{Z}(u - 3)$. Since $u - 3 = -u(t - 1)$ and $N_{K|\mathbb{Q}}(t - 1) = (-1)^3 f(1) = -7$, then Q_7 is the principal ideal generated by $t - 1$. Hence Q_7^2 is the principal ideal generated by $(t - 1)^2$ and Q_7' is the principal ideal generated by $7/(t - 1)^2$. If t', t'' are the conjugates of t, then $t + t' + t'' = 0$, $t\,t'\,t'' = -9$ so $-7 = (t - 1)(t' - 1)(t'' - 1)$, $(t' - 1)(t'' - 1) = -9/t + t + 1$,

$$\frac{7}{(t - 1)^2} = \frac{1}{7}(t' - 1)^2(t'' - 1)^2 = \frac{(t^2 + t - 9)^2}{7t^2}$$

$$= \frac{[(t^2 + t - 9)u]^2}{63} = \frac{9(1 + t - 3u)^2}{63} = -(2 + t),$$

so $Q_7' = A(t + 2)$. We could also see this directly, noting that $Q_7' = \mathbb{Z} \cdot 7 \oplus \mathbb{Z}(t + 2) \oplus \mathbb{Z}(u - 2)$, $N_{K|\mathbb{Q}}(t + 2) = (-1)^3 f(-2) = -7$. Thus $7 \in A(t + 2)$ and $u(t + 2) = 3 + 2u = 2(u - 2) + 7$ therefore $4u(t + 2) = (u - 2) + 7(u - 2) + 2$, and $u - 2 = 4u(t + 2) - (u - 1) \cdot 7 \in A(t + 2)$. This shows again that $Q_7' = A(t + 2)$.

Over $\mathbb{F}_{11}$ we have, similarly, $X^3 - 3X + 9 \equiv (X + 1)^2(X - 2) \pmod{11}$ so $A \cdot 11 = Q_{11}^2 \cdot Q_{11}'$, where Q_{11}, Q_{11}' have inertial degree equal to 1.

Q_{11} is the kernel of the homomorphism $\psi_{11} \colon A \to \mathbb{F}_{11}$ such that $\psi_{11}(t) = -\bar{1}$. Similarly, $\psi_{11}' \colon A \to \mathbb{F}_{11}$ has kernel Q_{11}' and $\psi_{11}'(t) = \bar{2}$. Then $3\psi_{11}(u) = \bar{3} - \psi_{11}(t^2) = \bar{3} - \bar{1} = \bar{2}$, hence $\psi_{11}(u) = \bar{8}$. Similarly, $\psi_{11}'(u) = \bar{7}$. Thus

$$Q_{11} = \mathbb{Z} \cdot 11 \oplus \mathbb{Z}(t + 1) \oplus \mathbb{Z}(u + 3),$$

$$Q_{11}' = \mathbb{Z} \cdot 11 \oplus \mathbb{Z}(t - 2) \oplus \mathbb{Z}(u + 4).$$

Since $u + 3 = u(t + 1)$ and $N_{K|\mathbb{Q}}(t + 1) = (-1)^3 f(-1) = -11$ then $11 \in A(t + 1)$ therefore $Q_{11} = A(t + 1)$. Next, $N_{K|\mathbb{Q}}(t - 2) = (-1)^3 f(2) = -11$, so $11 \in A(t - 2)$. Also $u(t - 2) = 3 - 2u = -2(u + 4) + 11$,

$$5u(t - 2) = (u + 4) - 11(u + 4) + 11$$

hence $u + 4 \in A(t - 2)$ showing that $Q_{11}' = A(t - 2)$.

Now we describe the decomposition of $A \cdot 3$. The method indicated in **(1D)** cannot be applied to the prime number 3.

From the relations satisfied by t, u it follows that if $\psi \colon A \to \mathbb{F}_3$ is any homomorphism then $\psi(t) = \bar{t}$, $\psi(u) = \bar{u}$ satisfy $\bar{t}^2 = \bar{0}$, $\bar{u}^2 = \bar{u}$, $\bar{t}\bar{u} = \bar{0}$. The only possibilities ψ_3, ψ_3' are:

$$\psi_3(t) = \bar{0}, \psi_3(u) = \bar{0},$$

and

$$\psi_3'(t) = 0, \psi_3'(u) = \bar{1}.$$

If $Q_3 = \mathrm{Ker}\,(\psi_3)$, $Q_3' = \mathrm{Ker}\,(\psi_3')$ then Q_3, Q_3' have inertial degree 1. We have $Q_3 = \mathbb{Z} \cdot 3 \oplus \mathbb{Z}t \oplus \mathbb{Z}u$, $Q_3' = \mathbb{Z} \cdot 3 \oplus \mathbb{Z}t \oplus \mathbb{Z}(u-1)$. From $t = -(u-1)u$, and $3 = N_{K|\mathbb{Q}}(u)$ it follows that Q_3 is the principal ideal generated by u. Similarly, $3 = N_{K|\mathbb{Q}}(u-1)$, so Q_3' is the principal ideal generated by $u - 1$. Thus $At = Q_3 \cdot Q_3'$, and $A \cdot 3 = Atu = Q_3^2 \cdot Q_3'$.

Example 3. Now we discuss a classical example of Dedekind. Let $K = \mathbb{Q}(t)$, where t is a root of $f = X^3 + X^2 - 2X + 8$.

f is irreducible over $\mathbb{Q}$, because if $f = f_1 f_2$, with $f_1, f_2 \in \mathbb{Z}[X]$, $\deg\,(f_1) > 0$ $\deg\,(f_2) > 0$, then reducing the coefficients modulo 2 we would have $\bar{f} = X^3 + X^2 = X^2(X + \bar{1})$. The constant term of f_2 is congruent to 1 mod 2. Since it divides 8 it must be 1 or -1. But $f(1) \neq 0$, $f(-1) \neq 0$, so f is irreducible.

The discriminant of f is equal to $d = 4 - 4 \times 8 + 18 \times (-2) \times 8 - 4 \times (-2)^3 - 27 \times 8^2 = -2012 = -4 \times 503$ (see Chapter 2, Exercise **48**).

We shall determine a subring A_1 of A which contains properly $\mathbb{Z}[t]$. From

$$t^3 + t^2 - 2t + 8 = 0 \tag{3}$$

we have $1 + 1/t - 2/t^2 + 8/t^3 = 0$ and, multiplying by 8, $8 + 8/t - 16/t^2 + 64/t^3 = 0$. Letting $u = 4/t$ then

$$u^3 - u^2 + 2u + 8 = 0, \tag{4}$$

hence $u \in A$.

The $\mathbb{Z}$-module A_1 generated by $\{1, t, u\}$ is a subring of A. In fact, dividing (3) by t and (4) by u we have

$$\begin{cases} t^2 = 2 - t - 2u \\ u^2 = -2 - 2t + u \\ ut = 4 \end{cases}$$

These relations provide the multiplication table in A_1 and show that A_1 is a subring of A. Moreover $u \notin \mathbb{Z}[t]$ since $u = 1 - (1/2)t - (1/2)t^2$ (and the expression of u in terms of the $\mathbb{Q}$-basis $\{1, t, t^2\}$ is unique). Thus $\mathbb{Z}[t]$ is properly contained in A_1. If $d_1 = \mathrm{discr}_{K|\mathbb{Q}}(1,t,u)$ then $d = m^2 d_1$ with $1 < m^2$. Hence $m^2 = 4$ and $d_1 = -503$. Since 503 is prime then $A = A_1$ and the discriminant of K is $\delta_K = d_1 = -503$.

The only ramified prime is 503. In our discussion in (**1D**) we have

$$\alpha = \sqrt{\frac{\mathrm{discr}_{K|\mathbb{Q}}(1,t,t^2)}{\mathrm{discr}_{K|\mathbb{Q}}(1,t,u)}} = 2.$$

Let us study the decomposition of the primes 2 and 503 in the ring A.

From the relations satisfied by t, u we see that if $\psi\colon A \to \mathbb{F}_2$ is any homomorphism then $\psi(\bar{t}) = \bar{t}$, $\psi(u) = \bar{u}$ satisfy $\bar{t}^2 = \bar{t}$, $\bar{u}^2 = \bar{u}$, $\bar{t}\bar{u} = \bar{0}$. The only possibilites $\psi_2, \psi_2', \psi_2''$ are:

$$\psi_2(t) = \bar{0}, \qquad \psi_2(u) = \bar{0},$$
$$\psi_2'(t) = \bar{1}, \qquad \psi_2'(u) = \bar{0},$$
$$\psi_2''(t) = \bar{0}, \qquad \psi_2''(u) = \bar{1}.$$

If $Q_2 = \mathrm{Ker}\,(\psi_2)$, $Q_2' = \mathrm{Ker}\,(\psi_2')$, $Q_2'' = \mathrm{Ker}\,(\psi_2'')$ then Q_2, Q_2', Q_2'' have inertial degree 1 and $A \cdot 2 = Q_2 \cdot Q_2' \cdot Q_2''$. We have

$$Q_2 = \mathbb{Z} \cdot 2 \oplus \mathbb{Z}t \oplus \mathbb{Z}u,$$
$$Q_2' = \mathbb{Z} \cdot 2 \oplus \mathbb{Z}(t-1) \oplus \mathbb{Z}u,$$
$$Q_2'' = \mathbb{Z} \cdot 2 \oplus \mathbb{Z}t \oplus \mathbb{Z}(u-1).$$

Now we show that these prime ideals are principal. $N_{K|\mathbb{Q}}(t) = -8$ and similarly $N_{K|\mathbb{Q}}(u) = -8$ since $ut = 4$. t divides $N_{K|\mathbb{Q}}(t)$, hence the only prime ideals appearing in the decomposition of At are those which divide 2. But $t \notin Q_2'$ and $t \in Q_2$, $t \in Q_2''$. Similarly, $u \notin Q_2''$ but $u \in Q_2$, $u \in Q_2'$. So $A \cdot tu = A \cdot 4 = Q_2^2 \cdot Q_2'^2 \cdot Q_2''^2$. We show that Q_2^2 does not divide At. Otherwise either $At = Q_2^2 \cdot Q_2''$, thus $Au = Q_2'^2 \cdot Q_2''$ which is impossible; or $At = Q_2^2 \cdot Q_2''^2$, $Au = Q_2'^2$, then $8 = |N_{K|\mathbb{Q}}(u)| = N(Au) = (N(Q_2'))^2 = 4$, absurd. With the same argument, we see that Q_2^2 does not divide Au. Therefore $At = Q_2 \cdot Q_2''^2$ and $Au = Q_2 \cdot Q_2'^2$.

Let us note that if $a \in \mathbb{Q}$ and if t,t',t'' are conjugate over $\mathbb{Q}$ then

$$N_{K|\mathbb{Q}}(t - a) = (t - a)(t' - a)(t'' - a) = -8 + 2a - a^2 - a^3 = -f(a).$$

If $a \in \mathbb{Z}$ is odd then $t - a \in Q_2'$, $t - a \notin Q_2$, $t - a \notin Q_2''$ and $u - a \in Q_2''$, $u - a \notin Q_2$, $u - a \notin Q_2'$. Thus $N_{K|\mathbb{Q}}(t-1) = -8$, therefore the prime ideals dividing $A(t-1)$ must be among Q_2, Q_2', Q_2''. From the above we know that $A(t-1)$ must be a power of Q_2' and taking norms we conclude that $A(t-1) = Q_2'^3$. In the same manner, we see from $N_{K|\mathbb{Q}}(t+3) = -8 - 6 - 9 + 27 = 4$ that $A(t+3) = Q_2'^2$. Hence

$$Q_2' = A\left(\frac{t-1}{t+3}\right)$$

is a principal ideal. In terms of the integral basis, we may write $(t - 1)/(t + 3) = a + bt + cu$ where $a, b, c \in \mathbb{Z}$ are easily determined taking into account the multiplication table; namely $(t - 1)/(t + 3) = -5 + 3t + 2u$.

If $a \in \mathbb{Z}$ is even but not a multiple of 4 then $t - a \notin Q_2'$, and $t - a \in Q_2$, $t - a \in Q_2''$, $t - a \notin Q_2''^2$ (as may be seen from the norms). Similarly, $u - a \notin Q_2''$, and $u - a \in Q_2$, $u - a \in Q_2'$, $u - a \notin Q_2'^2$. Thus $N_{K|\mathbb{Q}}(t - 2) = -16$, $N_{K|\mathbb{Q}}(t + 2) = -8$ and the decomposition of $A(t - 2)$, $A(t + 2)$ is easily seen: $A(t - 2) = Q_2^3 \cdot Q_2''$, $A(t + 2) = Q_2^2 \cdot Q_2''$ (because of the norms). Hence

$$Q_2 = A\left(\frac{t - 2}{t + 2}\right) = A(2 - t - u).$$

Finally, $A \cdot 2 = Q_2 \cdot Q_2' \cdot Q_2''$, thus Q_2'' is also a principal ideal, namely

$$Q_2'' = A\left(\frac{2(t + 2)(t + 3)}{(t - 1)(t - 2)}\right) = A(-5 - 2t + u).$$

Now we study the decomposition of other primes p into prime ideals of A.

If $p = 3$ we have $A \cdot 3 = Q_3$ since $\bar{f} = X^3 + X^2 + X + \bar{2}$ is irreducible over $\mathbb{F}_3$.

Let $p = 5$. Then $\bar{f} = X^3 + X^2 + 3X + \bar{3} = (X + \bar{1})(X^2 + \bar{3})$ over $\mathbb{F}_5$. Hence $A \cdot 5 = Q_5 \cdot Q_5'$, where Q_5 has inertial degree 1 and Q_5' has inertial degree 2. If ψ_5, ψ_5' are homomorphisms from A with kernels Q_5, Q_5' respectively, then $\psi_5(t)\psi_5(u) = \bar{4}$ hence $\psi_5(u) = \bar{1}$ and by a computation already explained, we see that $Q_5 = \mathbb{Z} \cdot 5 \oplus \mathbb{Z}(t + 1) \oplus \mathbb{Z}(u - 1)$. Similarly, $Q_5' = \mathbb{Z} \cdot 5 \oplus \mathbb{Z}(t^2 + 3) \oplus \mathbb{Z}(u^2 + 2)$. From $N_{K|\mathbb{Q}}(t + 1) = (-1)^3 f(-1) = 10$ we see that the prime ideals dividing $t + 1$ are among those dividing $A \cdot 2$, $A \cdot 5$. We have seen that $t + 1 \in Q_2'$. Taking the norms into account, we must have $A(t + 1) = Q_2' \cdot Q_5$. So

$$Q_5 = A\left(\frac{(t + 1)(t + 3)}{t - 1}\right)$$

and this generator of Q_5 may be easily expressed in terms of the integral basis $\{1, t, u\}$. From $A \cdot 5 = Q_5 \cdot Q_5'$ we deduce also that Q_5' is a principal ideal generated by

$$\frac{5(t - 1)}{(t + 1)(t + 3)} \in A.$$

For the primes $p = 7$, $p = 11$, we see with some computation that $\bar{f}$ is irreducible over $\mathbb{F}_7$, respectively over $\mathbb{F}_{11}$. Hence $A \cdot 7 = Q_7$, $A \cdot 11 = Q_{11}$.

We conclude the study of this example noting the following facts.

For every integer $v \in A$ we have $A \neq \mathbb{Z} \oplus \mathbb{Z}v \oplus \mathbb{Z}v^2$. Indeed, $A \cdot 2 = Q_2 \cdot Q_2' \cdot Q_2''$, so the prime ideals Q_2, Q_2', Q_2'' have inertial degree equal to 1.

Thus $\psi_2(A) = \psi_2'(A) = \psi_2''(A) = \mathbb{F}_2$. If we had $A = \mathbb{Z}[v]$ then the homomorphisms would be determined by the image of v. The only possibilities are $\bar{0}, \bar{1} \in \mathbb{F}_2$, so there would only exist two homomorphisms from A onto $\mathbb{F}_2$, a contradiction.

This tells that the discriminant of K has an unessential factor, namely 2. Indeed $A \cdot 2 = Q_2 \cdot Q_2' \cdot Q_2''$, $N(A \cdot 2) = 2 < 3 = [K:\mathbb{Q}]$ (see Chapter 10 (7)).

The class number of K is 1; that is, every ideal of A is principal. It is enough to show that every prime ideal is principal. By Chapter 9, (M), in every class of ideals of K there exists an integral non-zero ideal J such that

$$N(J) \leqslant \left(\frac{4}{\pi}\right)^{r_2} \frac{n!}{n^n} \sqrt{|\delta_K|}.$$

In our case, $n = 3$, $r_2 \leqslant 1$, $|\delta_K| = 503$, hence $N(J) < 7$. Thus it suffices to prove that every prime ideal of A having norm less than 7 is principal. This has already been established.

Example 4. Let $K = \mathbb{Q}(t)$ be a field of degree 3 over $\mathbb{Q}$ (where t is an algebraic integer). Let $t_1 = t$, t_2, t_3 be the conjugates of t over $\mathbb{Q}$. We assume that $K_1 = \mathbb{Q}(t_1) = K$, $K_2 = \mathbb{Q}(t_2)$, $K_3 = \mathbb{Q}(t_3)$ are distinct fields. Let $K = K_1K_2 = K_1K_3 = K_2K_3$ (since $t_1 + t_2 + t_3 \in \mathbb{Z}$) so $K \mid \mathbb{Q}$ is a Galois extension of degree 6. We denote by A the ring of integers of K and by B_i the ring of integers of K_i ($i = 1,2,3$).

The Galois group of $K \mid \mathbb{Q}$ is the symmetric group on 3 letters $\mathfrak{K} = \mathfrak{S}_3$. Moreover $K = K_i(\sqrt{\delta})$ (for $i = 1,2,3$) where $\delta = \delta_{K_1} = \delta_{K_2} = \delta_{K_1}$. Indeed, $\mathbb{Z}[t_i] \subseteq B_i$, hence $d_i = \mathrm{discr}_{K_i|\mathbb{Q}}(t_i) = (t_1 - t_2)^2(t_1 - t_3)^2(t_2 - t_3)^2 = m_i^2\delta$ (with $m_i \in \mathbb{Z}$); thus $K \supseteq K_1(\sqrt{\delta}) = K_1(\sqrt{d_1}) \supseteq K_1$. But $\sqrt{d_1} \notin K_1$ hence $K = K_1(\sqrt{\delta})$. In fact, if $\sqrt{d_1} \in K_1$ from $[K_1:\mathbb{Q}] = 3$ it would follow that $\sqrt{d_1} \in \mathbb{Q}$; however, for the permutation

$$\sigma = \begin{pmatrix} t_1 & t_2 & t_3 \\ t_1 & t_3 & t_2 \end{pmatrix} \in \mathfrak{K}$$

we have $\sigma(\sqrt{d_1}) = (t_1 - t_3)(t_1 - t_2)(t_3 - t_2) = -\sqrt{d_1}$.

Let $L = \mathbb{Q}(\sqrt{\delta})$ thus $[K:L] = 3$, $[L:\mathbb{Q}] = 2$, $[K:K_i] = 2$, $[K_i:\mathbb{Q}] = 3$ (for $i = 1,2,3$). We denote by C the ring of integers of L.

The non-trivial subgroups of $\mathfrak{K} = \mathfrak{S}_3$ are $\mathfrak{A} = G(K \mid L)$, the alternating group on 3 letters, $\mathfrak{B}_i = G(K \mid K_i)$, group of order 2 generated by the transposition

$$\tau_i = \begin{pmatrix} t_i & t_j & t_k \\ t_i & t_k & t_j \end{pmatrix} \quad \text{(for } i = 1, 2, 3\text{)}.$$

Clearly $\mathfrak{A} \cap \mathfrak{B}_i = \{\varepsilon\}$, $\mathfrak{A}\mathfrak{B}_i = \mathfrak{K}$, $\mathfrak{A}$ is a normal subgroup of $\mathfrak{K}$, $\mathfrak{B}_1$, $\mathfrak{B}_2$, $\mathfrak{B}_3$ are conjugate subgroups.

We shall discuss all possible types of decomposition of an arbitrary prime number p in $K \mid \mathbb{Q}$.

The following notations will be used:

$$P = P_1, P_2, \ldots \text{ denote prime ideals of } A,$$
$$Q_{i1}, Q_{i2}, \ldots \quad \text{denote prime ideals of } B_i,$$
$$R_1, R_2, \ldots \quad \text{denote prime ideals of } C.$$

Case 1: p is unramified in $K \mid \mathbb{Q}$.

The inertial group and the inertial field of P in $K \mid \mathbb{Q}$ are respectively equal to $\mathscr{T}_P(K \mid \mathbb{Q}) = \{\varepsilon\}$, $T_P(K \mid \mathbb{Q}) = K$. The possibilities for the decomposition group of P in $K \mid \mathbb{Q}$ are the following:

(a) $\mathscr{Z}_P(K \mid \mathbb{Q}) = \{\varepsilon\}$,
(b) $\mathscr{Z}_P(K \mid \mathbb{Q}) = \mathfrak{A}$,
(c) $\mathscr{Z}_P(K \mid \mathbb{Q}) = \mathfrak{B}_i$ $(i = 1,2,3)$.

(a) In this case $Ap = P_1 P_2 P_3 P_4 P_5 P_6$, $B_i p = Q_{i1} Q_{i2} Q_{i3}$ $(i = 1,2,3)$, $Cp = R_1 R_2$, where the above prime ideals are distinct. Each prime ideal P_i, Q_{ij}, R_i has degree 1 over $\mathbb{Q}$.

In fact, since p is unramified, by the fundamental relation $n = efg$ we have:

(1) in the extension $K \mid \mathbb{Q}$: $n = g = 6$, $e = f = 1$,
(2) in the extension $K_i \mid \mathbb{Q}$: $n = g = 3$, $e = f = 1$,
(3) in the extension $L \mid \mathbb{Q}$: $n = g = 2$, $e = f = 1$.

(b) In this case $Ap = P_1 P_2$, $B_i p = Q_{i1}$ $(i = 1,2,3)$, $Cp = R_1 R_2$.

In fact, $g = (\mathfrak{K} : \mathfrak{A}) = 2$ so $Ap = P_1 P_2$. Also $\mathscr{Z}_P(K \mid K_i) = \mathscr{Z}_P(K \mid \mathbb{Q}) \cap \mathfrak{B}_i = \{\varepsilon\}$ hence $P \cap B_i$ decomposes into the product of two prime ideals of A, thus necessarily $A(P_1 \cap B_i) = P_1 P_2$. Therefore $B_i p$ must be a prime ideal of B_i, $B_i p = Q_{i1}$. On the other hand, by Chapter 11, **(E)**, $\mathscr{Z}_{P \cap C}(L \mid \mathbb{Q}) = \mathscr{Z}_P(K \mid \mathbb{Q})/\mathscr{Z}_P(K \mid L) = \mathfrak{A}/\mathfrak{A} = \{\varepsilon\}$, thus Cp is the product of two prime ideals, $Cp = R_1 R_2$.

(c) In this case $Ap = P_1 P_2 P_3$, $B_i p = Q_{i1} Q_{i2}$ $(i = 1, 2, 3)$, $Cp = R_1$. Moreover, $AQ_{i1} = P_1$, $f(P_1 \mid Q_{i1}) = 2$, $AQ_{i2} = P_2 P_3$, $f(Q_{i1} \mid \mathbb{Z}p) = 1$, $(Q_{i2} \mid \mathbb{Z}p) = 2$, $f(R_1 \mid \mathbb{Z}p) = 2$.

In fact, $g = (\mathfrak{K} : \mathfrak{B}_i) = 3$ so Ap is the product of 3 prime ideals. Since $\mathscr{Z}_P(K \mid K_i) = \mathscr{Z}_P(K \mid \mathbb{Q}) \cap \mathfrak{B}_i = \mathfrak{B}_i$ then $P \cap B_i$ (denoted by Q_{i1}) generates a prime ideal of A, that is $AQ_{i1} = P_1$. From the fundamental relation in $K \mid K_i$ we have $f(P_1 \mid Q_{i1}) = 2$. Let $Q_{i2} = P_2 \cap B_i$, so we know that $Q_{i2} \neq Q_{i1}$. Since P_2 is conjugate to P_1 by some $\sigma_2 \in \mathfrak{K}$ then $\sigma_2 \notin \mathfrak{B}_i$ and $\mathscr{Z}_{P_2}(K \mid \mathbb{Q}) = \sigma_2^{-1} \mathscr{Z}_P(K \mid \mathbb{Q}) \sigma_2 = \mathfrak{B}_j$ $(j \neq i)$. Hence $\mathscr{Z}_{P_2}(K \mid K_i) = \mathscr{Z}_{P_2}(K \mid \mathbb{Q}) \cap \mathfrak{B}_i = \{\varepsilon\}$, thus $P_2 \cap B_i = Q_{i2}$ decomposes into the product of two prime ideals of A (which are distinct from P_1), hence $AQ_{i2} = P_2P_3$.

Since $f(P_2 \mid \mathbb{Z}p) = f(P_1 \mid \mathbb{Z}p) = f(P_1 \mid Q_{i1}) \cdot f(Q_{i1} \mid \mathbb{Z}p) = 2$ and $f(P_2 \mid Q_{i2}) = f(P_3 \mid Q_{i2})$ cannot be 2 then $f(Q_{i2} \mid \mathbb{Z}p) = 2$.

From $\mathscr{Z}_P(K \mid L) = \mathscr{Z}_P(K \mid \mathbb{Q}) \cap \mathfrak{A} = \mathfrak{B}_i \cap \mathfrak{A} = \{\varepsilon\}$ it follows that if $R_1 = P \cap C$ then AR_1 is the product of 3 prime ideals of A, thus necessarily $AR_1 = P_1P_2P_3$, $Cp = R_1$ and $f(R_1 \mid \mathbb{Z}p) = 2$.

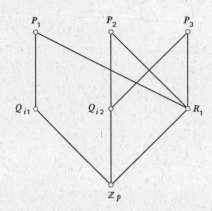

Case 2: p is ramified in $K \mid \mathbb{Q}$.

The possibilities for the inertial group and the decomposition group of P in $K \mid \mathbb{Q}$ are the following:

(a) $\mathscr{T}_P(K \mid \mathbb{Q}) = \mathscr{Z}_P(K \mid \mathbb{Q}) = \mathfrak{K}$
(b) $\mathscr{T}_P(K \mid \mathbb{Q}) = \mathfrak{A}$, $\mathscr{Z}_P(K \mid \mathbb{Q}) = \mathfrak{K}$
(c) $\mathscr{T}_P(K \mid \mathbb{Q}) = \mathscr{Z}_P(K \mid \mathbb{Q}) = \mathfrak{A}$
(d) $\mathscr{T}_P(K \mid \mathbb{Q}) = \mathscr{Z}_P(K \mid \mathbb{Q}) = \mathfrak{B}_i$ (for $i = 1, 2, 3$).

It suffices to recall that $\mathscr{T}_P(K \mid \mathbb{Q}) \neq \{\varepsilon\}$ and that $\mathscr{T}_P(K \mid \mathbb{Q})$ is a normal subgroup of $\mathscr{T}_P(K \mid \mathbb{Q})$.

(a) This case may only happen when $p = 3$. Then $A \cdot 3 = P_1^6$, $B_i \cdot 3 = Q_{i1}^3$, $C \cdot 3 = R_1^2$.

Let $\mathscr{V}_1$ be the first ramification group of P in $K \mid \mathbb{Q}$. $\mathscr{V}_1$ is a normal subgroup of $\mathscr{T} = \mathscr{T}_P(K \mid \mathbb{Q})$ and $\mathscr{T}/\mathscr{V}_1$ is a cyclic group (Chapter 11, Theorem 2); thus $\mathscr{V}_1 \neq \{\varepsilon\}$ since $\mathfrak{K} = \mathfrak{S}_3$. But the order of $\mathscr{V}_1$ is a power of p and $\#\mathfrak{S}_3 = 6$ thus $\mathscr{V}_1 \neq \mathfrak{K}$. Hence $\mathscr{V}_1 = \mathfrak{A}$, it has order 3, so necessarily $p = 3$.

From $\mathscr{T}_P(K \mid \mathbb{Q}) = \mathscr{Z}_P(K \mid \mathbb{Q}) = \mathfrak{K}$ it follows that the decomposition number and inertial degree of P in $K \mid \mathbb{Q}$ are equal to 1, hence $A \cdot 3 = P_1^6$.

By transitivity of the decomposition number and inertial degree, those of $P_1 \cap B_i$, $P_1 \cap C$ are also equal to 1, hence $B_i \cdot 3 = Q_{i1}^3$, $C \cdot 3 = R_1^2$.

The case in question may actually arise; for example when $K_1 = \mathbb{Q}(\sqrt[3]{3})$ (the reader should verify this statement).

(b) In this case $Ap = P_1^3$, $B_i p = Q_{i1}^3$ $(i = 1, 2, 3)$, $Cp = R_1$.

We have $\mathscr{T}_P(K \mid L) = \mathscr{L}_P(K \mid L) = \mathfrak{A}$, therefore if $P_1 \cap C = R_1$ then the decomposition number of P_1 in $K \mid L$ is 1, the inertial degree is also 1, so by the fundamental relation R_1 is totally ramified, that is $AR_1 = P_1^3$. Also $\mathscr{T}_{R_1}(L \mid \mathbb{Q}) = \mathscr{T}_P(K \mid \mathbb{Q})/\mathscr{T}_P(K \mid L) = \{\varepsilon\}$ hence the inertial degree $f(R_1 \mid \mathbb{Z}p) = 2$. Thus $f(P_1 \mid \mathbb{Z}p) = 2$ and by the fundamental relation for $K \mid \mathbb{Q}$, P_1 is the only prime ideal of A dividing $\mathbb{Z}p$; that is, $Ap = P_1^3$. Then there is only one prime ideal in B_i dividing $B_i p$, say Q_{i1}; by the fundamental relation $B_i p = Q_{i1}^3$ because $f(Q_{i1} \mid \mathbb{Z}p)$ divides the degree 3 and the inertial degree $f(P_1 \mid \mathbb{Z}p) = 2$.

(c) In this case $p \neq 2$ and $Ap = P_1^3 P_2^3$, $B_i p = Q_{i1}^3$ $(i = 1, 2, 3)$, $Cp = R_1 R_2$.

We have $\mathscr{L}_P(K \mid \mathbb{Q}) = \mathfrak{A}$, hence $\mathscr{L}_P(K \mid L) = \mathfrak{A}$ and $\mathscr{L}_P(L \mid \mathbb{Q}) = \{\varepsilon\}$. Thus $Cp = R_1 R_2$ and $AR_1 = P_1^3$, $AR_2 = P_2^3$ hence $Ap = P_1^3 P_2^3$. Also $\mathscr{L}_P(K \mid K_i) = \mathfrak{A} \cap \mathfrak{B}_i = \{\varepsilon\}$ thus if $Q_{i1} = P_1 \cap B_i$ then AQ_{i1} is the product of two distinct prime ideals, thus necessarily $AQ_{i1} = P_1 P_2$. Hence $B_i p$ has only one prime factor Q_{i1}. But the inertial degrees of P_1, P_2 in $K \mid \mathbb{Q}$ are equal to 1, so $f(Q_{i1} \mid \mathbb{Z}p) = 1$ and by the fundamental relation $B_i p = Q_{i1}^3$.

We show now that $p \neq 2$. If $p = 2$ let $\mathscr{V}_1$ be the first ramification group of P_1 in $K \mid \mathbb{Q}$; by Chapter 11, Theorem 2, $\#(\mathscr{T}_{P_1}(K \mid \mathbb{Q})/\mathscr{V}_1)$ divides $\#(\mathbb{F}_2^{\cdot}) = 1$, so $\mathscr{V}_1 = \mathfrak{A}$ has order 3. But $\#(\mathscr{V}_1)$ is a power of p, thus $p = 3$, a contradiction.

(d) In this case $Ap = P_1^2 P_2^2 P_3^2$, $B_i p = Q_{i1} Q_{i2}^2$ $(i = 1, 2, 3)$, $Cp = R_1^2$.

In fact, $g = (\mathfrak{K} : \mathfrak{B}_i) = 3$ so Ap has 3 different prime factors. Since p is ramified then $Ap = P_1^2 P_2^2 P_3^2$. Let $Q_{i1} = B_i \cap P_1$. Since $\mathscr{L}_P(K \mid K_i) = \mathfrak{B}_i$ then $AQ_{i1} = P_1^2$. On the other hand $\mathscr{L}_{P_2}(K . \mathbb{Q}) = \mathfrak{B}_j$ $(j \neq i)$ hence $\mathscr{L}_{P_2}(K \mid K_i) = \{\varepsilon\}$. If $Q_{i2} = B_i \cap P_2$ $(\neq Q_{i1})$ then AQ_{i2} is the product of two different prime ideals; that is, $AQ_{i2} = P_2 P_3$ and therefore $B_i p = Q_{i1} Q_{i2}^2$. From $\mathscr{L}_P(K \mid L) = \mathfrak{B}_i \cap \mathfrak{A} = \{\varepsilon\}$, if $R_1 = P_1 \cap C$ then AR_1 has three. different prime factors; that is, $AR_1 = P_1 P_2 P_3$ and so $Cp = R_1^2$.

From this discussion, we deduce:

If $A . 2 = P_1 P_2 P_3 P_4 P_5 P_6$ then 2 is unramified in $K_1 \mid \mathbb{Q}$; however, it divides the discriminant of every primitive integral element b of K_1. Thus 2 is an unessential factor of the discriminant δ_{K_1}.

Indeed, 2 is unramified in $K \mid \mathbb{Q}$, hence also in $K_1 \mid \mathbb{Q}$. On the other hand, if $b = b_1, b_2, b_3$ are the conjugates of b over $\mathbb{Q}$, then

$$\operatorname{discr}_{K_1 \mid \mathbb{Q}}(b) = (b_1 - b_2)^2 (b_1 - b_3)^2 (b_2 - b_3)^2.$$

But $N(P_1) = 2$ (since P_1 has inertial degree 1), so b_1, b_2, b_3 are not all in different residue classes modulo P_1. Hence $P_1{}^2$ divides $\mathrm{discr}_{K_1|\mathbb{Q}}(b) \in \mathbb{Z}$, so 4 divides $\mathrm{discr}_{K_1|\mathbb{Q}}(b)$.

3. BIQUADRATIC FIELDS

Example 5. Let $K = \mathbb{Q}(\sqrt{7}, i)$, let A be the ring of integers of K, B the ring of integers of $\mathbb{Q}(\sqrt{7})$, C the ring of integers of $\mathbb{Q}(i)$.

$K \mid \mathbb{Q}$ is a Galois extension, its Galois group is $\{\varepsilon, \sigma, \tau, \sigma\tau\}$ where $\sigma(\sqrt{7}) = \sqrt{7}$, $\sigma(i) = -i$, $\tau(\sqrt{7}) = -\sqrt{7}$, $\tau(i) = i$ and $\sigma\tau(\sqrt{7}) = -\sqrt{7}$, $\sigma\tau(i) = -i$. Thus the fixed field of $\{\varepsilon, \sigma\}$ is $\mathbb{Q}(\sqrt{7})$, while the fixed field of $\{\varepsilon, \tau\}$ is $\mathbb{Q}(i)$.

Every element of K may be written uniquely in the form

$$x = a + b\sqrt{7} + ci + d\sqrt{7}i, \text{ with } a,b,c,d \in \mathbb{Z}. \text{ Then}$$

$$\sigma(x) = a + b\sqrt{7} - ci - d\sqrt{7}i,$$

$$\tau(x) = a - b\sqrt{7} + ci - d\sqrt{7}i,$$

$$\sigma\tau(x) = a - b\sqrt{7} - ci + d\sqrt{7}i.$$

If $x \in A$ then $\mathrm{Tr}_{K|\mathbb{Q}(\sqrt{7})}(x) = x + \sigma(x) \in B$, $N_{K|\mathbb{Q}(\sqrt{7})}(x) = x \cdot \sigma(x) \in B$ and similarly $\mathrm{Tr}_{K|\mathbb{Q}(i)}(x) = x + \tau(x) \in C$, $N_{K|\mathbb{Q}(i)}(x) = x \cdot \tau(x) \in C$. We express these conditions in terms of the coefficients of x:

$$\begin{cases} 2a + 2b\sqrt{7} \in B, \\ (a + b\sqrt{7})^2 + (c + d\sqrt{7})^2 = [(a^2 + c^2) + 7(b^2 + d^2)] + 2(ab + cd)\sqrt{7} \in B \\ 2a + 2ci \in C \\ (a + ci)^2 - 7(b + di)^2 = [(a^2 - c^2) - 7(b^2 - d^2)] + 2(ac - 7bd)i \in C. \end{cases}$$

Taking into account that $B = \mathbb{Z} \oplus \mathbb{Z}\sqrt{7}$, $C = \mathbb{Z} \oplus \mathbb{Z}i$, then $2a \in \mathbb{Z}$, $2b \in \mathbb{Z}$, $2c \in \mathbb{Z}$, $(a^2 + c^2) + 7(b^2 + d^2) \in \mathbb{Z}$, $2(ab + cd) \in \mathbb{Z}$, $(a^2 - c^2) - 7(b^2 - d^2) \in \mathbb{Z}$, $2(ac - 7bd) \in \mathbb{Z}$. From these relations we deduce: $2c^2 + 14b^2 \in \mathbb{Z}$, $2a^2 + 14d^2 \in \mathbb{Z}$. Letting $a = (1/2)a'$, $b = (1/2)b'$, $c = (1/2)c'$ with $a',b',c' \in \mathbb{Z}$, then $12b^2 = 3b'^2 \in \mathbb{Z}$, so $(c'^2 + b'^2)/2 \in \mathbb{Z}$, and therefore b',c' have the same parity. Since $14d^2$ has a denominator at most equal to 2 then $d = (1/2)d'$, $d' \in \mathbb{Z}$, and again $12d^2 = 3d'^2 \in \mathbb{Z}$ so $(a'^2 + d'^2)/2 \in \mathbb{Z}$ and a',d' have the same parity. But

$$\frac{a'}{2} + \frac{d'}{2}\sqrt{7}\,i = \frac{a' - d'}{2} + d'\left(\frac{1 + \sqrt{7}\,i}{2}\right) \quad \text{with} \quad \frac{a' - d'}{2} \in \mathbb{Z},$$

Similarly

$$\frac{b'}{2}\sqrt{7} + \frac{c'}{2}i = \frac{b' - c'}{2}\sqrt{7} + c'\frac{\sqrt{7} + i}{2} \quad \text{with} \quad \frac{b' - c'}{2} \in \mathbb{Z}.$$

Thus every element $x \in A$ is a linear combination with coefficients in $\mathbb{Z}$ of 1, $\sqrt{7}$, $(\sqrt{7} + i)/2$, $(1 + \sqrt{7}i)/2$, and since these elements are linearly independent, they constitute an integral basis of $K \mid \mathbb{Q}$.

Let us note here that $t = (\sqrt{7} + i)/2$ is a primitive integral element of K. Its minimal polynomial is easily computed and equal to $f = X^4 - 3X^2 + 4$. Its discriminant is $d = 16 \times (-3)^4 \times 4 - 128 \times (-3)^2 \times 4^2 + 256 \times 4^3 = 4^3 \times 7^2$ (see Chapter 2, Exercise **48**). The fact that d has square factors does not allow us to decide at once whether $\mathbb{Z}[t]$ is equal to A. But computing the discriminant of the integral basis 1, $\sqrt{7}$, $(\sqrt{7} + i)/2$, $(1 + \sqrt{7}i)/2$ we arrive at

$$
\delta_K = \begin{pmatrix} 1 & \sqrt{7} & \dfrac{\sqrt{7} + i}{2} & \dfrac{1 + \sqrt{7}\,i}{2} \\[2mm] 1 & -\sqrt{7} & \dfrac{-\sqrt{2} + i}{2} & \dfrac{1 - \sqrt{7}\,i}{2} \\[2mm] 1 & \sqrt{7} & \dfrac{\sqrt{7} - i}{2} & \dfrac{1 - \sqrt{7}\,i}{2} \\[2mm] 1 & -\sqrt{7} & \dfrac{-\sqrt{7} - i}{2} & \dfrac{1 + \sqrt{7}\,i}{2} \end{pmatrix}^2 = 4^2 \times 7^2.
$$

Thus $\mathbb{Z}[t] \neq A$. In the notation of (**1D**) $\alpha = 2$. The only ramified prime ideals are 2, 7.

Since $f \equiv (X^2 + 2)^2 \pmod 7$ and $X^2 + \bar{2}$ is irreducible over $\mathbb{F}_7$ it follows that $A \cdot 7 = Q_7{}^2$ where Q_7 has inertial degree 2.

Let ψ_7 be the homomorphism from A onto $\mathbb{F}_{7^2}$ having kernel equal to Q_7. From $\psi_7(7) = 0$ it follows that $\psi_7(\sqrt{7}) = 0$. From $i^2 = -1$ it follows that $\psi_7(i)$ is the square root of -1 over $\mathbb{F}_7$ ($X^2 + \bar{1}$ is irreducible over $\mathbb{F}_7$ because -1 is not a square modulo 7, see Chapter 4, (**H**)). So $\psi_7(t) = (1/2)\sqrt{-1} = \gamma \in \mathbb{F}_{7^2}$. Finally $\psi_7[(1 + \sqrt{7}i)/2] = \psi_7(1/2) = \bar{4} \in \mathbb{F}_7$.

If $x = a + b\sqrt{7} + ct + du \in Q_7$, with $u = (1 + \sqrt{7}i)/2$ then $\bar{a} + \bar{c}\gamma + \bar{d}\bar{4} = \bar{0}$, and since $\gamma \notin \mathbb{F}_7$ then $\bar{a} + \bar{4}\bar{d} = \bar{0}$, $\bar{c} = \bar{0}$, so there exist integers $l, m \in \mathbb{Z}$ such that $a = 7l - 4d$, $c = 7m$ and $x = 7l + b\sqrt{7} + 7mt + d(u - 4)$. This shows that $Q_7 = \mathbb{Z} \cdot 7 \oplus \mathbb{Z}\sqrt{7} \oplus \mathbb{Z} \cdot 7t \oplus \mathbb{Z}(u - 2)$. Now we conclude that Q_7 is the principal ideal generated by $\sqrt{7}$, because $u - 4 = -[(\sqrt{7} - i)/2]\sqrt{7}$ and $(\sqrt{7} - i)/2 \in A$, being a conjugate of t.

We now consider the decomposition of 2. We need to describe the possible homomorphisms ψ from A onto a field extension of $\mathbb{F}_2$. From $[\psi(\sqrt{7})]^2 = \psi(7) = \psi(1) = \bar{1}$ it follows that $\psi(\sqrt{7}) = \bar{1}$. Similarly $[\psi(i)]^2 = \psi(-1) = \bar{1}$, hence $\psi(i) = \bar{1}$.

Let $t' = (-\sqrt{7} + i)/2$ so $t + t' = i$, $tt' = -2$, therefore, $\psi(t) + \psi(t') = \bar{1}$, $\psi(t)\psi(t') = \bar{0}$. Let $u' = (1 - \sqrt{7}i)/2$ so $u + u' = 1$, $uu' = 2$, hence $\psi(u) + \psi(u') = \bar{1}$, $\psi(u)\psi(u') = \bar{0}$. From $tu = \dfrac{\sqrt{7} + i}{2} \cdot \dfrac{1 + \sqrt{7}i}{2} = 2i$ and $t'u' = \dfrac{-\sqrt{7} + i}{2} \cdot \dfrac{1 - \sqrt{7}i}{2} = 2i$ we have $\psi(t)\psi(u) = \bar{0}$, $\psi(t')\psi(u') = \bar{0}$. Then $\psi(t)$, $\psi(u)$ are either $\bar{0}$ or $\bar{1}$ and there are only two possible homomorphisms ψ_2, ψ_2', namely:

$$\psi_2(\sqrt{7}) = \bar{1}, \qquad \psi_2(t) = \bar{0}, \qquad \psi_2(u) = \bar{1}.$$
$$\psi_2'(\sqrt{7}) = \bar{1}, \qquad \psi_2'(t) = \bar{1}, \qquad \psi_2'(u) = \bar{0}.$$

Letting Q_2, Q_2' be their kernels, then Q_2, Q_2' have inertial degree equal to 1. But $K \mid \mathbb{Q}$ is a Galois extension, so the ramification indices of Q_2, Q_2' are equal, hence $A \cdot 2 = Q_2^2 \cdot Q_2'^2$. Now, it is easily seen that

$$Q_2 = \mathbb{Z} \cdot 2 \oplus \mathbb{Z}(1 - \sqrt{7}) \oplus \mathbb{Z}t \oplus \mathbb{Z}(u - 1),$$
$$Q_2' = \mathbb{Z} \cdot 2 \oplus \mathbb{Z}(1 - \sqrt{7}) \oplus \mathbb{Z}(t - 1) \oplus \mathbb{Z}u.$$

But $2 = -iut \in At$, $u - 1 = (1 + \sqrt{7}i)/2 - 1 = (-1 + \sqrt{7}i)/2 = it \in At$. Thus Q_2 is the ideal generated by $1 - \sqrt{7}$ and t. We note that $N_{K \mid \mathbb{Q}}(1 - \sqrt{7}) = (1 - \sqrt{7})^2(1 + \sqrt{7})^2 = 36$ and

$$N_{K \mid \mathbb{Q}}(t) = \left(\frac{\sqrt{7} + i}{2}\right) \cdot \left(\frac{\sqrt{7} - i}{2}\right) \cdot \left(\frac{-\sqrt{7} + i}{2}\right) \cdot \left(\frac{-\sqrt{7} - i}{2}\right) = 4,$$

hence $t \notin A(1 - \sqrt{7})$. It is easily seen that $1 - \sqrt{7} \notin At$ (for if $1 - \sqrt{7} = xt$ with $x \in A$, and if we express x in terms of the integral basis, we arrive at an impossibility).

For Q_2' we observe that $u - 1 = it$ implies that $u = i(t - i) \in A(t - i)$, $2 = -iut = t(t - i) \in A(t - i)$ thus Q_2' is the ideal generated by $1 - \sqrt{7}$, $t - i$. We note also that

$$N_{K \mid \mathbb{Q}}(t - i) = \left(\frac{\sqrt{7} - i}{2}\right) \cdot \left(\frac{\sqrt{7} + i}{2}\right) \cdot \left(\frac{-\sqrt{7} - i}{2}\right) \cdot \left(\frac{-\sqrt{7} + i}{2}\right) = 4$$

hence $t - i \notin A(1 - \sqrt{7})$ and it is also easily seen that $1 - \sqrt{7} \notin A(t - i)$.

Let us now compute the decomposition of 3. $f = X^4 - 3X^2 + 4 \equiv X^4 + 1 = (X^2 + X + 2)(X^2 - X + 2)$ (mod 3), these factors being irreducible over $\mathbb{F}_3$. Then $A \cdot 3 = Q_3 \cdot Q_3'$, where Q_3, Q_3' have inertial degree equal to 2. Let ψ be a homomorphism from A onto a field extension of $\mathbb{F}_3$. Then $\psi(\sqrt{7})^2 = \psi(7) = \bar{1}$ so $\psi(\sqrt{7}) = \bar{1}$ or $\psi(\sqrt{7}) = \bar{2}$. From $\psi(i)^2 = \psi(-1) = -\bar{1}$ and the fact that -1 is not a square module 3, then $\psi(i) = \gamma$, where $\gamma \in \mathbb{F}_{3^2}$ is a root of $X^2 + \bar{1}$. Then $\psi(t) = -(\psi(\sqrt{7}) + \gamma)$, $\psi(u) = -(1 + \psi(\sqrt{7})\gamma)$ and we have the following homomorphisms ψ_3, ψ_3' defined by:

$$\psi_3(\sqrt{7}) = \bar{1}, \ \psi_3(t) = -(\gamma + \bar{1}), \ \psi_3(u) = -(\gamma + \bar{1}).$$
$$\psi_3'(\sqrt{7}) = -\bar{1}, \ \psi_3'(t) = -(\gamma + \bar{2}), \ \psi_3'(u) = \gamma + \bar{2}.$$

If Q_3 is the kernel of ψ_3 and Q_3' is the kernel of ψ_3' then by a computation already explained,

$$Q_3 = \mathbb{Z} . 3 \oplus \mathbb{Z}(\sqrt{7} - 1) \oplus \mathbb{Z} . 3t \oplus \mathbb{Z}(u - t).$$
$$Q_3' = \mathbb{Z} . 3 \oplus \mathbb{Z}(\sqrt{7} + 1) \oplus \mathbb{Z} . 3t \oplus \mathbb{Z}(u + t).$$

Since $(\sqrt{7} - 1)\left(\dfrac{\sqrt{7} + i}{2}\right) = \dfrac{\sqrt{7} + \sqrt{7}i - \sqrt{7} - i}{2} = 3 + (u - t)$ then Q_3 is the ideal generated by 3 and $\sqrt{7} - 1$. Moreover, $3 \notin A(\sqrt{7} - 1)$ and $\sqrt{7} - 1 \notin A . 3$, as one sees taking norms. We deduce also that Q_3' is the ideal generated by 3 and $\sqrt{7} + 1$, since Q_3' is conjugate to Q_3.

4. CYCLOTOMIC FIELDS

Example 6. Let p be a prime number, $m = p^k > 2$ (so if $p = 2$ then $k \geqslant 2$), let ζ be a primitive mth root of unity, $K = \mathbb{Q}(\zeta)$, A the ring of integers of K.

We have already indicated in Chapter 10, **(3A)** and **(3B)**, the decomposition into prime ideals of Aq, where q is any prime number. In particular, if $q \neq p$ we have seen that $Aq = Q_1 \ldots Q_g$ where the prime ideals Q_i are distinct and have norm f, equal to the order of $\bar{q}$ in the multiplicative group $P(m)$.

Now we shall determine explicitly the ring A and compute the discriminant $\delta_{\mathbb{Q}(\zeta)}$. Afterwards, we shall find generators for the non-zero prime ideals of A.

We begin noting that if $\mu = \varphi(p^k)$ then $\{1, \zeta, \ldots, \zeta^{\mu-1}\}$ is a $\mathbb{Q}$-basis of K and of course $\mathbb{Z}[\zeta] \subseteq A$. We shall prove:

4A. $A = \mathbb{Z}[\zeta]$, so $\{1, \zeta, \ldots, \zeta^{\mu-1}\}$ *is an integral basis. The discriminant is* $\delta_{\mathbb{Q}(\zeta)} = (-1)^{\varphi(p^k)/2} p^{p^{k-1}(k(p-1)-1)}$.
Proof: First we compute the discriminant; $d = \operatorname{discr}_{K|\mathbb{Q}}(1, \zeta, \ldots, \zeta^{\mu-1})$. By Chapter 2, **(11)**,

$$d = (-1)^{(1/2)\mu(\mu-1)} N_{K|\mathbb{Q}}(\Phi_m'(\zeta)).$$

Following Chapter 6, **(2)** we have:

$$X^m - 1 = (X^{m/p} - 1)\Phi_m,$$

hence

$$mX^{m-1} = \frac{m}{p} X^{(m/p)-1}\Phi_m + (X^{m/p} - 1)\Phi_m,$$

so

$$m\zeta^{m-1} = (\zeta^{m/p} - 1)\Phi_m'(\zeta).$$

Considering the conjugates of this expression and taking their product, we arrive at

$$m^\mu \left(\prod_{\bar{a}\in P(m)} \zeta^a \right)^{m-1} = \prod_{\bar{a}\in P(m)} (\zeta^{a(m/p)} - 1) \prod_{\bar{a}\in P(m)} \Phi_m'(\zeta^a).$$

But

$$\prod_{\bar{a}\in P(m)} \zeta^a = N_{K|\mathbb{Q}}(\zeta) = (-1)^\mu \cdot 1 = (-1)^\mu,$$

$$\prod_{\bar{a}\in P(m)} (\zeta^{a(m/p)} - 1) = (-1)^\mu \prod_{\bar{a}\in P(m)} (1 - \zeta^{ap^{k-1}}) = (-1)^\mu p^{p^{k-1}}$$

because $\eta = \zeta^{p^{k-1}}$ is a primitive pth root of unity, $\prod_{\bar{c}\in P(p)}(1 - \zeta^c) = \Phi_p(1) = p$ and a system of representatives of the prime residue classes modulo $m = p^k$ gives rise to p^{k-1} systems of representatives of prime residue classes modulo p. Thus

$$m^\mu(-1)^{\mu(m-1)} = (-1)^\mu p^{p^{k-1}} N_{K|\mathbb{Q}}(\Phi_m'(\zeta)),$$

and so

$$d = (-1)^{(1/2)\mu(\mu-1)+\mu(m-2)} p^{p^{k-1}(k(p-1)-1)}.$$

Since $\mu(m - 2)$ is always even and $(1/2)\mu(\mu - 1) \equiv (1/2)\mu$ (mod 2) then

$$d = (-1)^{(1/2)\mu} p^{p^{k-1}(k(p-1)-1)}.$$

In order to show that $\mathbb{Z}[\zeta] = A$ we consider an arbitrary element $x \in A$. It may be written in unique way in the form

$$x = x_0 + x_1\zeta + x_2\zeta^2 + \cdots + x_{\mu-1}\zeta^{\mu-1}$$

where each $x_i \in \mathbb{Q}$. It is our purpose to show that in fact each x_i is an integer. We shall prove that if q is a prime number and $x \in Aq$ then also $x_i \in \mathbb{Z}q$ ($i = 0, 1, \ldots, \mu - 1$). If this is established we are able to conclude that each $x_i \in \mathbb{Z}$. In fact, let $x_i = a_i/b_i$ with $a_i, b_i \in \mathbb{Z}$ relatively prime integers. Let $l = lcm(b_0, b_1, \ldots, b_{\mu-1})$ and $l = b_i l_i'$. Assume that there exists a prime number q such that q^r divides l, q^{r+1} does not divide l (with $r \geqslant 1$), let j be an index such that q^r is the highest power of q dividing b_j; so q does not divide l_j'. From $l = q^r l'$ we have

$$lx = \sum_{i=0}^{\mu-1} (a_i l_i')\zeta^i = (l'x)q^r \in Aq^r.$$

If we show that q divides each coefficient $a_i l_i'$, in particular q divides $a_j l_j'$; but q does not divide l_j', hence q divides a_j, a contradiction.

Considering the conjugates in $K \mid \mathbb{Q}$ we obtain the μ relations

$$\sigma_i(x) = \sum_{j=0}^{\mu-1} x_j \sigma_i(\zeta^j) \qquad (i = 1, \ldots, \mu).$$

Thus $(x_0, \ldots, x_{\mu-1})$ is a solution of the system of linear equations with coefficients $\sigma_i(\zeta^j) \in A$ and determinant whose square is

$$(\det(\sigma_i(\zeta^j)))^2 = \operatorname{discr}_{K \mid \mathbb{Q}}(1, \zeta, \ldots, \zeta^{\mu-1}) = d.$$

Let α_j be the determinant of the matrix obtained from $(\sigma_i(\zeta_j))$ by replacing the jth column by that formed with elements $\sigma_1(x), \ldots, \sigma_\mu(x) \in A$. Thus $\alpha_j \in A$. By Cramer's rule $dx_j = \alpha_j \sqrt{d} \in A \cap \mathbb{Q} = \mathbb{Z}$ (for $j = 0, 1, \ldots, \mu - 1$).

If $q \neq p$ is a prime number such that $x \in Aq$ then $x = qy$, $y \in A$, $x_j = qy_j$ so $dx_j = dy_j q$, with $dy_j \in \mathbb{Z}$. Thus q divides $dx_j = d(a_j/b_j)$ $(a_j, b_j \in \mathbb{Z})$, hence q divides da_j. But $q \neq p$, hence q does not divide d, so q divides a_j; that is, q divides x_j (for $j = 0, 1, \ldots, \mu - 1$).

Now we shall prove that if $x \in Ap$ then p divides also x_j (for every $j = 0, 1, \ldots, \mu - 1$). Let $h = x_0 + x_1 X + \cdots + x_{\mu-1} X^{\mu-1}$ so if $\xi = 1 - \zeta$ then

$$x = h(\zeta) = h(1 - \xi)$$

$$= h(1) - \xi \cdot h'(1) + \xi^2 \cdot \frac{h''(1)}{2!} - \xi^3 \cdot \frac{h'''(1)}{3!} + \cdots$$

$$+ (-1)^{\mu-1} \xi^{\mu-1} \cdot \frac{h^{(\mu-1)}(1)}{(\mu - 1)!}.$$

The coefficients $h^{(k)}(1)/(k!)$ are integers which may be easily computed:

$$\frac{h^{(\mu-1)}(1)}{(\mu - 1)!} = x_{\mu-1},$$

$$\frac{h^{(\mu-2)}(1)}{(\mu - 2)!} = x_{\mu-2} + (\mu - 1)x_{\mu-1},$$

$$\frac{h^{(\mu-3)}(1)}{(\mu - 3)!} = x_{\mu-3} + (\mu - 2)x_{\mu-2} + \binom{\mu - 1}{2} x_{\mu-1}.$$

.

.

.

$$\frac{h''(1)}{2!} = x_2 + \binom{3}{2} x_3 + \cdots + \binom{k}{2} x_k + \cdots + \binom{\mu - 1}{2} x_{\mu-1},$$

$$\frac{h'(1)}{1!} = x_1 + 2x_2 + 3x_3 + \cdots + (\mu - 1)x_{\mu-1},$$

$$h(1) = x_0 + x_1 + x_2 + \cdots + x_{\mu-1}.$$

Since $p = u\xi^\mu$ and p divides x then ξ divides x, hence ξ divides $h(1)$. Thus $h(1) \in A\xi \cap \mathbb{Z} = Ap$, so p divides $h(1)$. But $\mu > 1$, thus ξ divides $p/\xi \in A$, which divides

$$-\frac{x - h(1)}{\xi} = h'(1) - \xi \frac{h''(1)}{2!} + \cdots + (-1)^{\mu-1}\xi^{\mu-2} \frac{h^{(\mu-1)}(1)}{(\mu - 1)!}.$$

Thus ξ divides $h'(1)$ and so p divides $h'(1)$. We may continue in this manner showing successively that p divides $h''(1)/(2!) \cdot h'''(1)/(3!), \ldots,$ $[h^{\mu-1}(1)]/(\mu - 1)!$. Taking into account the values of these elements, we deduce that p divides $x_{\mu-1}$, p divides $x_{\mu-2} + (\mu - 1)x_{\mu-1}$, hence p divides $x_{\mu-2}$. Similarly, p divides $x_{\mu-3}, \ldots, p$ divides x_0.

Thus we have established that $A = \mathbb{Z}[\zeta]$ and therefore the discriminant of K is $\delta_K = d = (-1)^{(1/2)\mu} p^{p^{k-1}(k(p-1)-1)}$. ∎

Now we shall describe explicitly the non-zero prime ideals of A. If P divides Ap then $P = A\xi$, so it is a principal ideal.

Let us consider now any prime ideal dividing Aq, with $q \neq p$. Let $\Phi_m \equiv h_1 h_2 \ldots h_g \pmod{q}$ where $h_1, \ldots, h_g \in \mathbb{Z}[X]$ are such that their images $\bar{h}_1, \ldots, \bar{h}_g \in \mathbb{F}_q[X]$ are irreducible polynomials. We note that the polynomials $h_1, \ldots, h_g$ must be distinct. As we know, if Q_i is the kernel of the homomorphism $\psi_i : A \to \mathbb{F}_{q^f}$ then $\psi_i(\zeta) = \theta_i$ is a root of h_i. Thus Q_i contains the ideal generated by q and $h_i(\zeta)$. But if $j \neq i$ then $\psi_i(h_j(\zeta)) = \bar{h}_j(\theta_i) \neq 0$ since the polynomials $h_1, \ldots, h_g$ are distinct. Thus if a prime ideal Q contains $(q, h_i(\zeta))$ then it contains Aq, hence Q is equal to some of the prime ideals $Q_1, \ldots, Q_g$. But Q_j does not contain $h_i(\zeta)$ when $j \neq i$, so $Q = Q_i$. It follows that

$$\bigcap_{i=1}^{g} (q, h_i(\zeta)) = \prod_{i=1}^{g} (q, h_i(\zeta))$$

(because these ideals are pairwise relatively prime). Hence,

$$Aq \subseteq \bigcap_{i=1}^{g} (q, h_i(\zeta)) = \prod_{i=1}^{g} (q, h_i(\zeta)) \subseteq \prod_{i=1}^{g} Q_i = Aq,$$

so

$$\prod_{i=1}^{g} (q, h_i(\zeta)) = \prod_{i=1}^{g} Q_i \quad \text{and} \quad (q, h_i(\zeta)) = Q_i$$

for every $i = 1, \ldots, g$.

Example 7. We consider now the cyclotomic field $K = \mathbb{Q}(\zeta)$ generated by a mth root of unity ζ, where $m > 2$ is any integer. We may assume that if m is even then 4 divides m. Indeed, if $m = 2m'$, where m' is odd, if ζ' is a primitive m'th root of unity then $\zeta'^m = 1$, so $\zeta' \in \mathbb{Q}(\zeta)$. On the other hand, $(\zeta^{m'})^2 = 1$, so $\zeta^{m'} = 1$, or $\zeta^{m'} = -1$, and in this case $(-\zeta)^{m'} = -\zeta^{m'} = 1$.

So ζ or $-\zeta$ belongs to $\mathbb{Q}(\zeta')$. We denote by s the number of distinct prime factors of m.

Let A be the ring of integers of K. We shall prove:

4B.   (1) *A prime number is ramified in $\mathbb{Q}(\zeta)$ if and only if it divides m.*

(2) $\delta_{\mathbb{Q}(\zeta)} = (-1)^{(1/2)\varphi(m)s} \dfrac{m^{\varphi(m)}}{\prod\limits_{q\,|\,m} q^{\varphi(m)/(q-1)}}.$

(3) $A = \mathbb{Z}[\zeta]$.

(4) Φ_m *is irreducible over* $\mathbb{Q}$ (see Chapter 2, Exercise **39**).

(5) *If p is a prime number, if $m = p^k m'$ with $k \geqslant 0$ and $\gcd(p,m') = 1$, if f is the order of p modulo m' in the group $P(m')$ then $Ap = (P_1 \ldots P_g)^{\varphi(p^k)}$ where $fg = \varphi(m')$.*

Proof: Let s be the number of distinct prime factors of m. The proof will be done by induction on s. In the previous example we have already established these facts when $s = 1$. We now assume that these assertions hold for $s - 1$.

Let p be a prime number dividing m, let $m = p^k m'$, where $k > 0$, $\gcd(p,m') = 1$. Then there exist integers a,b such that $1 = ap^k + bm'$ and $\zeta = \zeta^{ap^k} \cdot \zeta^{bm'}$, with $(\zeta^{ap^k})^{m'} = 1$, $(\zeta^{bm'})^{p^k} = 1$. Thus if ξ is a primitive m'th root of unity and η is a primitive p^k th root of unity, then ζ^{ap^k} is a power of ξ, $\zeta^{bm'}$ is a power of η and we conclude that $\mathbb{Q}(\zeta) = \mathbb{Q}(\xi) \cdot \mathbb{Q}(\eta)$.

By induction on s, a prime number q is ramified in $\mathbb{Q}(\xi)$ if and only if q divides m', and q is ramified in $\mathbb{Q}(\eta)$ if and only if q divides p^k. By Chapter 10, Theorem 1, the discriminants $\delta_{\mathbb{Q}(\xi)}$ and $\delta_{\mathbb{Q}(\eta)}$ are relatively prime. By Chapter 10, (**7Q**) and assertion (4) for m' we have

$$\delta_{\mathbb{Q}(\zeta)} = \delta_{\mathbb{Q}(\xi)}^{\varphi(p^k)} \cdot \delta_{\mathbb{Q}(\eta)}^{\varphi(m')}.$$

Hence q is ramified in $\mathbb{Q}(\zeta)$ exactly when it divides $\delta_{\mathbb{Q}(\zeta)}$; that is, q divides $\delta_{\mathbb{Q}(\xi)}$ or $\delta_{\mathbb{Q}(\eta)}$. This means that q divides m' or p^k, or in other words, q divides m, proving (1).

To show (2) we just use the previous relation for $\delta_{\mathbb{Q}(\zeta)}$:

$$\delta_{\mathbb{Q}(\zeta)} = \delta_{\mathbb{Q}(\xi)}^{\varphi(p^k)} \cdot \delta_{\mathbb{Q}(\eta)}^{\varphi(m')}$$

$$= (-1)^{(s-1)[\varphi(m')/2]\delta^{(p^k)}} \times$$

$$\times \frac{m'^{\varphi(m')\varphi(p^k)}}{\prod\limits_{q\,|\,m'} q^{[\varphi(m')/(q-1)]\varphi(p^k)}} \cdot (-1)^{[\varphi(p^k)/2]\cdot\varphi(m')} \cdot \frac{p^{k\varphi(p^k)\varphi(m')}}{p^{[\varphi(p^k)/(p-1)]\varphi(m')}}$$

$$= (-1)^{[\varphi(m)/2]s} \frac{m^{\varphi(m)}}{\prod\limits_{q\,|\,m} q^{\varphi(m)/(q-1)}}.$$

(3) also follows from Chapter 10, **(7Q)**: if B is the ring of integers of $\mathbb{Q}(\xi)$ and C is the ring of integers of $\mathbb{Q}(\eta)$ then $B = \mathbb{Z}[\xi]$, $C = \mathbb{Z}[\eta]$ and so $A = B \cdot C = \mathbb{Z}[\zeta]$.

(4) is again a consequence of Chapter 10, **(7Q)**:

$$[\mathbb{Q}(\zeta): \mathbb{Q}] = [\mathbb{Q}(\xi): \mathbb{Q}] \cdot [\mathbb{Q}(\eta): \mathbb{Q}] = \varphi(m') \cdot \varphi(p^k) = \varphi(m),$$

so Φ_m is irreducible over $\mathbb{Q}$.

Finally, we show (5). Let $Ap = (P_1 \ldots P_{g'})^{e'}$ where each prime ideal P_i has inertial degree f' over $\mathbb{Q}$. We know that $e' f' g' = [\mathbb{Q}(\zeta): \mathbb{Q}] = \varphi(m) = \varphi(p^k) \cdot \varphi(m')$. By the previous example, $Cp = P^{\varphi(p^k)}$, hence the ramification index e' is at least equal to $\varphi(p^k)$. Since p is unramified in $\mathbb{Q}(\xi)$ (by induction and (1)) then $Bp = Q_1 \ldots Q_g$, Q_i has inertial degree f, $fg = [\mathbb{Q}(\xi): \mathbb{Q}] = \varphi(m')$ and f is the order of p modulo m' in $P(m')$, by induction. Altogether, we have the inequalities $e' f' g' = \varphi(p^k) \cdot \varphi(m') \leqslant e'(fg) \leqslant e' f' g'$. Noting that $f \leqslant f'$, $g \leqslant g'$, then we have $f = f'$, $g = g'$, $e' = \varphi(p^k)$ and so $Ap = (P_1 \ldots P_g)^{\varphi(p^k)}$, $fg = \varphi(m')$ and f is the order of p modulo m' in the group $P(m')$. ∎

5. BINOMIAL EXTENSIONS

Example 8.*   Let $f = X^p - a \in \mathbb{Z}[X]$, where p is a prime number, $p \neq 2$ and $a = q_1 q_2 \ldots q_r$, each q_i being a prime number different from p.

The roots of $f = X^p - a$ are $t, t\zeta, \ldots, t\zeta^{p-1}$ where ζ is a primitive pth root of unity, $t^p = a$. Thus

$$f = \prod_{i=0}^{p-1} (X - t\zeta^i).$$

Now we show that f is irreducible over $\mathbb{Q}$. It has no linear factor, otherwise there would exist a rational number whose pth power is equal to a.

If the minimal polynomial $g \in \mathbb{Q}[X]$ of t has degree less than p, it is of the form

$$g = \prod_{j=1}^{k} (X - t\zeta^{i_j}) \in \mathbb{Z}[X]$$

with $\quad 1 < k < p \quad$ and $\quad 0 = i_1 < i_2 < \cdots < i_k \leqslant p - 1$.   Hence, $t(\zeta^{i_1} + \zeta^{i_2} + \cdots + \zeta^{i_k}) \in \mathbb{Q}$ and $t \in \mathbb{Q}(\zeta)$. But $\mathbb{Q}(t\zeta^{i_2})$ and $\mathbb{Q}(t)$ are conjugate over $\mathbb{Q}$, hence the Galois groups $G(\mathbb{Q}(\zeta) \mid \mathbb{Q}(t\zeta^{i_2}))$ and $G(\mathbb{Q}(\zeta) \mid \mathbb{Q}(t))$ are conjugate subgroups of $G(\mathbb{Q}(\zeta) \mid \mathbb{Q})$. Since this is an Abelian group then $\mathbb{Q}(t) = \mathbb{Q}(t\zeta^{i_2})$ hence $\zeta^{i_2} \in \mathbb{Q}(t)$; taking j such that $ji_2 \equiv 1 \pmod{p}$ we have $\zeta \in \mathbb{Q}(t)$, that is $\mathbb{Q}(\zeta) = \mathbb{Q}(t)$. Thus g has degree $p - 1$ and f would have a linear factor over $\mathbb{Q}$, which is impossible.

* See: Gautheron, V. and Flexor, M. "Un Exemple de Détermination des Entiers d'un Corps de Nombres," *Bull. Soc. Math. France*, **93**, 1969, 3–13.

It follows that $K = \mathbb{Q}(t)$ has degree p over $\mathbb{Q}$. Let A be the ring of integers of K. A contains $\mathbb{Z}[t]$ and we know that $A, \mathbb{Z}[t]$ are free Abelian additive groups of rank p.

5A. *We have* $Ap \subset \mathbb{Z}[t]$ *and the Abelian group* $A/\mathbb{Z}[t]$ *is isomorphic to* $(\mathbb{Z}/\mathbb{Z}p)^j$ *where* $0 \leqslant j < p$.

Proof: By Chapter 6, **(B)**, $\mathbb{Z}[t] \subseteq A \subseteq [1/f'(t)]\mathbb{Z}[t]$, where $f'(t) = pt^{p-1}$. Then $Apt^{p-1} \subseteq \mathbb{Z}[t]$, hence multiplying by t, $Apa \subseteq \mathbb{Z}[t]t \subseteq \mathbb{Z}[t]$. Let us note that $\mathbb{Z}[t]$ is not contained in Ap, because $1 \notin Ap$ (otherwise $1/p \in A \cap \mathbb{Q} = \mathbb{Z}$).

In order to show that $Ap \subseteq \mathbb{Z}[t]$ it suffices to prove that if $\psi: A \to A/At$ is the canonical homomorphism then $\psi(\mathbb{Z}) = \psi(A)$. In fact, this means that given any element of A there exists an element of $\mathbb{Z}$ in the same residue class modulo At, so $A = \mathbb{Z} + At$. Repeating, $A = \mathbb{Z} + (\mathbb{Z} + At)t = \mathbb{Z} + At^2$ and in this manner $A = \mathbb{Z} + At^p = \mathbb{Z} + Aa$ hence $Ap = \mathbb{Z}p + Aap \subseteq \mathbb{Z}[t]$.

We shall compare the decomposition into prime ideals of $\mathbb{Z}a = \mathbb{Z}q_1 \cdot \mathbb{Z}q_2 \ldots \mathbb{Z}q_r$ and At. Let $At = Q_1^{l_1} \cdot Q_2^{l_2} \ldots Q_s^{l_s}$, where each Q_i is a prime ideal of A and $1 \leqslant l_i$. We shall prove that $s = r$, $l_i = 1$ and $Aq_i = Q_i^p$ for every index $i = 1, \ldots, r$.

Indeed, $Aa = (At)^p = Q_1^{pl_1} \ldots Q_s^{pl_s}$; on the other hand $Aa = Aq_1 \ldots Aq_r$. Hence we have $r \leqslant s$, because of the uniqueness of the decomposition into prime ideals. Each Q_i divides one (and only one) ideal Aq_j. For example, the decomposition into prime ideals of Aq_1 is (after renumbering) of the type $Aq_1 = Q_1^{h_1} \cdot Q_2^{h_2} \ldots Q_k^{h_k}$, where the exponents h_j satisfy the fundamental relation $\Sigma_{j=1}^{k} f_{Q_j}(K \mid \mathbb{Q}) \cdot h_j = p$. So $pl_1 = h_1$ and from $h_1 \leqslant p$ it follows that $h_1 = p$ and $l_1 = 1$. Therefore, by the fundamental relation $k = 1$, so $Aq_1 = Q_1^p$. By the same token $Aq_i = Q_i^p$ for $i = 1, \ldots, r$ and we conclude that $s = r$.

So, each prime ideal q_i is totally ramified in $K \mid \mathbb{Q}$ and therefore $A/Q_i \cong \mathbb{Z}/\mathbb{Z}q_i$ $(i = 1, \ldots, r)$.

Now we have $A/At \cong \Pi_{i=1}^{r} A/Q_i \cong \Pi_{i=1}^{r} \mathbb{Z}/\mathbb{Z}q_i \cong \mathbb{Z}/\mathbb{Z}a$. If $\psi: A \to A/At$ denotes the canonical homomorphism then $\psi(\mathbb{Z})$ is a subring of the ring $A/At \to \mathbb{Z}/\mathbb{Z}a$. Since this ring has no non-trivial subring then $\psi(\mathbb{Z}) = \psi(A)$ as we intended to show.

From $Ap \subset \mathbb{Z}[t] \subseteq A$ it follows that $A/\mathbb{Z}[t] \cong (A/Ap)/(\mathbb{Z}[t]/Ap)$ and since A/Ap is a vector space of dimension p over $\mathbb{F}_p$ (see Chapter 10, **(1M)**) and $\mathbb{Z}[t]/Ap \neq 0$, then $A/\mathbb{Z}[t]$ has dimension j, $0 \leqslant j < p$. ∎

5B. *The discriminant of* $K \mid \mathbb{Q}$ *is* $\delta_K = \mathbb{Z}p^{p-2j}a^{p-1}$.

Proof: The discriminant of t is

$$d = \text{discr}_{K|\mathbb{Q}}(1, t, \ldots, t^{p-1}) = (-1)^{(1/2)p(p-1)}N_{K|\mathbb{Q}}(f'(t)).$$

But $f'(t) = pt^{p-1}$. The conjugates of t are $t, t\zeta, \ldots, t\zeta^{p-1}$ (where ζ is a primitive pth root of unity), so

$$N_{K|\mathbb{Q}}(f'(t)) = p^p t^{p(p-1)} \prod_{i=0}^{p-1} \zeta^i = p^p a^{p-1},$$

and therefore $|d| = p^p a^{p-1}$.

We have seen that $\mathbb{Z}[t]$ is contained in A and the Abelian group $A/\mathbb{Z}[t]$ is a vector space of dimension j over $\mathbb{F}_p$. By Chapter 6, (L), there exists an integral basis $\{x_1, \ldots, x_p\}$ of the Abelian group A and integers $f_1, \ldots, f_p$ such that if $y_i = f_i x_i$ ($i = 1, \ldots, p$) then $\{y_1, \ldots, y_p\}$ is a basis of the Abelian group $\mathbb{Z}[t]$. Then $\mathbb{Z}d = (f_1 \ldots f_p)^2 \delta_K$. But

$$\mathbb{F}_p^j \cong A/\mathbb{Z}[t] \cong \overset{p}{\underset{i=1}{\oplus}} (\mathbb{Z}x_i/\mathbb{Z}f_i x_i) \cong \overset{p}{\underset{i=1}{\oplus}} \mathbb{Z}/\mathbb{Z}f_i,$$

thus $\Pi_{i=1}^p f_i = p^j$. We conclude that $\delta_K = \mathbb{Z}p^{p-2j} a^{p-1}$. ∎

Since p is odd then $2j < p$. From this result, we know that the only ramified primes are p and $q_1, \ldots, q_r$. As we saw in (5A), each prime q_i is totally ramified in $K | \mathbb{Q}$. Now we study the ramification of p.

5C. *The following facts are equivalent:*

(1) $A = \mathbb{Z}[t]$

(2) $j = 0$

(3) *p is totally ramified (that is $Ap = P^p$ where P is a prime ideal of A).*

If these conditions are not satisfied then $Ap = P_1^{l_1} \ldots P_s^{l_s}$, with $l_i < p$, $2j = \Sigma_{i=1}^s f_i$, where f_i is the inertial degree of P_i over p.

Proof: The equivalence of (1) and (2) has been seen in (5A).

(1) → (3).   Let $A = \mathbb{Z}[t]$. By reducing modulo p the coefficients of $f = X^p - a$, we obtain $\bar{f} = X^p - \bar{a} = (X - \bar{a})^p$, because $\bar{a} \in \mathbb{F}_p$ hence $\bar{a}^p = \bar{a}$.

From (1D) we know that there exists only one prime ideal P of A above p; that is, $Ap = P^e$; if f is the inertial degree then $ef = p$. But p is ramified in $K | \mathbb{Q}$, hence $e > 1$, so $e = p, f = 1$ and p is therefore totally ramified.

(3) → (2).   Let $Ap = P^p$ so the inertial degree of P over p is equal to 1. Since the ramification index of P over p is the characteristic of the residue class field, it follows from Chapter 10, (7I) that the exponent of the different at P is $s_P \geqslant e_P = p$; that is, P^p divides the different Δ_K. Taking norms, we conclude from Chapter 10, (7J) that $p^p = N(P^p)$ divides $N(\Delta_K) = \delta_K = \mathbb{Z}p^{p-2j} a^{p-1}$. Hence $j = 0$.

Now we assume that p is not totally ramified in $K | \mathbb{Q}$, so $Ap = P_1^{l_1} \ldots P_s^{l_s}$, with each $l_i < p$ and $\Sigma_{i=1}^s l_i f_i = p$, where f_i is the inertial degree of P_i in

$K \mid \mathbb{Q}$. Hence p does not divide l_i and therefore by Chapter 10, **(7I)** the exponent of the different at each P_i is $s_i = l_i - 1$.

We have seen in the proof of **(5A)** that $At = Q_1 Q_2 \ldots Q_r$, where each prime ideal Q_i of A is totally ramified over q_i; that is, $Aq_i = Q_i{}^p$. Since $p \neq q_i$, by Chapter 10, **(7I)** the exponent of the different Δ_k at Q_i is equal to $p - 1$. Since $P_1, \ldots, P_s$ and $Q_1, \ldots, Q_r$ are the only ramified prime ideals, then

$$\Delta_K = P_1^{l_1 - 1} \ldots P_s^{l_s - 1} \cdot Q_1^{p-1} \ldots Q_r^{p-1}.$$

Taking norms we have

$$\delta_K = N_{K|\mathbb{Q}}(\Delta_K) = \mathbb{Z} p^{\sum_1^s f_i(l_i - 1)} \cdot (q_1 \ldots q_r)^{p-1}.$$

But $\delta_K = \mathbb{Z} p^{p-2j} a^{p-1}$ and comparing $p - 2j = \sum_1^s f_i l_i - \sum_1^s f_i$, so $2j = \sum_1^s f_i$. ∎

5D. *If p is not totally ramified then $j = 1$, $Ap = P_1^e P_2^{p-e}$, the inertial degrees of P_1, P_2 are equal to 1.*

Proof: We shall find a new relation between j and the inertial degrees f_1, $\ldots, f_s$. This will be done considering the conductor of A into $\mathbb{Z}[t]$ and localizing above p.

Let S be the multiplicative set complement of $\mathbb{Z}p$ in $\mathbb{Z}$, let $\mathbb{Z}' = S^{-1}\mathbb{Z}$, $\mathbb{Z}[t]' = S^{-1}\mathbb{Z}[t]$, $A' = S^{-1}A$. The ring $\mathbb{Z}'$ has only one non-zero prime ideal which is $\mathbb{Z}'p$ and $\mathbb{Z}'/\mathbb{Z}'p \cong \mathbb{Z}/\mathbb{Z}p = \mathbb{F}_p$ (by Chapter 10, **(4H)**). Reducing the coefficients of $f = X^p - a$ modulo p we obtain $\bar{f} = X^p - \bar{a} = (X - \bar{a})^p$ (since $\bar{a}^p = \bar{a}$ in $\mathbb{F}_p$). By **(1D)** there exists only one prime ideal $\underline{P}$ of $\mathbb{Z}[t]$ such that $\underline{P} \cap \mathbb{Z} = \mathbb{Z}p$ and $\mathbb{Z}[t]p = \underline{P}^p$. Hence the inertial degree of $\underline{P}$ over p is equal to 1. It follows that $\mathbb{Z}[t]'\underline{P}$ is the only non-zero prime ideal of $\mathbb{Z}[t]'$ and $\mathbb{Z}[t]'/\mathbb{Z}[t]'\underline{P} \cong \mathbb{Z}[t]/\underline{P} = \mathbb{F}_p$. We note also that $\underline{P}$ is the ideal of $\mathbb{Z}[t]$ generated by $t - a$ and p.

Since p is not totally ramified then $A \neq \mathbb{Z}[t]$. Let F_t be the conductor of A in $\mathbb{Z}[t]$. Thus $F_t = Af'(t)\Delta_K^{-1}$. We shall show that $A'F_t = \mathbb{Z}[t]'\underline{P}$. By Chapter 10, **(7H)**

$$\Delta(A' \mid \mathbb{Z}') = A' \cdot \Delta_K = A'(Af'(t) \cdot F_t^{-1}) = A'p \cdot A't^{p-1} \cdot (A'F_t)^{-1}$$

$$= A'p \cdot (A'F_t)^{-1} = \prod_1^s A'P_i^{l_i} \cdot (A'F_t)^{-1}.$$

On the other hand, since p is not totally ramified, each $l_i < p$, so the exponent of the different at P_i is $l_i - 1$, hence $\Delta(A' \mid \mathbb{Z}') = \prod_1^s A'P_i^{l_i - 1}$. Comparing these expressions $A'F_t = \prod_1^s A'P_i = \bigcap_1^s A'P_i$. But by Chapter 10, **(7M)**, $F_t \subseteq \mathbb{Z}[t]$ so $A'F_t \subseteq \mathbb{Z}[t]'$ and so

$$A'F_t = A'F_t \cap \mathbb{Z}[t]' = \bigcap_1^s (A'P_i \cap \mathbb{Z}[t]') = \mathbb{Z}[t]'\underline{P}$$

because each prime ideal $A'P_i \cap \mathbb{Z}[t]'$ lies over $\mathbb{Z}'p$, therefore $A'P_i \cap \mathbb{Z}[t]' = \mathbb{Z}[t]'\underline{P}$.

We shall prove that $j = (\Sigma_1^s f_i) - 1$, where f_i is the inertial degree of P_i. We have $A'/A'F_t \cong \Pi_1^s A'/A'P_i$, hence $\#(A'/A'F_t) = p^{\Sigma_i s_i f_i}$, because $A'/A'P_i = A/P_i$; $\mathbb{Z}[t]'/A'F_t = \mathbb{Z}[t]'/\mathbb{Z}[t]'\underline{P} = \mathbb{F}_p$, hence $\#(\mathbb{Z}[t]'/A'F_t) = p$. So the Abelian groups $A'/\mathbb{Z}[t]' \cong (A'/A'F_t)/(\mathbb{Z}[t]'/A'F_t)$ has $p^{\Sigma_i s_i f_i - 1}$ elements. But $A'/\mathbb{Z}[t]' \cong A/\mathbb{Z}[t] \cong F_p{}^j$ and so $j = \Sigma_1^s f_i - 1$.

From $2j = \Sigma_1^s f_i$ we deduce that $2j = j + 1$, hence $j = 1$, $\Sigma_1^s f_i = 2$ and $s \leqslant 2$.

If we had $s = 1$ then $f_1 = 2$ and from the fundamental relation $e_1 f_1 = p$, so p would be even, which is not true. Thus $s = 2$ and $f_1 = f_2 = 1$, $e_1 + e_2 = p$ and we have concluded the proof. ∎

It remains to indicate conditions to decide whether $j = 0$ or $j = 1$.

5E. *The following conditions are equivalent:*
  (1) $A = \mathbb{Z}[t]$
  (2) $a^{p-1} \not\equiv 1 \pmod{p^2}$.
  If $A \neq \mathbb{Z}[t]$ then

$$\left\{ 1, t, \ldots, t^{p-2}, \frac{1 + ta^{p-2} + t^2 a^{p-3} + \cdots + t^{p-1}}{p} \right\}$$

is an integral basis.
Proof:

(1) → (2). By (**5C**) we know that $Ap = P^p$. In the proof of (**5D**) we have seen that P is generated by the elements $t - a$, p. Since $p \in P^p$ then $t - a \notin P^2$. Thus $A(t - a) = PJ$ where P does not divide the ideal J of A. Taking norms $|N_{K|\mathbb{Q}}(t - a)| = N(A(t - a)) = N(P) \cdot N(J) = p \cdot N(J)$. Since P is the only prime ideal of A whose norm is a power of p, from the decomposition of J into a product of prime ideals, it follows that p does not divide $N(J)$.

But the minimal polynomial of $t - a$ over $\mathbb{Q}$ is $(X + a)^p - a$ (this polynomial is irreducible because $\mathbb{Q}(t - a) = \mathbb{Q}(t)$). So $|N_{K|\mathbb{Q}}(t - a)| = a^p - a = a(a^{p-1} - 1)$. We conclude that $a(a^{p-1} - 1) \not\equiv 0 \pmod{p^2}$ and since p does not divide a, $a^{p-1} \not\equiv 1 \pmod{p^2}$.

(2) → (1). Now we assume that $A \neq \mathbb{Z}[t]$, hence $j \neq 0$; therefore $j = 1$ and $Ap = P_1^e P_2^{p-e}$, with $P_1 \cap \mathbb{Z}[t] = P_2 \cap \mathbb{Z}[t] = \underline{P}$, the only non-zero prime ideal of $\mathbb{Z}[t]$. Since $t - a \in \underline{P}$ then $A(t - a)$ is a multiple of P_1 and of P_2, so we may write $A(t - a) = P_1^{m_1} P_2^{m_2} J$ where J is an ideal of A not containing p and $m_1 \geqslant 1$, $m_2 \geqslant 1$. Taking norms we have

$$|a(a^{p-1} - 1)| = N(A(t - a)) = p^{m_1 + m_2} \cdot N(J),$$

with $m_1 + m_2 \geqslant 2$, hence $a^{p-1} \equiv 1 \pmod{p^2}$ since p does not divide a.

Now we shall prove that when $A \neq \mathbb{Z}[t]$ there exists an integral basis of the type indicated. Let us begin showing that

$$z = (1 + ta^{p-2} + t^2 a^{p-3} + \cdots + t^{p-1})/p$$

belongs to A. We have

$$(t^{p-1})^p - 1 = (t^{p-1} - 1)(1 + t^{p-1} + t^{2(p-1)} + \cdots + t^{(p-1)^2})$$

$$= (t^{p-1} - 1)(1 + t^{p-1} + at^{p-2} + a^2 t^{p-3} + \cdots + a^{p-2}t)$$

$$= (t^{p-1} - 1)pz.$$

From $(t^{p-1})^p = a^{p-1}$, we deduce that $pzt^{p-1} = a^{p-1} - 1 + pz$, and therefore $p^p z^p a^{p-1} = (a^{p-1} - 1 + pz)^p$. Thus z is a root of the polynomial $p^p a^{p-1} X^p - (a^{p-1} - 1 + pX)^p$ which has leading coefficient $p^p(a^{p-1} - 1) \in \mathbb{Z}$. The coefficient of X^i $(0 \leqslant i < p)$ is equal to $c_i = \begin{pmatrix} p \\ p - i \end{pmatrix} p^i (a^{p-1} - 1)^{p-i}$. But $a^{p-1} \equiv 1 \pmod{p^2}$ because $A \neq \mathbb{Z}[t]$, hence for $i = 1, \ldots, p - 1$

$$\begin{pmatrix} p \\ p - i \end{pmatrix} p^i (a^{p-1} - 1)^{p-i} = p^{1+i+2(p-i)}b \text{ (with } b \in \mathbb{Z});$$

therefore, $1 + i + 2(p - i) = 2p + 1 - i \geqslant p$; showing that $p^p(a^{p-1} - 1)$ divides c_i. Similarly, $p^p(a^{p-1} - 1)$ divides $c_0 = (a^{p-1} - 1)^p$ because $(a^{p-1} - 1)^{p-1} = p^{2(p-1)}b$, $b \in \mathbb{Z}$ and $p \leqslant 2(p - 1)$. Hence z is an algebraic integer.

Since $1, t, \ldots, t^{p-2}, t^{p-1}$ are linearly independent over $\mathbb{Z}$, then the same holds obviously for the elements $1, t, \ldots, t^{p-2}, z$. It remains to show that these elements generate A. We have $A/\mathbb{Z}[t] \cong \mathbb{Z}/\mathbb{Z}p$ (since $j = 1$), hence $A/\mathbb{Z}[t]$ is a cyclic Abelian additive group, generated by every element different from 0. Since $z \in A$, $z \notin \mathbb{Z}[t]$ then $A = \mathbb{Z}z + \mathbb{Z}[t]$. Finally t^{p-1} belongs to the Abelian group generated by $1, t, \ldots, t^{p-2}, z$ because

$$t^{p-1} = -1 - a^{p-2}t - a^{p-3}t^2 - \cdots - at^{p-2} + pz. \quad \blacksquare$$

We conclude with the following observation. The results indicated above do not hold when some square divides a. For example, let $f = X^3 - 4$, $K = \mathbb{Q}(t)$, where $t^3 = 4$. The discriminant of t is $d = -27 \times 16$. Letting $u = 2/t$, from $t^3 - 4 = 0$ we have $2 - 8/t^3 = 0$, so $u^3 - 2 = 0$, therefore $u \in A$. The $\mathbb{Z}$-module A_1 generated by $1, t, u$ is a subring of A, because $t^2 - 4/t = 0$ implies $t^2 = 2u$. Similarly, $u^2 = t$ and $ut = 2$. Since $\{1, t, t^2\}$ is a $\mathbb{Q}$-basis of K and $u = t^2/2$ then $u \notin \mathbb{Z}[t]$. Hence $\mathbb{Z}[t] \neq A$. We have $3u = (3/2)t^2 \notin A$, thus pA is not contained in $\mathbb{Z}[t]$.

Let $d_1 = \operatorname{discr}_{K|\mathbb{Q}}(1, t, u)$. From the expression of $1, t, t^2$ in terms of $1, t, u$ it follows that $d = 4d_1$, hence $d_1 = -27 \times 4$. Since $A_1 \subseteq A$ then $d_1 = m^2 \delta_K$, $m^2 \geqslant 1$. It follows that $|\delta_K| \neq 3^{3-2j} \times 4^2$.

The prime ideals of A dividing 3 are the kernels of the homomorphisms $\psi: A \to \mathbb{F}_3$ such that $\bar{u}^2 = \bar{t}$, $\bar{t}^2 = 2\bar{u}$, $\bar{u}\bar{t} = \bar{2}$. The only possibility is $\bar{t} = \bar{1}$, $\bar{u} = \bar{2}$. So there is only one prime ideal P dividing 3 and $A \cdot 3 = P^3$. Thus 3 is totally ramified, yet $\mathbb{Z}[t] \neq A$.

6. RELATIVE CYCLOTOMIC EXTENSIONS

Example 9. Let p be a prime number, K an algebraic number field containing a primitive pth root of unity ζ and A the ring of algebraic integers of K. Let $a \in K$ be an element which is not the pth power of an element of K and consider the polynomial $F = X^p - a \in K[X]$. Let t be a root of F, $L = K(t)$ and B be the ring of integers of L. We shall determine the decomposition of the prime ideals of A in $L \mid K$.

First we note that $L \mid K$ is a Galois extension. In fact, the roots of F are $t, t\zeta, \ldots, t\zeta^{p-1}$ and they belong to L.

Next, we show that $[L: K] = p$. Indeed, by hypothesis the minimal polynomial G of t over K cannot be linear (since a is not the pth power of an element of K). Thus $G = \Pi_{j=1}^{k}(X - t\zeta^{i_j})$ with $0 = i_1 < i_2 < \cdots < i_k \leqslant p - 1$ and $1 < k \leqslant p$. It follows that $t(\Sigma_{j=1}^{k}\zeta^{i_j}) \in K$. If $k < p$ then $\Sigma_{j=1}^{k}\zeta^{i_j} \neq 0$ so $t \in K(\zeta) = K$ against the hypothesis about a. Hence $k = p$; that is, G has degree p, $G = F$ and $[L: K] = p$.

We deduce that the Galois group $G(L \mid K)$ is the cyclic group with p elements.

Let $\underline{P}$ be any non-zero prime ideal of A, then $B\underline{P} = \Pi_{i=1}^{g}P_i^{e}$ and all the prime ideals P_i have the same inertial degree f over K. From the fundamental relation $p = efg$ we deduce the following possibilities:

$\quad$ (1) $e = p, f = g = 1$: $B\underline{P} = P^p$ $\qquad$ $\underline{P}$ is totally ramified,
$\quad$ (2) $f = p, e = g = 1$: $B\underline{P} = P$ $\qquad$ $\underline{P}$ is inert,
$\quad$ (3) $g = p, e = f = 1$: $B\underline{P} = P_1 \ldots P_p$ $\qquad$ $\underline{P}$ is decomposed.

Now we shall indicate which case holds for any given prime ideal $\underline{P}$ of A.

6A. *Let $Aa = \underline{P}^h J$, where J is an ideal of A not multiple of $\underline{P}$ and $h > 0$ is an integer not a multiple of p. Then $\underline{P}$ is ramified.*

Proof: We may assume without loss of generality that $\underline{P}$ divides Aa, but $\underline{P}^2$ does not divide Aa. Indeed, let $b \in \underline{P}$, $b \notin \underline{P}^2$, let l, l' be integers such that $lh + l'p = 1$. Taking $a' = a^l b^{pl'}$ then $X^p - a'$ generates the same field extension $L \mid K$. In fact, if $t'^p = a' = a^l b^{pl'} = t^{pl} b^{pl'}$ then $t' = \zeta^i t^l b^{l'} \in L$ (for some pth root of unity ζ^i); conversely $a'^h = a^{lh} b^{pl'h} = a(b^h a^{-1})^{l'p}$, hence $a = a'^h (ab^{-h})^{l'p}$ and in the same way we see that $t \in K(t')$. From the choice of b we deduce that the exact power of $\underline{P}$ which divides $Aa' = Aa^l \cdot Ab^{pl'}$ is $\underline{P}^{hl+pl'} = \underline{P}$.

Thus $Aa = \underline{P} \cdot \underline{J}$ where $\underline{J}$ is not a multiple of $\underline{P}$. Now let P be the ideal of B generated by $B\underline{P}$ and Bt; that is, $P = gcd(B\underline{P}, Bt)$. Then

$$P^p = gcd(B\underline{P}^p, Ba) = gcd(B\underline{P}, B\underline{P} \cdot B\underline{J}) = B\underline{P}.$$

It follows from the fundamental relation that P is necessarily a prime ideal of B and $\underline{P}$ is totally ramified in $L \mid K$. $\blacksquare$

In the second case, $\underline{P}$ is a prime ideal of A such that $Aa = \underline{P}^h J$, where J is an ideal of A not multiple of $\underline{P}$ and p divides the integer $h \geqslant 0$. Actually, we may assume without loss of generality that $\underline{P}$ does not divide Aa. Indeed, we choose an element $b \in \underline{P}$, $b \notin \underline{P}^2$; if $h > 0$ and if $a' = a(b^{-h/p})^p$ then $X^p - a'$ generates the same field extension $L \mid K$ (because a root t' of this equation satisfies $t'^p = a' = t^p(b^{-h/p})^p$, so $t' \in K(t)$ and conversely $t \in K(t')$). Moreover, the exact power of $\underline{P}$ dividing Aa' is now equal to $h - (h/p)p = 0$. In particular $gcd(Aa, \underline{P}) = A$.

6B. *Assume that $\underline{P}$ does not divide Aa neither Ap.*

(1) *If the congruence $X^p \equiv a \pmod{\underline{P}}$ has a solution in A then $\underline{P}$ decomposes in $L \mid K$.*

(2) *If the above congruence has no solution in A then $\underline{P}$ is inert in $L \mid K$.*

Proof: Let us assume that there exists $x \in A$ such that $x^p - a \in \underline{P}$. We have $x^p - a = (x - t)(x - \zeta t) \ldots (x - \zeta^{p-1}t)$.

Let $P_i = B\underline{P} + B(x - \zeta^{i-1}t) = gcd(B\underline{P}, B(x - \zeta^{i-1}t))$ (for $i = 1, \ldots, p$). These ideals of B are conjugate over K.

We have $P_1 P_2 \ldots P_p = B\underline{P}$ because every element of $P_1 P_2 \ldots P_p$ is a sum of elements of the form

$$\prod_{i=1}^{p} (y_i + z_i(x - \zeta^{i-1}t)) = y + z(x^p - a) \in B\underline{P}$$

(where $y_i, y \in B\underline{P}$, $z_i, z \in B$). Hence each P_i is different from B (for these ideals are conjugate, so if some $P_i = B$ then $P_1 = \cdots = P_p = B$ and $P_1 P_2 \ldots P_p = B$, contradiction). It follows that $A \neq P_i \cap A \supseteq \underline{P}$ and therefore $P_i \cap A = \underline{P}$ for $i = 1, 2, \ldots, p$.

Taking into account the fundamental relation, from $B\underline{P} = P_1 P_2 \ldots P_p$, with each $P_i \neq B$, we see that each P_i must already be a prime ideal of B.

It remains to show that these ideals P_i are all distinct. By Galois theory, the decomposition group of P_i in $L \mid K$ has order 1 or p, so either $P_1 = P_2 = \cdots = P_p$ or these ideals are all distinct. In the first case, we would have $x - t, x - t\zeta \in P_1 = \cdots = P_p$, hence $t(1 - \zeta) \in P_1$; since $gcd(Aa, \underline{P}) = A$ then $gcd(Bt, P_1) = B$ so there exist elements $y \in P_1$, $z \in B$ such that $1 = y + zt$ and $1 - \zeta = y(1 - \zeta) + zt(1 - \zeta) \in P_1 \cap A = \underline{P}$. Now $p = u(1 - \zeta)^{p-1}$ where u is a unit, so $p \in \underline{P}$, that is $\underline{P}$ divides Ap against the hypothesis. We conclude that $\underline{P}$ decomposes in $L \mid K$.

(3) Let us assume now that $\underline{P}$ is not inert in $L \mid K$. So either $B\underline{P} = P_1{}^p$ or $B\underline{P} = P_1 P_2 \ldots P_p$. At any rate, P_1 is a prime ideal of B with inertial degree over A equal to 1, thus $B/P_1 \cong A/\underline{P}$ and $N_{L|K}(P_1) = \underline{P}$. In particular, there exists $x \in A$ such that $x - t \in P_1$. So P_1 divides $B(x - t)$ and therefore $\underline{P} = N_{L|K}(P_1)$ divides $N_{L|K}(B(x - t)) = A(x^p - a)$ because $N_{L|K}(x - t) = \prod_{i=1}^{p}(x - \zeta^{i-1}t) = x^p - a$. This means that the congruence $X^p \equiv a \pmod{\underline{P}}$ has the solution $x \in A$. ∎

It remains now to study the case where $\underline{P}$ divides Ap but does not divide Aa. Since $p = u(1 - \zeta)^{p-1}$ then $\underline{P}$ divides $A(1 - \zeta)$. We write $A(1 - \zeta) = \underline{P}^m \cdot \underline{J}$ where $\underline{P}$ does not divide the ideal $\underline{J}$ and $m \geqslant 1$.

6C. *With these notations, we have:*

(1) *If the congruence $X^p \equiv a \pmod{\underline{P}^{mp+1}}$ has a solution in A then $\underline{P}$ decomposes in L.*

(2) *If the above congruence has no solution in A, but the congruence $X^p \equiv a \pmod{\underline{P}^{mp}}$ has a solution in A, then $\underline{P}$ is inert in L.*

(3) *If the congruence $X^p \equiv a \pmod{\underline{P}^{mp}}$ has no solution in A, then $\underline{P}$ is ramified in L.*

Proof:

(1) We first show that the congruence $X^p \equiv a \pmod{\underline{P}^{mp+1}}$ has a solution if and only if $B\underline{P} = P_1 P_2 \ldots P_p$, where $P_1, \ldots, P_p$ are distinct prime ideals of B. This will establish (1).

In fact, if $\underline{P}$ decomposes in $L \mid K$ then the inertial degree of each P_i is equal to 1. Hence $P_i{}^s \cap A \neq \underline{P}^{s-1}$ for $s \geqslant 1$; otherwise $P_i{}^s$ divides $(B\underline{P})^{s-1} = B\underline{P}^{s-1}$ and the exponent of P_i in the decomposition of $B\underline{P}$ into prime ideals could not be equal to 1. Now we show that $P_i{}^s \cap A = \underline{P}^s$ for every $s \geqslant 1$; this is true for $s = 1$ and if it is true for $s - 1$ then $\underline{P}^s \subseteq P_i{}^s \cap A \subseteq P^{s-1} \cap A = \underline{P}^{s-1}$ with $P_i{}^s \cap A \neq \underline{P}^{s-1}$, hence also $P_i{}^s \cap A = \underline{P}^s$. We deduce that $A/\underline{P}^s$ is a subring of $B/P_i{}^s$. But these rings have the same number of elements, namely $N(\underline{P})^s = N(P_i)^s$, so $B/P_i{}^s = A/\underline{P}^s$.

Thus there exists $x \in A$ such that $t \equiv x \pmod{P_1^{mp+1}}$; that is, P_1^{mp+1} divides $B(x - t)$. Taking norms and noting that $N_{L|K}(x - t) = x^p - a$, we deduce that $\underline{P}^{mp+1}$ divides $x^p - a$, so the congruence $X^p \equiv a \pmod{\underline{P}^{mp+1}}$ has a solution in A.

Conversely, let $x \in A$ be such that $x^p \equiv a \pmod{\underline{P}^{mp+1}}$. Let $u \in \underline{P}^{-m}$, $u \notin \underline{P}^{-m+1}$, hence $Au \cdot \underline{P}^m = \underline{I}$ is an integral ideal of A. Consider $v = u(x - t)$. This element is a root of

$$(X - ux)^p - u^p a = X^p - \binom{p}{1} u x X^{p-1} + \binom{p}{2} u^2 x^2 X^{p-2} - \cdots$$

$$+ \binom{p}{p-1} u^{p-1} x^{p-1} X - u^p(x^p - a).$$

The coefficients of this polynomial belong to A as we show now. We know that $\underline{P}^{m(p-1)}$ divides $A(1 - \zeta)^{p-1} = Ap$. For every j such that $1 \leqslant j \leqslant p - 1$ we have $m(p - 1) - mj \geqslant 0$, hence $\binom{p}{j} u^j x^j \in A$. Moreover, $x^p - a \in \underline{P}^{mp+1}$, thus $u^p(x^p - a) \in \underline{P}$.

Therefore, $v \in B$ and the same holds for the conjugates $u(x - \zeta^{i-1}t)$ (for $i = 0, 1, \ldots, p - 1$). We consider the integral ideals

$$P_i = B\underline{P} + Bu(x - \zeta^{i-1}t) = gcd(B\underline{P}, Bu(x - \zeta^{i-1}t)) \text{ (for } i = 0, 1, \ldots, p - 1),$$

conjugate over K. We have $P_1 P_2 \ldots P_p = B\underline{P}$, because every element of which are $P_1 P_2 \ldots P_p$ is a sum of elements of the form

$$\prod_{i=1}^{p} (y_i + z_i u(x - \zeta^{i-1}t)) = y + zu^p(x^p - a) \in B\underline{P}$$

where $y_i, y \in B\underline{P}$, $z_i, z \in B$. Hence each ideal P_i is different from B (since these ideals are conjugate and $P_1 P_2 \ldots P_p = B\underline{P}$). Therefore $P_i \cap A = \underline{P}$ for $i = 1, \ldots, p$.

Taking into account the fundamental relation, from $B\underline{P} = P_1 P_2 \ldots P_p$, with each $P_i \neq B$, we see that each P_i must already be a prime ideal of B.

It remains to show that all these conjugate ideals P_i are distinct. Considering the decomposition group of P_1 in $L \mid K$, we see that either all ideals P_i are distinct or $P_1 = P_2 = \cdots = P_p$ (because the Galois group of $L \mid K$ has order p). In the latter case $u(x - t)$ and $u(x - \zeta t)$ belong to P_1 so $ut(1 - \zeta) \in P_1$ and taking the pth power $u^p a(1 - \zeta)^p \in P_1 \cap A = P$. Now, the exact power of P dividing $A(u^p a(1 - \zeta)^p)$ is $\underline{P}^{-mp+mp}$, so $\underline{P}$ would not divide this ideal, a contradiction.

(2) Now we assume that the congruence $X^p \equiv a \pmod{\underline{P}^{mp}}$ has a solution $x \in A$. We choose $u \in P^{-m}$, $u \notin P^{-m+1}$ and $v = u(x - t)$. As before, the minimal polynomial of v over K is $g = (X - ux)^p - u^p a$, hence $v \in B$ because $x^p - a \in \underline{P}^{mp}$. The different of the element v in $L \mid K$ is $g'(v) = p(ut)^{p-1}$, so $gcd(P, Bg'(v)) = B$ for every prime ideal P dividing $B\underline{P}$. Indeed, if P contains $g'(v)$ then $P \cap A = \underline{P}$ contains $g'(v)^p = p^p u^{p(p-1)} a^{p-1}$. But this is not true, because the exact power of $\underline{P}$ dividing this element is $\underline{P}^{p(p-1)m - p(p-1)m}$. We conclude from Chapter 10, (7N) that $\Delta_{L|K}$ does not divide any of the prime ideals P such that $P \cap A = \underline{P}$. This shows that $\underline{P}$ is unramified in $L \mid K$.

If we assume also that the congruence $X^p \equiv a \pmod{\underline{P}^{mr+1}}$ has no solution in A, then from (1) we deduce that $\underline{P}$ is not decomposed. Therefore, $\underline{P}$ is inert in $L \mid K$.

(3) We assume that the congruence $X^p \equiv a \pmod{\underline{P}^{mp}}$ has no solution in A.

However, the congruence $X^p \equiv a \pmod{\underline{P}}$ has a solution in A. Indeed, $A/\underline{P}$ is a finite field with p^r elements and the multiplicative group of $A/\underline{P}$ is cyclic. Let ω be a generator, so there exists an integer α such that ω^α is the residue class of a modulo $\underline{P}$. Since p, $p^r - 1$ are relatively prime, let c, $c' \in \mathbb{Z}$ be such that $cp + c'(p^r - 1) = 1$. Hence $\alpha \equiv c\alpha p \pmod{p^r - 1}$. Taking $\beta = c\alpha$ we have $\omega^{\beta p} = \omega^\alpha$ and if $x \in A$ has residue class modulo $\underline{P}$ equal to ω^β then $x^p \equiv a \pmod{P}$.

Let $l \geqslant 1$ be the largest integer such that the congruence $X^p \equiv a \pmod{\underline{P}^l}$ has a solution in A. We show that l is not a multiple of p. For suppose that $l = ph$, that $x \in A$ is a solution of $X^p \equiv a \pmod{\underline{P}^{ph}}$ and that $X^p \equiv a \pmod{\underline{P}^{ph+1}}$ has no solution in A. Let $y \in \underline{P}^h$, $y \notin \underline{P}^{h+1}$, hence $y^p \in \underline{P}^{ph}$, $y^p \notin \underline{P}^{ph+1}$. Similarly, by hypothesis $a - x^p \in \underline{P}^{ph}$, $a - x^p \notin \underline{P}^{ph+1}$. But $\underline{P}^{ph}/\underline{P}^{ph+1}$ is a vector space of dimension 1 over $A/\underline{P}$ (see Chapter 8, (A)), hence there exists $z \in A$ such that $zy^p \equiv a - x^p \pmod{\underline{P}^{ph+1}}$. Since $A/\underline{P}$ is a finite field of characteristic p, the raising to the pth power is an automorphism, so there exists $w \in A$ such that $x \equiv w^p \pmod{P}$. Therefore, $z^p + y^p w^p \equiv a \pmod{\underline{P}^{ph+1}}$. But then $x + yw \in A$ would satisfy the congruence $(x + yw)^p \equiv a \pmod{\underline{P}^{ph+1}}$, which is a contradiction. Indeed, $(x + yw)^p \equiv x^p + y^p w^p \pmod{\underline{P}^{ph+1}}$ since the exact power of $\underline{P}$ dividing each of the terms $\binom{p}{i} x^{p-i} y^i w^i$ is at least $m(p - 1) + ih \geqslant ph + 1$ (because $h \leqslant m - 1$, $1 \leqslant i \leqslant p - 1$).

This shows that l is not a multiple of p. We write $l = ph + k$ (with $1 \leqslant k < p$ and $h \leqslant m - 1$).

Let $u \in \underline{P}^h$, $u \notin \underline{P}^{-h+1}$ and $x^p \equiv a \pmod{\underline{P}^l}$, and consider $v = u(x - t)$, which is a root of $(X - ux)^p - u^p a$. As before, we deduce that $v \in B$ and $\underline{P}^k$ is the exact power of $\underline{P}$ dividing $u^p(x^p - a) = N_{L|K}(v)$. Therefore $v \notin B\underline{P}$, otherwise $B\underline{P}$) divides Bv, hence $\underline{P}^p = N_{L|K}(B\underline{P})$ divides $A \cdot N_{L|K}(v)$, which is false since $k < p$.

Let $P = B\underline{P} + Bv = \gcd(B\underline{P}, Bv)$. So $B\underline{P} \neq P$. Also $P \neq B$. This may be seen by considering the ideals $P_1 = P, P_2, \ldots, P_p$ which are the conjugates of P in $L \mid K$. Then $P_1 P_2 \ldots P_p = B\underline{P}$, by the argument already used: If $y_i \in B\underline{P}$, $z_i \in B$ then

$$\prod_{i=1}^{p} (y_i + z_i u(x - \zeta^{i-1}t)) = y + zu^p(x^p - a) \in B\underline{P}$$

(with $y \in B\underline{P}$, $z \in B$). Since the ideals P_i are all equal or all distinct, $P \neq B$.

We conclude that $B\underline{P}$ is not a prime ideal of B, hence $\underline{P}$ is not inert in $L \mid K$. By (1) $\underline{P}$ is not decomposed, hence $\underline{P}$ is ramified in $L \mid K$. ∎

From the study of the decomposition of the prime ideals of A in $L \mid K$ we may infer the following result about the relative discriminant:

6D. *The relative discriminant $\delta_{L|K}$ is the unit ideal of A if and only if the following conditions hold:*
   (1) *Aa is the pth power of an ideal of A.*

(2) *If gcd(Aa,Ap) = A then there exists $x \in A$ which satisfies the congruence*
$X^p \equiv a \pmod{A(1 - \zeta)^p}$.

Proof: We assume that $\delta_{L|K} = A$, so by Chapter 10, Theorem 1, no prime
ideal of A is ramified in $L \mid K$.

Let $\underline{P}$ be a prime ideal of A dividing Aa, and let $Aa = \underline{P}^m \cdot \underline{J}$, where $\underline{J}$
is an ideal of A not multiple of $\underline{P}$. By (6A) m is a multiple of p. This being
true for every prime ideal $\underline{P}$ dividing Aa, it follows that Aa is the pth power
of an ideal of A.

Now we assume that $gcd(Aa,Ap) = A$ and let $A(1 - \zeta) = \Pi_{i=1}^{r} \underline{P}_i^{m_i}$ be
the decomposition of $A(1 - \zeta)$ into prime ideals of A. Thus $\underline{P}_i$ divides Ap,
hence $\underline{P}_i$ does not divide Aa. Since $\underline{P}_i$ is not ramified, by (6C) there exists
$x_i \in A$ such that $x_i^p \equiv a \pmod{\underline{P}_i^{m_i p}}$. By the Chinese remainder theorem,
there exists $x \in A$ such that $x \equiv x_i \pmod{\underline{P}_i^{m_i}}$ for $i = 1, \ldots, r$, so $x^p \equiv a$
$\pmod{A(1 - \zeta)^p}$.

Conversely, we assume now that conditions (1), (2) hold and we shall
show that every prime ideal $\underline{P}$ of A is unramified in $L \mid K$. If $Aa = \underline{P}^m \cdot \underline{J}$,
where $\underline{J}$ is an ideal of A and $m > 0$, then by (1) m is a multiple of p. Replacing a by another element a', we may assume without loss of generality
that $\underline{P}$ does not divide Aa.

If $\underline{P}$ does not divide Ap then by (6B) $\underline{P}$ is not ramified in $L \mid K$. If $\underline{P}$
divides Ap then necessarily Ap does not divide Aa (because $\underline{P}$ does not divide
Aa). Since $Ap = A(1 - \zeta)^{p-1}$ then $\underline{P}$ divides $A(1 - \zeta)$. Let $A(1 - \zeta) =$
$\underline{P}^m \cdot \underline{J}$, with $m \geqslant 1$, where $\underline{J}$ is an ideal of A, not a multiple of $\underline{P}$. By (2) there
exists $x \in A$ such that $x^p - a \in A(1 - \zeta)^p \subseteq \underline{P}^{mp}$. By (6C) it follows that
$\underline{P}$ is not ramified. By Chapter 10, Theorem 1, $\delta_{L|K}$ is the unit ideal. ∎

In Chapter 9, (K) we showed that if K is any algebraic number field then
its absolute discriminant $\delta_{K|\mathbb{Q}}$ is different from the unit ideal. In the preceding
result we saw that this fact need no more be true for the relative discriminant.

In class field theory, it is important to consider field extensions with
relative discriminant equal to 1.

7. THE CLASS NUMBER OF QUADRATIC EXTENSIONS

In this section we indicate a method to compute the class number of
quadratic extensions; we follow the exposition of Hasse [8]. The procedure
is appropriate for small values of the discriminant. There exist more efficient
methods using the theory of characters and transcendental arguments; the
interested reader should consult the literature (see [10], [3]).

Let $K = \mathbb{Q}(\sqrt{d})$ where d is a square-free integer, let $\delta = \delta_K$ be the discriminant, A the ring of integers of K. We recall that if $d \equiv 2, 3 \pmod{4}$
then $\delta = 4d$, if $d \equiv 1 \pmod{4}$ then $\delta = d$.

Also, if $d \equiv 2, 3 \pmod 4$ then $\{1, \sqrt{d}\}$ is an integral basis of K, if $d \equiv 1$ (mod 4) then $\{1, (1 + \sqrt{d})/2\}$ is an integral basis. In this latter case the elements of A are of the form $(a + b\sqrt{d})/2$, where $a, b \in \mathbb{Z}$, $a \equiv b \pmod 2$. In all cases, the elements of A are of the form $(u + v\sqrt{d})/2$ with $u, v \in \mathbb{Z}$, $u^2 \equiv v^2 d \pmod 4$.

Let $\mathscr{D}$ be the set of prime numbers p which are decomposed in K, that is $Ap = PP'$, with P, P' distinct prime ideals of A.

Let $\mathscr{I}$ be the set of prime numbers q which are inert in K, that is $Aq = Q$, where Q is a prime ideal of A.

Let $\mathscr{R}$ be the set of prime numbers r which are ramified in K; that is, $Ar = R^2$, where R is a prime ideal of A.

A non-zero integral ideal I of A is said to be *normalized* when its norm satisfies the following conditions:

(1) $N(I) = \prod_{r \in \mathscr{R}} r^{e_r} \prod_{p \in \mathscr{D}} p^{e_p}$ with $e_p = 0$ or 1, $e_p \geqslant 0$.

(2) If $\delta > 0$ then $N(I) \leqslant (1/2)\sqrt{|\delta|}$; if $\delta < 0$ then $N(I) \leqslant (2/\pi)\sqrt{|\delta|}$.

A non-zero integral ideal I of A is *primitive* when there exists no integer $m \in \mathbb{Z}$, $m \neq 1$, such that Am divides I.

Let $\mathscr{N}$ be the set of normalized primitive ideals of A.

7A. *Every non-zero fractional ideal of K is equivalent to some ideal belonging to $\mathscr{N}$.*

Proof: From Chapter 9, **(L)** it follows that every non-zero fractional ideal of K is equivalent to an ideal J such that $N(J) \leqslant [2^{r_1+r_2}/\text{vol}(D)]\sqrt{|\delta|}$, where D is a symmetric convex body in $\mathbb{R}^2$ such that $D \subseteq D_1$ and D_1 is defined as follows. If $\delta > 0$ then $r_1 = 2$, $r_2 = 0$ and $D_1 = \{(\xi_1, \xi_2) \mid |\xi_1| \, |\xi_2| \leqslant 1\}$. If $\delta < 0$ then $r_1 = 0$, $r_2 = 1$ and $D_1 = \{(\xi_1, \xi_2) \mid \xi_1^2 + \xi_2^2 \leqslant 1\}$.

Thus if $\delta > 0$ we may choose D to be the square with vertices $(2,0)$, $(0,2)$, $(-2,0)$, $(0,-2)$; therefore vol $(D) = 8$, hence $N(J) \leqslant (1/2)\sqrt{|\delta|}$.

If $\delta < 0$ we take $D = D_1$ hence vol$(D) = \pi$ and $N(J) \leqslant (2/\pi)\sqrt{|\delta|}$.

Noting that $Ar = R^2$ for every $r \in \mathscr{R}$, $Aq = Q$ for every $q \in \mathscr{I}$, $A = PP'$ for every $p \in \mathscr{D}$ are principal ideals, then the integral ideal J is equivalent to an integral ideal I such that

$$I = \prod_{r \in \mathscr{R}} R^{e_r} \cdot \prod_{p \in \mathscr{D}} P^{e_p} P'^{e_p'}$$

with $e_r = 0$ or 1, $e_p, e_p' \geqslant 0$ and $e_p = 0$ or $e_p' = 0$. Thus I is a normalized ideal. But I is also primitive, as it follows from the limitation on the exponents $e_r(r \in \mathscr{R})$, $e_p(p \in \mathscr{D})$. $\blacksquare$

Let $N(\mathscr{N})$ denote the set of integers $N(I)$, where $I \in \mathscr{N}$. Thus if $m \in N(\mathscr{N})$ then

$$m = \prod_{r \in \mathscr{R}} r^{e_r} \prod_{p \in \mathscr{D}} p^{e_p}$$

with $e_r = 0$ or 1, $e_p \geqslant 0$ and $m \leqslant (1/2)\sqrt{|\delta|}$, when $\delta > 0$ or $m \leqslant (2/\pi)\sqrt{|\delta|}$ when $\delta < 0$.

Given the field K it is possible to determine all integers $m \in N(\mathcal{N})$ by the computation of a finite number of Legendre symbols.

Once the set of integers $N(\mathcal{N})$ is known, the question is to find all possible ideals in $\mathcal{N}$ and to decide when $I, I' \in \mathcal{N}$ are equivalent.

If $m \in N(\mathcal{N})$, let $k_m = \#\{p \in \mathcal{D} \mid p \mid m\}$; then there exists 2^{k_m} ideals $I \in \mathcal{N}$ such that $N(I) = m$. Indeed, if p^{e_p} is the exact power of $p \in \mathcal{D}$ dividing m, if $Ap = PP'$, we may choose an ideal $I \in \mathcal{N}$ such that P^{e_p} divides I or such that P'^{e_p} divides I. This gives rise to 2^{k_m} possible ideals $I \in \mathcal{N}$ such that $N(I) = m$.

7B. *Let* $m \in N(\mathcal{N})$ *and* $x = (u + v\sqrt{d})/2$, *with* $u, v \in \mathbb{Z}$, $u \equiv v \pmod 2$. *The following conditions are equivalent:*
(1) $N(Ax) = m$.
(2) $(u^2 - v^2 d)/4 = \pm m$ *and* $\gcd(u/2, v/2) = 1$ *when* $d \equiv 2, 3 \pmod 4$, *or* $\gcd(u - v)/2, v) = 1$ *when* $d \equiv 1 \pmod 4$.
Proof:

$1 \to 2$. $m = N(Ax) = |N_{K|\mathbb{Q}}(x)| = |(u^2 - v^2 d)/4|$. Moreover, since $m \in N$ $(\mathcal{N})$ then $\gcd(u/2, v/2) = 1$ when $d \equiv 2, 3 \pmod 4$ or $\gcd((u - v)/2, v) = 1$ when $d \equiv 1 \pmod 4$.

$2 \to 1$. Let $I = Ax$ so $N(I) = m$ and I is a primitive ideal, as follows from the hypothesis on u, v. ∎

We need therefore to find for which integers $m \in N(\mathcal{N})$, m or $-m$ admits a primitive representation as indicated in (2) of (**7B**).

Of course, if $d < 0$ then necessarily $-m$ has no such representation.

If $d \equiv 2, 3 \pmod 4$ then we have to consider representations of the type $\pm m = a^2 - b^2 d$, with $\gcd(a,b) = 1$.

As a corollary, we deduce:

7C. *Assume that the class number of K is $h = 1$ and that $m \in N(\mathcal{N})$.*
If $d < 0$ then m admits a primitive representation.
If $d > 0$ and the fundamental unit ε of K has norm $N(\varepsilon) = 1$ then m or $-m$ admits a primitive representation.
If $d > 0$ and $N(\varepsilon) = -1$ then m and $-m$ admit both a primitive representation.
Proof: In view of (**7B**) and the hypothesis that $h = 1$, for every $m \in N(\mathcal{N})$, m or $-m$ admits a primitive representation. If $d < 0$, $-m$ has no primitive representation, as already said.

If $d > 0$ and $N(\varepsilon) = -1$, if m (or $-m$) has a primitive representation, so has also $-m$ (resp. m). ∎

Summarizing, to determine the class number of K we proceed as follows:

Step 1: to determine the set $\mathcal{N}$ of normalized primitive ideals and the set $N(\mathcal{N})$ of integers which are their norms.

Step 2: for every pair of ideals $I, J \in \mathcal{N}$, to decide whether I, J are in the same class; it is equivalent to decide, for every pair of ideals $I, J \in \mathcal{N}$ whether $I . J$ is a principal ideal.

Now we shall consider numerical examples.

Case 1: $d > 0$.
Let n be the largest integer such that $n \leqslant (1/2)\sqrt{|\delta|}$.

(a) $\underline{n = 1}$. The fields with discriminant $\delta = 5, 8, 12, 13$ are the only ones such that $n = 1$. In this case $\mathcal{N}$ contains only one element and therefore $h = 1$.

(b) $\underline{n = 2}$. The fields with discriminant $\delta = 17, 21, 24, 28, 29, 33$ are the only ones such that $n = 2$. Let us consider some of these values.

If $\delta = 17$ then $N(\mathcal{N}) = \{1, 2\}$ since $17 \equiv 1 \pmod 8$ so $A . 2 = P_2 . P_2'$. But $-2 = (3^2 - 17 . 1^2)/4$ is a primitive representation of -2, so $P_2 = Ax$, $x = (3 + \sqrt{17})/2$, $P_2' = Ax'$, $x' = (3 - \sqrt{17})/2$. Thus $h = 1$.

If $\delta = 24$ then $d = 6$ and $N(\mathcal{N}) = \{1, 2\}$, since 2 is ramified, so $A . 2 = R_2^2$. But $-2 = 2^2 - 6 . 1^2$ is a primitive representation of -2, hence $R_2 = Ax$, $x = 2 + \sqrt{6}$. Thus $h = 1$.

(c) $\underline{n = 3}$. The fields with discriminant $\delta = 37, 40, 41, 44, 53, 56, 57, 60, 61$ are the only ones such that $n = 3$. We consider one of these values.

If $\delta = 40$ then $d = 10$ and $N(\mathcal{N}) = \{1, 2, 3\}$ since 2 is ramified and $\left(\dfrac{10}{3}\right) = \left(\dfrac{1}{3}\right) = 1$; hence $A . 2 = R_2^2$ and $A . 3 = P_3 . P_3'$.

But ± 2 have no primitive representation, otherwise $\pm 2 = a^2 - 10b^2$ (with $a, b \in \mathbb{Z}$) hence $a^2 = 10b^2 \pm 2$; this is impossible since the last digit of a square is not equal to 2 or to 8. For the same reason ± 3 has no primitive representation.

Finally, $-2 . 3 = 2^2 - 10 . 1^2$, hence $R_2 . P_3$ is principal and so is $R_2 . P_3'$. Therefore $h = 2$.

It should be noted that when the discriminant is positive, the question of primitive representation of a prime number may not be as immediate to answer as in the above example. In general, it requires the theory of the Hilbert symbol.

From the tables of class numbers of real quadratic fields, one sees that there exist 142 square free integers d, $2 \leqslant d < 500$, such that the class number of $\mathbb{Q}(\sqrt{d})$ is equal to 1. It is not known whether there exists an infinite number or real quadratic fields with class number 1.

Case 2: $d < 0$.

Let n be the largest integer such that $n \leqslant (\pi/2)\sqrt{|\delta|}$.

(a) $\underline{n = 1.}$ The fields with discriminant $\delta = -3, -4, -7, -8$ are the only ones such that $n = 1$. In this case $\mathcal{N}$ contains only one element and therefore $h = 1$.

(b) $\underline{n = 2.}$ The fields with discriminant $\delta = -11, -15, -19, -20$ are the only ones such that $n = 2$.

We consider some of these values.

If $\delta = -11$ then $N(\mathcal{N}) = \{1\}$ since $-11 \equiv 5 \pmod 8$ so $A . 2$ is a prime ideal. Hence $h = 1$.

If $\delta = -15$ then $N(\mathcal{N}) = \{1,2\}$ since $-15 \equiv 1 \pmod 8$, so $A . 2 = P_2 . P_2'$. Now $2 \neq (u^2 + 15v^2)/4$ for $u,v \in \mathbb{Z}$, $gcd(u,v) = 1$. Hence P_2, P_2' are not principal ideals; but $P_2 . P_2'$ is principal, hence $h = 2$.

(c) $\underline{n = 3.}$ The fields with discriminant $\delta = -23, -24, -31, -35, -39$ are the only ones such that $n = 3$. We consider some of these values.

If $\delta = -31$ then $N(\mathcal{N}) = \{1,2\}$, since $-31 \equiv 1 \pmod 8$ so $A . 2 = P_2 . P_2'$; on the other hand $\left(\dfrac{-31}{3}\right) = -1$ hence 3 is inert in $\mathbb{Q}(-\sqrt{31})$ and $3 \notin N(\mathcal{N})$. Since $2 \neq (u^2 + 31v^2)/4$ for $u,v \in \mathbb{Z}$, $gcd(u,v) = 1$ then P_2, P_2' are not principal ideals. Similarly $2^2 \neq (u^2 + 31v^2)/4$ for $u,v \in \mathbb{Z}$, but $2^3 = (1^2 + 31 . 1^2)/4$, hence P_2^2 is not a principal ideal, but P_2^3 is a principal ideal. From $P_2 . P_2' = A . 2$ it follows that $P_2', P_2'^2$ are not principal ideals, but $P_2'^3$ is a principal ideal; actually $P_2^2 P_2'^{-1} = P_2^3 . (A . 2)^{-1}$ is a principal ideal, so P_2' is equivalent to P_2^2. We conclude that $h = 3$ and the class of the ideal P_2 is a generator of the class group.

The question of determination of the imaginary quadratic fields with class number 1 may be tackled as follows.

If $\mathbb{Q}(\sqrt{d})$, $d < 0$, has class number 1 then $\mathcal{N}$ consists only of the unit ideal. This is true when $|\delta| < 7$, so let $|\delta| \geqslant 7$. If $I \in \mathcal{N}$ then there exists a prime ideal P dividing I such that $N(P) = p \leqslant (2/\pi)\sqrt{|\delta|}$. But if $u, v \in \mathbb{Z}$, $v \neq 0$ (with $v \equiv 0 \pmod 2$ when $d \equiv 2, 3 \pmod 4$) then $(u^2 - dv^2)/4 \geqslant |\delta|/4 > (2/\pi)|\delta| \geqslant p$ when $|\delta| \geqslant 7$, hence P would not be principal, and therefore $h > 1$.

On the other hand, if $h = 1$ then $\delta = -4, -8$ or $\delta = -p$, where p is a prime number, $p \equiv 3 \pmod 4$.

This follows from the theory of genera, which we shall not develop here. Precisely, it is possible to prove:

If $d < 0$ *then the class number h of* $\mathbb{Q}(\sqrt{d})$ *divides* 2^{t-1}, *where t is the number of distinct prime factors of the discriminant δ.*

Thus if $\delta \neq -4, -8$ and $d = -p$, where p is prime number, $p \equiv 1 \pmod 4$ or if d has at least two distinct prime factors, then δ has at least two distinct prime factors, hence 2 divides h, so $h \neq 1$.

If $\delta \neq -3, -4, -7, -8$, in order that $\mathcal{N}$ consists only of the unit ideal it is necessary and sufficient that $-p \equiv 5 \pmod 8$ and if q is an odd prime number, $q \leqslant n$ (largest integer such that $n \leqslant (2/\pi)\sqrt{|\delta|}$) then $\left(\dfrac{-p}{q}\right) = -1$ (this means that q is inert).

Let $n = 2$. Then $-20 \leqslant \delta \leqslant -11$. From $\delta = -p$, $p \equiv 3 \pmod 4$, then $\delta \in \{-11, -19\}$. But 2 should be inert, thus $\delta \equiv 5 \pmod 8$, so $\delta = -11$, -19 satisfy the required conditions.

Let $n = 3$. Then $-39 \leqslant \delta \leqslant -23$. From $\delta = -p$, $p \equiv 3 \pmod 4$ then $\delta \in \{-23, -31\}$. From $\delta \equiv 5 \pmod 8$ then $-23, -31$ do not satisfy the required conditions.

Let $n = 4$. Then $-59 \leqslant \delta \leqslant -40$. Again $\delta \in \{-41, -43, -47, -53\}$. From $\delta \equiv 5 \pmod 8$ it follows that $\delta \in \{-43\}$. Also $\left(\dfrac{-43}{3}\right) = -1$, hence $\delta = -43$ satisfies the required conditions.

Let $n = 5$. Then $-88 \leqslant \delta \leqslant -62$ so $\delta \in \{-67, -71, -73, -79, -83\}$. But $\delta \equiv 5 \pmod 8$, so $\delta \in \{-67, -83\}$. From $\left(\dfrac{-67}{3}\right) = -1$, $\left(\dfrac{-67}{5}\right) = -1$ and $\left(\dfrac{-83}{3}\right) = 1$ we deduce that $\delta = -67$ is the only possibility.

If $n = 6$ then $-120 \leqslant \delta \leqslant -89$ so $\delta \in \{-89, -97, -101, -103, -107, -109, -113, -117\}$. From $\delta \equiv 5 \pmod 8$ it follows that $\delta \in \{-107\}$. But $\left(\dfrac{-107}{3}\right) = 1$ so no value of δ is possible.

If $n = 7$ then $-157 \leqslant \delta \leqslant -121$, so $\delta \in \{-131, -137, -139, -149, -151, -157\}$. From $\delta \equiv 5 \pmod 8$ it follows that $\delta \in \{-131, -139\}$. But $\left(\dfrac{-131}{3}\right) = 1$ and $\left(\dfrac{-139}{5}\right) = 1$ so no value of δ is possible.

If $n = 8$ then $-199 \leqslant \delta \leqslant -158$ then $\delta \in \{-163, -167, -173, -181,$ $-191, -193, -197, -199\}$. From $\delta \equiv 5 \pmod 8$ it follows that $\delta \in \{-163\}$. Since $(-163/3) = -1$, $\left(\dfrac{-163}{5}\right) = -1$, $\left(\dfrac{-163}{7}\right) = -1$ then $\delta = -163$ is the only possibility.

Altogether, we have shown that if $-200 \leqslant \delta < 0$ and $\mathbb{Q}(\sqrt{d})$ has class number 1 then $\delta = -3, -4, -7, -8, -11, -19, -43, -163$. As we have mentioned in Chapter 5, (2), Stark has shown that these are the only imaginary quadratic fields with class number equal to 1.

EXERCISES

1. Let $K_1 = \mathbb{Q}(t)$, $K_2 = \mathbb{Q}(u)$, $K_3 = \mathbb{Q}(v)$, where t, u, v satisfy the equations:

$$t^3 - 18t - 6 = 0,$$
$$u^3 - 36u - 78 = 0,$$
$$v^3 - 54v - 150 = 0.$$

Show that these fields have the same discriminant $22356 = 2^2 \times 3^5 \times 23$.

2. Let K_1, K_2, K_3 be the fields of the previous exercise. Show that they are distinct, by considering the decomposition of the prime numbers 5,11.

3. Let $g = X^3 - 7X - 7$.
(a) Show that g is irreducible over $\mathbb{Q}$.
Let t be a root of g, $K = \mathbb{Q}(t)$.
(b) Find the discriminant of K and the ring of integers A of K.
(c) Determine the decomposition into prime ideals of the ideals of A generated by 2, 3, 5, 7, 11.
(d) Does there exist an unessential factor of the discriminant?

4. Let $K = \mathbb{Q}(\sqrt[3]{175})$.
(a) Show that $\{1, \sqrt[3]{175}, \sqrt[3]{245}\}$ is an integral basis, and the discriminant of K is $\delta = -3^3 \times 5^2 \times 7^2$.
(b) Show that 3,5,7 are completely ramified in $K \mid \mathbb{Q}$.
(c) Show that δ has no unessential factor and find the decomposition of 2 into prime ideals of A.
(d) Show that A has no integral basis of the form $\{1, x, x^2\}$.

5. Determine the ring of integers and the discriminant of the field $K = \mathbb{Q}(t)$, where $t^3 - t - 4 = 0$. Which prime numbers are ramified in K? Determine the decomposition into prime ideals of A of the ideals generated by 2, 3, 5. Determine whether there is an unessential factor of the discriminant.

6. Let a, b be square-free positive integers, $gcd(a, b) = 1$. Let $c = ab$, when $a^2 \equiv b^2 \pmod 9$ or $c = 3ab$ when $a^2 \equiv b^2 \pmod 9$. Show that the discriminant of $\mathbb{Q}(\sqrt[3]{ab^2})$ is $\delta = -3c^2$.

7. Determine the ring of integers, the discriminant and the decomposition of prime numbers in the field $\mathbb{Q}(\sqrt 2, i)$.

8. Determine the ring of integers, the discriminant and the decomposition of prime numbers in the field $\mathbb{Q}(\zeta_{12})$, where ζ_{12} is a primitive 12th root of unity.

9. Let ζ_7 be a primitive 7th root of unity. Determine the minimal polynomial of $\zeta_7 + \zeta_7^{-1}$ and show that the discriminant of $\mathbb{Q}(\zeta_7 + \zeta_7^{-1})$ is equal to 49.

10. Let q be a prime number not dividing n. In order that there exists $x \in \mathbb{Z}$ such that $\Phi_n(x) \equiv 0 \pmod q$ it is necessary and sufficient that $q \equiv 1 \pmod n$. In this case the solutions are the integers x such that $x^n \equiv 1 \pmod q$. The number of pairwise incongruent solutions modulo q is $\varphi(n)$.

Hint: Use the results about the decomposition of primes in cyclotomic fields.

11. Let q be a prime factor of n, let $n = q^a n_1$, n_1 not divisible by q. Show that there exists an integer x such that $\Phi_n(x) \equiv 0 \pmod q$ if and only if $q \equiv 1 \pmod{n_1}$. In this case, the solutions are the integers x such that $x^{n_1} \equiv 1 \pmod q$. The number of pairwise incongruent solutions modulo q is $\varphi(n_1)$.

Hint: Use the results about the decomposition of primes in cyclotomic fields.

12. Determine the integers x such that $\Phi_{20}(x) \equiv 0 \pmod{41}$.

13. Show that there exists no integer x such that $\Phi_{15}(x) \equiv 0 \pmod 5$.

14. Prove the following particular case of Dirichlet's theorem: for every natural number n there exists infinitely many prime numbers which are congruent to 1 modulo n.

Hint: Let m be the product of all primes congruent to 1 modulo $n > 2$; show that every prime factor of $\Phi_n(nm)$ is congruent to 1 modulo n, and note that this is impossible.

15. Let $m = p^k > 2$, p a prime number, $k \geqslant 1$; let ζ be a primitive mth root of unity. Show that if s is any integer, such that $1 \leqslant s \leqslant m, gcd(s,m) = 1$ then

$$v_s = \sqrt{\frac{1 - \zeta^s}{1 - \zeta} \cdot \frac{1 - \zeta^{-s}}{1 - \zeta^{-1}}}$$

is a real unit of $\mathbb{Q}(\zeta)$ (see Chapter 9, **2**).

16. Let $m = p_1^{k_1} \ldots p_r^{k_r}$ with $r > 1$, p_i prime numbers, $k_i \geqslant 1$ and let ζ be a primitive mth root of unity. Show that if s is any integer such that $1 \leqslant s \leqslant m$, $gcd(s,m) = 1$ then $v_s = \sqrt{(1 - \zeta^s)(1 - \zeta^{-s})}$ is a real unit of $\mathbb{Q}(\zeta)$ (see Chapter 9, **(2)** and the previous exercise).

17. Let K be the algebraic number field of example 9 and σ a generator of the Galois group of $\mathbb{Q}(\zeta) \mid \mathbb{Q}$, $\sigma(\zeta) = \zeta^s$, where $1 \leqslant s \leqslant p - 1$, $gcd(s,p) = 1$. Show:

(a) $K \mid \mathbb{Q}$ is a Galois extension if and only if there exists $r, 1 \leqslant r \leqslant p - 1$, such that $\sigma(a)/a^r$ is the pth power of an element of $\mathbb{Q}(\zeta)$.

(b) $K \mid \mathbb{Q}$ is an Abelian extension if and only if $\sigma(a)/a^s$ is the pth power of an element of $\mathbb{Q}(\zeta)$.

(c) If $K \mid \mathbb{Q}$ is a Galois extension then $K = \mathbb{Q}(\zeta) \cdot L$, where $L \mid \mathbb{Q}$ is an extension of degree p.

18. Let $n > 2$ and let h be an integer, $gcd(h,n) = 1$. Let ζ be a primitive nth root of unity. Show:

(a) $\mathbb{Q}(\cos (2\pi h/n)) \mid \mathbb{Q}$ has degree $\varphi(n)/2$.

(b) If $n \neq 4$ then

$$\mathbb{Q}(\sin (2\pi h/n)) \mid \mathbb{Q} \text{ has degree} \begin{cases} \varphi(n) & \text{when } gcd(n,8) < 4 \\ \varphi(n)/4 & \text{when } gcd(n,8) = 4 \\ \varphi(n)/2 & \text{when } gcd(n,8) > 4 \end{cases}$$

(c) If $n > 4$ then

$$\mathbb{Q}(\tan (2\pi h/n)) \mid \mathbb{Q} \text{ has degree} \begin{cases} \varphi(n) & \text{when } gcd(n,8) < 4 \\ \varphi(n)/2 & \text{when } gcd(n,8) = 4 \\ \varphi(n)/4 & \text{when } gcd(n,8) > 4 \end{cases}$$

19. Let ζ be a primitive nth root of unity. Show that $\mathbb{Q}(\zeta + \zeta^{-1}) \mid \mathbb{Q}$ has degree $\varphi(n)/2$.

20. Let $K \mid \mathbb{Q}$ be a real quadratic extension, let ε_0 be a fundamental unit of K, having norm equal to 1. Let x be an algebraic integer of K such that $N_{K\mid\mathbb{Q}}(x) < 0$ and $Ax = J^2$ (where J is an ideal of the ring of integers of K). Show that J is not a principal ideal.

21. Show that the class number of $\mathbb{Q}(\sqrt{34})$ is $h = 2$.

Hint: By a result in Chapter 9, reduce to the consideration of the principal ideals generated by 2, 3, 5; study their prime ideal decompositions in $\mathbb{Q}(\sqrt{34})$ and by the previous exercise, show that the ideal generated by 3 and $1 - \sqrt{34}$ is not a principal ideal, and also that the ideal generated by $5 - \sqrt{34}$ is not principal; conclude showing that these ideals are equivalent.

22. Show that the class number of $\mathbb{Q}(\sqrt{21})$ is $h = 1$.

23. Show that the class number of $\mathbb{Q}(\sqrt{37})$ is $h = 1$.

24. Show that the class number of $\mathbb{Q}(\sqrt{65})$ is $h = 2$.

25. Show that the class number of $\mathbb{Q}(\sqrt{-19})$ is $h = 1$.

26. Show that the class number of $\mathbb{Q}(\sqrt{-163})$ is $h = 1$.

27. Show that the class number of $\mathbb{Q}(\sqrt{-23})$ is $h = 3$.

28. Show that the class number of $\mathbb{Q}(\sqrt{-14})$ is $h = 4$.

29. Show that the class number of $\mathbb{Q}(\sqrt{-127})$ is $h = 5$.

30. Show that the class number of $\mathbb{Q}(\sqrt{-39})$ is $h = 4$.

31. Let p be a prime number, $p \equiv 1 \pmod 4$. Show that the ideal class group of $\mathbb{Q}(-p)$ has an element of order 2.

32. Let n be an integer, $n \geqslant 1$, let p be a prime number such that $p \equiv 2 \pmod 3$ and $p > 3^n$. Show that the ideal class group of $\mathbb{Q}(\sqrt{-p})$ has an element of order greater than n.

Bibliography

1. Artin, E. *Theory of Algebraic Numbers*, 1959. Göttingen: Göttingen Math. Institute.

2. Bachmann, G. *p-adic Numbers and Valuation Theory*, 1964. New York: Academic Press.

3. Borevich, Z. I. and Shafarevich, I. R. *Number Theory*, 1966. New York: Academic Press.

4. Bourbaki, N. *Algèbre*, Ch. V (*Corps Commutatifs*), 1950. Paris: Hermann.

5. Cohn, H. *A Second Course in Number Theory*, 1961. New York: John Wiley.

6. Dickson, L. E. and Vandiver, H. S. *et al. Algebraic Numbers, I, II*, 1923–1928. Bronx: Chelsea.

7. Hardy, G. H. and Wright, E. M. *An Introduction to the Theory of Numbers*, 1954. Oxford: Clarendon.

8. Hasse, H. *Zahlentheorie*, 2nd ed., 1963. Berlin: Akademie Verlag.

9. Hasse, H. *Vorlesungen uber Zahlentheorie*, 2nd ed., 1964. Berlin: Springer.

10. Hasse, H. *Über die Klassenzahl Abelscher Zahlkörper*, 1952. Berlin: Akademie Verlag.

11. Hecke, E. *Vorlesungen über die Theorie der Algebraischen Zahlen*, 1948 (reprint). New York: Chelsea.

12. Hilbert, D. *Gesammelte Abhandlungen* (*Zahlentheorie*), 1965. Bronx: Chelsea.

13. Holzer, L. *Zahlentheorie*, 3 vols., 1958. Leipzig: Teubner.

14. Lang, S. *Algebra*, 1965. Reading: Addison-Wesley.

15. Lang, S. *Algebraic Number Theory*, 1970. Reading: Addison-Wesley.

16. LeVeque, W. J. *Topics in Number Theory*, 1956. Reading: Addison-Wesley.

17. McCarthy, P. J. *Algebraic Extensions of Fields*, 1966. Waltham: Blaisdell.

18. Nagell, T. *Introduction to Number Theory*, 1951. New York: John Wiley.

19. Pollard, H. *The Theory of Algebraic Numbers*, 1950. Buffalo: Math. Assoc. of America.
20. Samuel, P. *Théorie Algébrique des Nombres*, 1967. Paris: Hermann.
21. Serre, J. P. *Corps Locaux*, 1962. Paris: Hermann.
22. Weber, H. *Lehrbuch der Algebra*, 1961 (reprint). New York: Chelsea.
23. Weyl, H. *Algebraic Theory of Numbers*, 1940. Princeton: Princeton Univ. Press.

Notation Index

Name Index

Artin, 34
Bézout, 6, 86
Blaschke, 137
Cashwell, 47
Chevalley, 24
Cramer, 72
Dedekind, 17, 104, 108, 202, 255
Dirichlet, 55, 66, 124, 135, 148, 159, 288
Eisenstein, 12, 19, 226
Euler, 15, 28, 45, 51, 66, 85, 205
Everett, 47
Fermat, 29, 45, 124, 126
Fibonacci, 8
Flexor, 270
Frobenius, 16
Gauss, 9, 52, 53, 58, 78, 159
Gautheron, 270
Gelfond, 13
Golod, 125
Hasse, 281
Heilbronn, 81
Hensel, 216
Hermite, 13, 147
Hilbert, 13, 113, 221, 285
Hurwitz, 108, 113
Jacobi, 61, 62
Kronecker, 61, 66, 219, 233, 243
Krull, 104, 108
Kummer, 124

Lagrange, 29, 86
Landau, 64
Legendre, 50
Lehmer, 82
Liouville, 44
Lindemann, 13
Linfoot, 81
Lucas, 8
Mangoldt, 44
Matusita, 104, 108
Mersenne, 46
Minkowski, 135, 138
Möbius, 43
Newton, 18, 21, 245
Noether, 104, 108
Ore, 213
Pell, 157
Schneider, 13
Shafarevič, 125
Speiser, 229
Stark, 82
Steinitz, 12
Stickelberger, 97
Stirling, 146
Sylow, 35
Warning, 24
Weber, 233, 243
Wilson, 40
Zorn, 113

Alphabetical Index

DATE DUE

1976 MAA Basic Library List